Universitext

Claude Sabbah

Isomonodromic Deformations and Frobenius Manifolds

An Introduction

With 10 Figures

Professor Claude Sabbah
CNRS, Centre de Mathématiques Laurent Schwartz
École Polytechnique
F-91128 Palaiseau Cedex
France

Mathematics Subject Classification (2000): 14F05, 32A10, 32G20, 32G34, 32S40, 34M25, 34M35, 34M50, 53D45, 33E30, 34E05

British Library Cataloguing in Publication Data
A catalogue record for this book is available from the British Library

Library of Congress Control Number: 2067939825

ISBN 978-1-84800-053-7   ISBN 978-1-84800-054-4 (eBook)
DOI 10.1007/978-1-84800-054-4

EDP Sciences ISBN 978-2-7598-0047-6
Translation from the French language edition:
Déformations isomonodromiques et variétés de Frobenius by Claude Sabbah
Copyright 2007 EDP Sciences, CNRS Editions, France.
http://www.edpsciences.org/
http://www.cnrseditions.fr/
All Rights Reserved

Printed on acid-free paper

9 8 7 6 5 4 3 2 1

springer.com

To Andrey Bolibrukh

Contents

Preface

Despite a somewhat esoteric title, this book deals with a classic subject, namely that of linear differential equations in the complex domain. The prototypes of such equations are the linear homogeneous equations (with respect to the complex variable t and the unknown function $u(t)$)

$$\frac{du}{dt} = \frac{\alpha}{t} u(t) \quad (\alpha \in \mathbb{C}), \quad \frac{du}{dt} = \frac{1}{t^2} u(t).$$

The solutions of the first equation are the "multivalued" functions $t \mapsto ct^\alpha$ ($\alpha \in \mathbb{C}$, $c \in \mathbb{C}$) and those of the second equation are the functions $t \mapsto c\exp(-1/t)$. On the other hand, the "multivalued" function $\log$ is a solution of the inhomogeneous linear equation

$$\frac{du}{dt} = \frac{1}{t},$$

or, if one wants to continue with homogeneous equations as we do in this book, of the equation of order 2:

$$t\frac{d^2u}{dt^2} + \frac{du}{dt} = 0.$$

Thus, the solutions of a differential equation with respect to the variable t, having polynomial or rational fractions as coefficients, are, in general, transcendental functions. Needless to say, other families of equations, such as the hypergeometric equations or the Bessel equations, are also celebrated.

Once these facts are understood, the question of knowing if it is necessary to explicitly solve the equations to obtain interesting properties of their solutions can be stated. In other words, one wants to know which properties of the solutions only depend in an algebraic way on (hence are in principle computable from) the coefficients of the equation, and which are those which need transcendental manipulations.

Following this reasoning to its end leads one to develop the theory of differential equations in the complex domain with the tools of algebraic or complex analytic geometry (i.e., the theory of complex algebraic equations). One is thus led to treat *systems* of linear differential equations, which depend on many variables. The algebraic geometry also invites us to consider the global properties of such systems, that is, to consider systems defined on *algebraic* or complex *analytic* manifolds.

The differential equations that we will consider in this book will be named *integrable connections on a vector bundle*. Our *Drosophila melanogaster* (fruit fly) will be the complex projective line, more commonly called the "Riemann sphere" and denoted by $\mathbb{P}^1(\mathbb{C})$ or $\mathbb{P}^1$, and will be the subject of some experiments concerning connections: analysis of singularities and deformations.

The theory of isomonodromic deformations serves as a machine to produce systems of *nonlinear* (partial) differential equations in the complex domain, starting from an equation or from a system of *linear* differential equations of one complex variable. It provides at the same time a procedure (far from being explicit in general) to solve them, as well as remarkable properties of the solutions of these systems (among others, the property usually called the "Painlevé property"). If, at the beginning, the main object of interest was the deformation of linear differential *equations* of a complex variable with polynomial coefficients, it has now been realized that the deformation theory of linear systems of many differential equations can shed light on this question, thanks to the use of tools coming from algebraic or differential geometry, such as vector bundles, connections, and the like.

For a long time (and such remains the case), this method serve specialists in dynamical systems and physicists who analyze the nonlinear equations produced by integrable dynamical systems; to exhibit these equations as isomonodromy equations is, in a way, a linearization of the initial problem. From this point of view, the Painlevé equations have played a prototypic role, beginning with the article by R. Fuchs [Fuc07] (followed by those of R. Garnier) who showed how the sixth one can be written as an isomonodromy equation, thus avoiding the strict framework of the search for new transcendental functions.

A nice application of this theory is the introduction of the notion of a Frobenius structure on a manifold. If this notion had clearly emerged from the articles of Kyoji Saito on the unfoldings of singularities of holomorphic functions, it has been extensively developed by Boris Dubrovin, who used motivations coming from physics, opening new perspectives on, and establishing a new link between, mathematical domains which are apparently not related (singularities, quantum cohomology, mirror symmetry).

My aim to keep this text a moderate length and level of complexity, as well as my lack of knowledge on more recent advances, led me to limit the number of themes, and to refer to the foundational article of B. Dubrovin [Dub96], or to the book of Y. Manin [Man99a], for further investigation of other topics.

Chapter 0, although slightly long, can be skipped by any reader having a basic knowledge of complex algebraic geometry; it can serve as a reference for notation. It presents the concepts referred to in the book concerning sheaf theory, vector bundles, holomorphic and meromorphic connections, and locally constant sheaves. The results are classic and exist, although scattered, in the literature.

The same considerations apply to Chapter I, although it can be more difficult to find a reference for the rigidity theorem of trivial vector bundles in elementary books on algebraic geometry. We restrict ourselves to bundles on the Riemann sphere, minimizing the knowledge needed of algebraic geometry. In this chapter, we do not give the proof of the finiteness theorem for the cohomology of a vector bundle on a compact Riemann surface, for which good references exist; we only need it for the Riemann sphere.

With Chapters II and III begins the study of linear systems of differential equations of a complex variable and their deformations. The type of singular points is analyzed there. Here also we do not give the proof of two theorems of analysis, inasmuch as the techniques needed, although very accessible, go too much beyond the scope of this book.

One of the fundamental objects attached to a differential equation or, more generally, to an integrable connection on a vector bundle, is the *group of monodromy transformations* in its natural representation, reflecting the "multivaluedness" of the solutions of this equation or connection. The *Riemann-Hilbert correspondence*—at least when the singularities of the equation are regular—expresses that this group contains the complete information on the differential equation. Thus, one of the classic problems of the theory consists of, given a differential equation, computing its monodromy group. Let us also mention another object, the *differential Galois group*—which we will not use in this book—that has the advantage of being defined algebraically from the equation.

We will not deal with this problem in this book, and one will not find explicit computations of such groups. As indicated above, we rather try to express the properties of the solutions of the equation in terms of algebraic objects, here the (meromorphic) vector bundle with connection. In this meromorphic bundle exist *lattices* (i.e., holomorphic bundles), which correspond to the various equivalent ways to write the differential system.

To find the simplest way to present a differential system up to meromorphic equivalence is the subject of the *Riemann-Hilbert problem* (in the case of regular singularities) or of *Birkhoff's problem*. In all cases, it is a matter of writing the system as a connection on the trivial bundle. Chapter IV expounds on some techniques used in the resolution of the Riemann-Hilbert or Birkhoff's problem. One will find in the works of A. Bolibrukh [AB94] and [Bol95] many more results.

Chapter V introduces the Fourier transform (which should possibly more accurately be called the Laplace transform) for systems of differential equations of one variable. It helps one in understanding the link between Schlesinger equations and the deformation equations for Birkhoff's problem, analyzed in Chapter VI. In the latter, the notion of isomonodromic deformation is explained in detail.

Chapter VII gives an axiomatic presentation of the notion of a Saito structure (as introduced by K. Saito) as well as that of a Frobenius structure (as introduced by B. Dubrovin, with its terminology). We show the equivalence between these notions, using the term "Frobenius-Saito structure". Many examples are given in order to exhibit various aspects of these structures. This chapter can serve as an introduction to the theory of K. Saito on the Frobenius-Saito structure associated with unfoldings of holomorphic functions with isolated singularities. The proofs of many results of this theory require techniques of algebraic geometry in dimension $\geqslant 1$, techniques which go beyond the scope of this book and would need another book (Hodge theory for the Gauss-Manin system).

This text, a much expanded version of my article [Sab98] on the same topic, stemmed from a series of graduate lectures that I gave at the universities of Paris VI, Bordeaux I and Strasbourg, and during a summer school on Frobenius manifolds at the CIRM (Luminy). Michèle Audin, Alexandru Dimca, Claudine Mitschi and Pierre Schapira gave me the opportunity to lecture on various parts of this text.

Many ideas, as well as their presentation, come directly from the articles of Bernard Malgrange, as well as from numerous conversations that we had. Many aspects of Frobenius manifolds would have remained obscure to me without the multiple discussions with Michèle Audin. I also had the pleasure of long discussions with Andrey Bolibrukh, who explained to me his work, particularly concerning the Riemann-Hilbert problem. Joseph Le Potier answered my electronic questions on bundles with good grace.

Various people helped me to improve the text, or pointed out a few mistakes: Gilles Bailly-Maitre, Alexandru Dimca, Antoine Douai, Claus Hertling, Adelino Paiva, Mathias Schulze and the anonymous referees.

I thank all of them.

The original (French) version of this book has been written within INTAS program no. 97-1644.

The English translation differs from the original French version only in the correction of various mistakes or inaccuracies, a list of which can be found on the author's web page `math.polytechnique.fr/~sabbah`.

Terminology and notation

Some words used in this book may have various meanings, depending on the context in which they appear. For instance:

- the *flatness* property (in differential geometry), the absence of curvature (for a metric, a connection...) is synonymous with integrability; on the other hand, in commutative algebra, it expresses good behaviour with respect to the tensor product;
- *torsion* is a geometric notion (for a connection on the tangent bundle), but is also an algebraic notion (for a module on a ring); one can consider the torsion of a flat connection, but a flat module on a integral ring has no torsion.

Analogously, the name "Frobenius" can refer to various distinct properties: Pfaff systems which share the integrability property in the sense of Frobenius are called foliations, although the notion of Frobenius manifold (or of Frobenius structure on a manifold) refers to the structure of Frobenius algebra on the tangent bundle of this manifold; nevertheless, the notion of integrability comes in the construction of such structures.

Make note of the distinction between the mathematicians L. Fuchs (Fuchs condition, Fuchsian equation, etc.) and R. Fuchs (isomonodromy and Painlevé equations), as well as K. Saito and M. Saito.

Regarding notation, I have tried to use and maintain throughout the text a simple rule.

The letters M, N, X only serve (in principle) for manifolds, the letter X often denoting the parameter space of a deformation, while the letter M rather denotes the space on which the deformed object lives.

The letters $\mathscr{E}, \mathscr{F}, \mathscr{G}$ serve for sheaves on a manifold, the letters E, F, G for the corresponding bundles, when meaningful, and the letters $\mathcal{E}, \mathcal{F}, \mathcal{G}$ for their germ at some point. The "meromorphic bundles" are denoted in general by the letter $\mathscr{M}$ (and sometimes $\mathscr{N}$), their germ at some point by $\mathcal{M}$ or $\mathcal{N}$. Lastly, in the algebraic framework, the module of global sections of a sheaf $\mathscr{E}, \mathscr{F}, \mathscr{G}, \mathscr{M}$ is denoted by $\mathbb{E}, \mathbb{F}, \mathbb{G}, \mathbb{M}$.

In a family parametrized by a space, the restriction of an object A for the value x^o of the parameter is denoted by A^o.

Finally, a white square on a white background $\square$ means the end of a proof, or its absence.

0

The language of fibre bundles

This chapter assembles the main concepts referred to in this book. It also establishes the notation used throughout the text.

We assume some familiarity with the language of sheaf theory.

The notion of vector bundle and of connection enables one to take up global problems of the theory of linear differential systems and of their singularities. We consider here only the holomorphic or meromorphic setting. We have tried to provide intrinsic statements, independent of the choices of coordinates or of basis. Therefore, the reader will not be surprised not to find the notion of fundamental matrix of solutions, of the Wronskian, etc. On the other hand, we do insist on the difference between the notion of a meromorphic differential system (where meromorphic base changes are allowed) and that of a lattice of such a system (where only holomorphic base changes are allowed).

1 Holomorphic functions on an open set of $\mathbb{C}^n$

The reader can refer to [GR65, Chap. I], [GH78, Chap. 0], [Kod86, Chap. 1] (and also to [Hör73, Chap. 1] or [LT97, Chap. 1]) for the elementary properties of holomorphic functions.

Let n be an integer $\geqslant 1$ and U be an open set of $\mathbb{C}^n$. The coordinates are denoted by $z_1, \ldots, z_n$, where $z_j = x_j + iy_j$ is the decomposition into real and imaginary parts.

Let $f : U \to \mathbb{C}$ be a function of class C^1. We put

$$\frac{\partial f}{\partial z_j} = \frac{1}{2}\left(\frac{\partial f}{\partial x_j} - i\frac{\partial f}{\partial y_j}\right), \qquad \frac{\partial f}{\partial \overline{z}_j} = \frac{1}{2}\left(\frac{\partial f}{\partial x_j} + i\frac{\partial f}{\partial y_j}\right).$$

A function f of class C^1 on U is said to be *holomorphic* if, for any $z \in U$, the Cauchy-Riemann equations

$$\frac{\partial f}{\partial \overline{z}_j}(z) = 0, \quad j = 1, \ldots, n,$$

are satisfied. We denote by $\mathscr{O}(U)$ the ring of holomorphic functions on U and by $\mathscr{O}_U$ the *sheaf* of holomorphic functions on U.

1.1 Theorem (The holomorphic functions are the analytic functions). *A function $f : U \to \mathbb{C}$ is holomorphic if and only if it is analytic, that is, can be expanded as a converging series of $(z_1 - z_1^o, \ldots, z_n - z_n^o)$ in the neighbourhood of any point $z^o \in U$.*

Proof. Analogous to the proof for functions of one variable. Let

$$\Delta(z^o, r) = \prod_{j=1}^{n} D(z_j^o, r_j)$$

be an open polydisc of polyradius $r = (r_1, \ldots, r_n) \in (\mathbb{R}_+^*)^n$, centered at $z^o \in U$ and contained in U. The result follows from the "Cauchy polyformula", that one shows by induction on n: for any $z \in \Delta(z^o, r)$ we have

$$f(z) = \frac{1}{(2i\pi)^n} \int_{C(z_1^o, r_1)} \cdots \int_{C(z_n^o, r_n)} \frac{f(w)}{(z_1 - z_1^o) \cdots (z_n - z_n^o)} \, dz_1 \wedge \cdots \wedge dz_n. \quad \square$$

Let us fix $z^o \in U$. The germ of the sheaf $\mathscr{O}_U$ at z^o, denoted by $\mathscr{O}_{U,z^o}$, can thus be identified with the ring of *converging series* with n variables, denoted by $\mathbb{C}\{z_1 - z_1^o, \ldots, z_n - z_n^o\}$.

1.2 Some properties.

- Any holomorphic mapping from an open set of $\mathbb{C}^n$ with values in $\mathbb{C}^p$ is a C^∞ mapping.
- Any bijective holomorphic mapping between two open sets of $\mathbb{C}^n$ is biholomorphic.
- Any nonconstant holomorphic function on a connected open set of $\mathbb{C}^n$ is an *open* mapping.
- We have at our disposal the implicit function theorem and the inverse function theorem for holomorphic mappings.
- Let U be an open set of $\mathbb{C}^n$, let $\varphi : U \to \mathbb{C}$ be a holomorphic function $\not\equiv 0$ and let $Z \subset U$ be the set defined by the equation $\varphi(z_1, \ldots, z_n) = 0$ in U. If f is any holomorphic function on $U \smallsetminus Z$ which is bounded in the neighbourhood of any point of Z, then f can be extended as a holomorphic function on U.

2 Complex analytic manifolds

The reader can refer to [GH78, Chap. 0] or [Kod86, Chap. 2] for more details or examples. Let M be a topological space. An *open covering* $\mathfrak{U}$ of M is a family $(U_i)_{i \in I}$ of open sets of M indexed by a finite set I, such that $\bigcup_i U_i = M$.

A topological space M is said to be *paracompact* if it is *Hausdorff* and if, for any open covering $\mathfrak{U}$ of M, there exists an open covering $\mathfrak{V}$ of M which is *locally finite* and *finer* than $\mathfrak{U}$, that is, any compact set cuts only a finite number of open sets of $\mathfrak{V}$ and any open set V_j of $\mathfrak{V}$ is contained in at least one open set U_i of $\mathfrak{U}$.

A *complex analytic manifold* M of dimension n is a paracompact topological space which admits an open covering $(U_i)_{i \in I}$ and *charts* $\varphi_i : U_i \to \mathbb{C}^n$, where each φ_i induces a homeomorphism from U_i to some open set Ω_i of $\mathbb{C}^n$, such that, for any $i, j \in I$, the change of chart

$$\varphi_i(U_i \cap U_j) \xrightarrow{\ \varphi_j \circ \varphi_i^{-1}\ } \varphi_j(U_i \cap U_j)$$

is a biholomorphism.

A *holomorphic function* on a complex analytic manifold is a function of class C^1 such that each $f \circ \varphi_i^{-1}$ is a holomorphic function on the open set $\varphi_i(U_i)$ of $\mathbb{C}^n$.

Any chart U_i admits then coordinates systems $(z_1, \ldots, z_n)$ coming from those of Ω_i.

A mapping between two complex analytic manifolds is said to be holomorphic if, for one (or any) choice of holomorphic local coordinates of the source and the target, the components of the mapping are holomorphic functions of coordinates.

A *complex analytic submanifold* N of M is a locally closed subset of M for which, in the neighbourhood of any point of N in M, there exist local coordinates $z_1, \ldots, z_n$ such that N is defined by the equations $z_1 = \cdots = z_p = 0$; one then says that p is the *(complex) codimension* of N in M. A *smooth hypersurface* is a *closed* submanifold of *codimension* 1.

A *closed analytic subset* M is a closed subset, which is *locally* defined as the set of zeroes of a family of analytic functions. We will have to use the following result (the reader can consult [GR65, Chap. II and III] for more details on analytic subsets):

2.1 Lemma (connectedness, see [GR65, Th. 2, p. 86]). *Let M be a connected complex analytic manifold. Then the complement in M of a closed analytic subset distinct from M is connected.* □

2.2 Examples.

(1) Any open set of $\mathbb{C}^n$ is a complex analytic manifold.
(2) Any covering space (cf. for instance [God71]) of a complex analytic manifold is equipped with a unique structure of complex analytic manifold for which the covering map is holomorphic. For this structure, any local section of the covering is a holomorphic mapping.
(3) The complex projective space $\mathbb{P}^n$ is the space of lines of $\mathbb{C}^{n+1}$: it is the quotient space of $\mathbb{C}^{n+1} \setminus \{0\}$ by the equivalence relation $v \sim \lambda v$ ($\lambda \in \mathbb{C}^*$). We denote by $[v] \in \mathbb{P}^n$ the line generated by the vector $v \in \mathbb{C}^{n+1} \setminus \{0\}$.

If $(Z_0, \ldots, Z_n)$ are the coordinates on $\mathbb{C}^{n+1}$, one covers $\mathbb{P}^n$ with the $n+1$ open sets $U_j = \{[v] \mid v_j \neq 0\}$ $(j = 0, \ldots, n)$. We set

$$\varphi_j : U_j \xrightarrow{\sim} \mathbb{C}^n$$

$$[Z] \longmapsto \left(\frac{Z_0}{Z_j}, \ldots, \frac{Z_n}{Z_j} \right).$$

We then get a structure of complex analytic manifold on $\mathbb{P}^n$: for $j \neq k$, let us denote by $(z_0, \ldots, \widehat{z_j}, \ldots, z_n)$ the coordinates on $\varphi_j(U_j)$ (the j-th coordinate does not appear) and similarly $(w_0, \ldots, \widehat{w_k}, \ldots, w_n)$ the coordinates on $\varphi_k(U_k)$; then $\varphi_j(U_j \cap U_k)$ is the open set $z_k \neq 0$ in $\mathbb{C}^n$ and $\varphi_k(U_j \cap U_k)$ is the open set $w_j \neq 0$; the change of chart is the isomorphism

$$(z_0, \ldots, \widehat{z_j}, \ldots, z_k, \ldots, z_n) \longmapsto \left(\frac{z_0}{z_k}, \ldots, \frac{1}{z_k}, \ldots, \widehat{1}, \ldots, \frac{z_n}{z_k} \right).$$

(4) A complex analytic manifold of dimension one is called a *Riemann surface*. Let us mention for instance:

- the projective line $\mathbb{P}^1$, covered by two charts U_0 and U_∞; the change of chart

$$\mathbb{C}^* \xrightarrow{\varphi_0^{-1}} U_0 \cap U_\infty \xrightarrow{\varphi_\infty} \mathbb{C}^*$$

  is defined by $z \mapsto 1/z$;
- the quotient space of $\mathbb{C}$ by a lattice $L = \mathbb{Z} \oplus \mathbb{Z}\tau$, with $\tau \in \mathbb{C} \smallsetminus \mathbb{R}$, naturally equipped with a structure of Riemann surface (elliptic curve).

2.3 Exercise (The Grassmannian, cf. [GH78, p. 193–194]). Show that the set of vector subspaces of dimension $r+1$ of $\mathbb{C}^{n+1}$ admits a structure of complex analytic manifold (called the *Grassmannian* of $r+1$-planes in $\mathbb{C}^{n+1}$). Remark that it is also the set of projective subspaces of dimension r of $\mathbb{P}^n$.

2.4 Remarks.

(1) The notion of C^∞ manifold is defined in the same way as above, by changing "holomorphic" with "C^∞". As any holomorphic mapping is C^∞, an analytic manifold can be equipped with a natural structure of C^∞ manifold and the notion of C^∞ function has a meaning on a complex analytic manifold. We denote by $\mathscr{C}_M^\infty$ the sheaf of germs of C^∞ functions on M.

(2) It will be convenient to generalize the notion of a chart of an analytic manifold. An *étale chart* $\varphi : U \to \mathbb{C}^n$ is a holomorphic mapping from an open set M to an open set in $\mathbb{C}^n$, which has maximal rank everywhere. The components $(\varphi_1, \ldots, \varphi_n)$ of φ form a coordinate system in the neighbourhood of any point of U. When there can be no confusion, we will also say that $(\varphi_1, \ldots, \varphi_n)$ is a coordinate system on U.

2.5 The structure sheaf. Let U be an open set of a complex analytic manifold and let $\mathscr{O}_M(U)$ denote the ring of holomorphic functions on U. For

$V \subset U$ there is a restriction morphism $\mathscr{O}_M(U) \to \mathscr{O}_M(V)$, which makes $\mathscr{O}_M$ a *presheaf*. This presheaf is in fact a *sheaf of commutative algebras with unit*: this is the structure sheaf of M.

Let $\varphi : M \to N$ be a holomorphic mapping. By composition, any holomorphic function g on an open set V of N induces a holomorphic function $g \circ \varphi$ on the open set $\varphi^{-1}(V)$. Whence a morphism of sheaves of algebras

$$\varphi^{-1}\mathscr{O}_N \longrightarrow \mathscr{O}_M$$

which makes $\mathscr{O}_M$ a *sheaf of modules* on $\varphi^{-1}\mathscr{O}_N$.

2.6 The constant sheaf. Let G be a group and let $\widetilde{G}$ be the presheaf on M defined by $\widetilde{G}(U) = G$ for any open set U of M, all the restriction mappings ρ_{VU} being equal to the identity $\mathrm{Id}_G : G \to G$. This presheaf *is not* a sheaf: if U has two connected components U_1 and U_2, we have $\widetilde{G}(U) \neq \widetilde{G}(U_1) \times \widetilde{G}(U_2)$.

The associated sheaf is *the constant sheaf with fibre G*, that we denote by G_M.

2.7 Exercise (The sections of the constant sheaf). Check that, if U is connected, we have $\Gamma(U, G_M) = G$ and that, in general, $\Gamma(U, G_M) = G^I$ if I is the set of connected components of U. Show that the restriction mappings are the identity on each factor of the product.

3 Holomorphic vector bundle

Let $\pi : E \to M$ be a holomorphic mapping between two complex analytic manifolds. We will say that π is a *vector fibration* of rank d, or also that π makes E a *vector bundle of rank d on M* if there exist an open covering $\mathfrak{U}$ of M (called *trivializing covering for E*) and, for any open set U of $\mathfrak{U}$, a *linear chart* φ_U which is biholomorphic, which makes the diagram

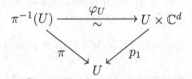

commute (p_1 denotes the first projection) and which satisfies the *linearity property*: for any pair (U, V) of open sets of the covering, the change of charts

$$\varphi_U \circ \varphi_V^{-1} : (U \cap V) \times \mathbb{C}^d \longrightarrow (U \cap V) \times \mathbb{C}^d$$

is *linear in the fibres of the first projection*, that is, is defined by a holomorphic mapping $\psi_{U,V} : U \cap V \to \mathrm{GL}_d(\mathbb{C})$, or also by a holomorphic section of the sheaf $\mathrm{GL}_d(\mathscr{O}_M)$ of germs of invertible matrices with holomorphic entries.

3.1 Exercise (The cocycle condition). Show that the mappings $\psi_{U,V}$ constructed in this way satisfy the following property (called the *cocycle condition*): for any triple U, V, W of open sets in $\mathfrak{U}$ and any $m \in U \cap V \cap W$,

$$(3.1)(*) \qquad \psi_{U,V}(m) \cdot \psi_{V,W}(m) \cdot \psi_{W,U}(m) = \mathrm{Id}_{\mathbb{C}^d}.$$

A *morphism* φ between two vector bundles E, E' on M is a holomorphic mapping from E to E' making the diagram

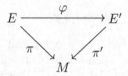

commute and inducing for any $m \in M$ a *linear* mapping $\varphi_m : \pi^{-1}(m) \to \pi'^{-1}(m)$. For instance, on a *trivializing open set* U for E, the linear chart φ_U considered above is an isomorphism from the restriction $E_{|U}$ to the *trivial bundle* of rank d. We denote by $\mathrm{L}(E, E')$ the space of morphisms from E to E'. It is a module over the ring $\Gamma(M, \mathcal{O}_M)$ of holomorphic functions on M.

3.2 Remark. The kernel and the cokernel of a morphism between vector bundles $\varphi : E \to E'$ are themselves vector bundles if and only if φ *has constant rank*. Therefore, the category of vector bundles on a manifold of dimension $\geqslant 1$ is not abelian[1]. Interpreting a vector bundle as a locally free sheaf of $\mathcal{O}_M$-modules, as we will do in paragraph 4, enables one to consider this category as a subcategory of the category of sheaves of $\mathcal{O}_M$-modules which *is* abelian: the kernel and the cokernel of φ do exist as sheaves of $\mathcal{O}_M$-modules, but these sheaves are possibly not be locally free.

3.3 Operations on vector bundles.

- Let $f : M \to N$ be a holomorphic mapping between two complex analytic manifolds and let $\pi : E \to N$ be a holomorphic vector bundle of rank d. The *fibre product*

$$E \underset{N}{\times} M = \{(e, m) \in E \times M \mid \pi(e) = f(m)\} \subset E \times M$$

equipped with the restriction of the projection to M is a vector bundle of rank d on M, denoted by f^*E, and called the *pullback of E by f* (consider the pullback of a covering of N trivializing for E and the corresponding linear charts).
When f is the inclusion of a point n in N, the pullback f^*E is a $\mathbb{C}$-vector space, called the *fibre of E at $n \in N$*.
When f is the inclusion of an open set U of N, the pullback of E to U is denoted by $E_{|U}$, as above.

---

[1] See §17 for elementary notions on categories.

- Let E (resp. E') be a vector bundle on an open set U (resp. U') of M. Giving an isomorphism of vector bundles $\varphi : E_{|U \cap U'} \to E'_{|U \cap U'}$ enables us to define a vector bundle F on $U \cup U'$ such that $F_{|U} \simeq E$ and $F_{|U'} \simeq E'$: the set F is that of equivalence classes with respect to the relation $\sim$ on the disjoint union $E \coprod E'$ such that $e \sim e'$ if $e' = \varphi(e)$ (the reader will check as an exercise that the projection to M makes F a holomorphic vector bundle).

- The usual operations on vector spaces can be extended to vector bundles: direct sum, tensor product, homomorphisms. Given two vector bundles E, E' of rank d, d' on M, define (exercise) the vector bundles $E \oplus E'$ (of rank $d + d'$), $E \otimes E'$ (of rank dd'), $\mathrm{L}(E, E')$ (of rank dd'), satisfying, for any $m \in M$,

$$(E \oplus E')_m = E_m \oplus E'_m$$
$$(E \otimes E')_m = E_m \otimes E'_m$$
$$\mathrm{L}(E, E')_m = \mathrm{L}(E_m, E'_m) \quad \text{(linear mappings)}.$$

4 Locally free sheaves of $\mathscr{O}_M$-modules

Let E be a vector bundle on M and let U be an open set of M. A *holomorphic section* of E on U is a holomorphic mapping $\sigma : U \to E$ which is a section of the projection π, i.e., which satisfies $\pi \circ \sigma = \mathrm{Id}_U$. The set $\mathscr{E}(U)$ of holomorphic sections is a module over $\mathscr{O}_M(U)$. The natural restriction mappings to open sets enables one to define a presheaf $\mathscr{E}$ on M, which is in fact a sheaf (exercise). It is a sheaf of $\mathscr{O}_M$-modules. Moreover, as E is locally isomorphic to a trivial bundle, this sheaf is locally isomorphic to the sheaf of holomorphic sections of the trivial bundle, which is nothing but $\mathscr{O}_M^d$: one says that $\mathscr{E}$ is *locally free of rank d on $\mathscr{O}_M$*.

We define in this way a *functor* from the category of vector bundles (the morphisms are the morphisms of vector bundles) to the category of locally free sheaves of $\mathscr{O}_M$-modules (a morphism between vector bundles gives rise, by composition, to a morphism between the sheaves of their holomorphic local sections).

4.1 Proposition (Equivalence between vector bundles and locally free sheaves). *The functor defined in this way is an equivalence[2] between the category of holomorphic vector bundles of rank d and that of locally free sheaves of $\mathscr{O}_M$-modules of rank d.*

Proof. Let us recall that a morphism $\varphi : \mathscr{F} \to \mathscr{G}$ of sheaves of $\mathscr{O}_M$-modules consists in giving, for any open set U of M, a $\mathscr{O}_M(U)$-linear mapping

---

[2] See §17 for elementary notions on categories.

$\Gamma(U, \mathscr{F}) \to \Gamma(U, \mathscr{G})$ such that, for any pair of open sets $V \subset U$, the restriction diagram commutes:

$$
\begin{array}{ccc}
\Gamma(U, \mathscr{F}) & \xrightarrow{\;\varphi(U)\;} & \Gamma(U, \mathscr{G}) \\
\downarrow & & \downarrow \\
\Gamma(V, \mathscr{F}) & \xrightarrow[\;\varphi(V)\;]{} & \Gamma(V, \mathscr{G})
\end{array}
$$

We denote by $\operatorname{Hom}_{\mathscr{O}_M}(\mathscr{F}, \mathscr{G})$ the space of $\mathscr{O}_M$-linear morphisms from $\mathscr{F}$ to $\mathscr{G}$. This is a module over the ring $\Gamma(M, \mathscr{O}_M)$ of holomorphic functions on M (multiplication by a holomorphic function).

4.2 Remark. If $\mathscr{F}$ is a sheaf of $\mathscr{O}_M$-modules, a section $\sigma \in \Gamma(U, \mathscr{F})$ of $\mathscr{F}$ on an open set U is nothing but a homomorphism $\varphi : \mathscr{O}_U \to \mathscr{F}_{|U}$; as such a homomorphism is determined by the value $\sigma = \varphi(\mathbf{1}) \in \Gamma(U, \mathscr{F})$ of the section $\mathbf{1} \in \Gamma(U, \mathscr{O}_M)$. We thus have a canonical isomorphism $\operatorname{Hom}_{\mathscr{O}_U}(\mathscr{O}_U, \mathscr{F}_{|U}) \xrightarrow{\;\sim\;} \Gamma(U, \mathscr{F})$.

Let us come back to the proposition. So, let E and E' be two holomorphic vector bundles on M and let $\mathscr{E}, \mathscr{E}'$ be the associated sheaves. Let us first show that the natural mapping

(4.3) $$L(E, E') \longrightarrow \operatorname{Hom}_{\mathscr{O}_M}(\mathscr{E}, \mathscr{E}')$$

is an isomorphism of $\Gamma(M, \mathscr{O}_M)$-modules.

Injectivity. Let $f : E \to E'$ be a morphism of vector bundles, whose image in $\operatorname{Hom}_{\mathscr{O}_M}(\mathscr{E}, \mathscr{E}')$ is zero. Let us pick $m \in M$ and let U be an open neighbourhood of m, which is trivializing for E and E'. We thus have $E_{|U} \simeq U \times \mathbb{C}^d$ and $E'_{|U} \simeq U \times \mathbb{C}^{d'}$. For any section $\sigma : U \to U \times \mathbb{C}^d$, the composition $f \circ \sigma : U \to U \times \mathbb{C}^{d'}$ vanishes. We deduce that the image by $f(m)$ of each vector of the canonical basis of $\mathbb{C}^d$ is zero, hence $f(m) : E_m \to E'_m$ is the zero linear map.

Surjectivity. Assume first that E and E' are trivializable, hence that $\mathscr{E}$ and $\mathscr{E}'$ are free ($\mathscr{E} \simeq \mathscr{O}_M^d$ and $\mathscr{E}' \simeq \mathscr{O}_M^{d'}$). A morphism $\varphi : \mathscr{O}_M^d \to \mathscr{O}_M^{d'}$ can be expressed, in the canonical bases, as a matrix (φ_{ij}) of size $d' \times d$, where $\varphi_{ij} : \mathscr{O}_M \to \mathscr{O}_M$ is a morphism. The sheaf $\mathscr{O}_M$ admits a canonical section, namely, the unit section $\mathbf{1}$, and $\varphi_{ij}(\mathbf{1})$ is a holomorphic function on M. Conversely, it is clear by linearity that $\varphi_{ij}(\mathbf{1})$ completely determines φ_{ij}. Therefore, giving the matrix $(\varphi_{ij}(\mathbf{1}))$ is equivalent to giving φ. This matrix defines a morphism between the trivial bundles

$$
\begin{aligned}
M \times \mathbb{C}^d & \longrightarrow M \times \mathbb{C}^{d'} \\
(m, v) & \longmapsto (m, \varphi(\mathbf{1})(m) \cdot v).
\end{aligned}
$$

In general, $\mathscr{E}$ and $\mathscr{E}'$ are locally free. If $\varphi : \mathscr{E} \to \mathscr{E}'$ is a morphism, one constructs a morphism of bundles $E_{|U} \to E'_{|U}$ after restricting to any trivializing open set. The injectivity shown above implies that these morphisms can be glued in a morphism $E \to E'$.

Essential surjectivity of the functor. It is a matter of verifying that any locally free $\mathscr{O}_M$-module $\mathscr{E}$ is isomorphic to the image of a bundle E by the functor. One first reconstructs the bundle E fibre by fibre from the sheaf $\mathscr{E}$, then one equips the disjoint union of fibres with the structure of an analytic manifold.

So, pick $m \in M$ and let $\mathscr{O}_{M,m}$ be the ring of germs at m of holomorphic functions on M. If $(z_1, \ldots, z_n)$ are holomorphic local coordinates for which m is the origin of coordinates, this ring is nothing but the ring of converging series $\mathbb{C}\{z_1, \ldots, z_n\}$. Let $\mathfrak{m}$ be the ideal of germs which vanish at m. If $\mathscr{E}_m$ denotes the germ of $\mathscr{E}$ at m, we set $E_m = \mathscr{E}_m / \mathfrak{m}\mathscr{E}_m$. This is a vector space of dimension d. We take for E the disjoint union of E_m, equipped with its natural projection to M. One can endow E with the structure of a complex analytic manifold by considering a trivializing covering for $\mathscr{E}$ (exercise: give details). $\square$

4.4 Operations on locally free $\mathscr{O}_M$-modules. We describe below the operations corresponding to that on vector bundles, seen in §3.3. These operations are more generally defined for sheaves of (not necessarily locally free) $\mathscr{O}_M$-modules. When the data are locally free, the result is so. The reader will check as an exercise the correspondence with §3.3.

- Let $f : M \to N$ be a holomorphic mapping and let $\mathscr{F}$ be a sheaf of $\mathscr{O}_N$-modules. Let $f^{-1}\mathscr{F}$ be the pullback sheaf (in the topological sense of sheaf theory; see for instance [God64, p. 121], where this operation is denoted by f^*). We see in particular that $f^{-1}\mathscr{O}_N$ is a sheaf of rings on M and that $\mathscr{O}_M$ is also a sheaf of $f^{-1}\mathscr{O}_N$-modules. The *pullback* as sheaf of $\mathscr{O}$-modules is

$$f^*\mathscr{F} := \mathscr{O}_M \underset{f^{-1}\mathscr{O}_N}{\otimes} f^{-1}\mathscr{F}$$

so that, by definition, we have $f^*\mathscr{O}_N = \mathscr{O}_M$. The operation f^* transforms a sheaf of $\mathscr{O}_N$-modules into a sheaf of $\mathscr{O}_M$-modules.
- Let $\mathscr{E}$ (resp. $\mathscr{E}'$) be a sheaf on an open set U (resp. U') of M. Giving an isomorphism of sheaves $\varphi : \mathscr{E}_{|U\cap U'} \xrightarrow{\sim} \mathscr{E}'_{|U\cap U'}$ enables one to define a sheaf $\mathscr{F}$ on $U\cup U'$ such that $\mathscr{F}_{|U} \simeq \mathscr{E}$ and $\mathscr{F}_{|U'} \simeq \mathscr{E}'$. Hence, if $\mathscr{E}$ (resp. $\mathscr{E}'$) is locally free on $\mathscr{O}_U$ (resp. $\mathscr{O}_{U'}$), the sheaf $\mathscr{F}$ is also $\mathscr{O}_{U\cup U'}$-locally free.
- The natural operations on modules can be extended to sheaves of $\mathscr{O}_M$-modules: direct sum, tensor product, homomorphisms.
- Let us recall the construction of the sheaf $\mathscr{H}om_{\mathscr{O}_M}(\mathscr{F}, \mathscr{G})$: we set, for any open set U of M,

$$\Gamma(U, \mathscr{H}om_{\mathscr{O}_M}(\mathscr{F}, \mathscr{G})) := \mathrm{Hom}_{\mathscr{O}_U}(\mathscr{F}_{|U}, \mathscr{G}_{|U})$$

where Hom has been defined in the proof of Proposition 4.1. One can check that there exist restriction mappings and that the presheaf obtained in this way is a sheaf. When $\mathscr{F}$ and $\mathscr{G}$ are locally free of ranks d and d' respectively, the sheaf $\mathscr{H}om_{\mathscr{O}_M}(\mathscr{F},\mathscr{G})$ is locally free of rank dd'.

One should be careful (see for instance [GM93, p. 104]) that in general the construction $\mathscr{H}om$ does not commute with taking germs in sheaf theory, but that, if $\mathscr{F},\mathscr{G}$ are locally free sheaves of $\mathscr{O}_M$-modules, we have, for any $m \in M$,

$$\mathscr{H}om_{\mathscr{O}_M}(\mathscr{F},\mathscr{G})_m = \mathscr{H}om_{\mathscr{O}_{M,m}}(\mathscr{F}_m,\mathscr{G}_m)$$

(this is a local property, so that, to prove it, one can assume $\mathscr{F} = \mathscr{O}_M^d$ and $\mathscr{G} = \mathscr{O}_M^{d'}$, where the result is clear).

4.5 Exercise (Subsheaves and subbundles). Prove that a homomorphism

$$\varphi : \mathscr{F} \longrightarrow \mathscr{G}$$

of locally free sheaves comes from an *injective* homomorphism between the corresponding bundles if and only if it is injective and if the quotient sheaf $\mathscr{G}/\varphi(\mathscr{F})$ is also locally free. We will then say that $\mathscr{F}$ is a subbundle of the locally free sheaf $\mathscr{G}$. Check that the homomorphism "multiplication by z" from $\mathscr{O}_{\mathbb{C}}$ into itself does not come from an injective morphism from the trivial bundle of rank one to itself.

5 Nonabelian cohomology

We refer to [Fre57] for the elementary properties of nonabelian cohomology (see also, for instance, [BV89a, Chap. II.1, p. 110–123]).

Let $\mathscr{G}$ be a sheaf of groups (not necessarily commutative). For instance, we will have to consider the sheaf $\mathrm{GL}_d(\mathscr{O}_M)$ whose sections on an open set of M are the invertible matrices of holomorphic functions on this open set.

Let $\mathfrak{U}$ be an open covering of M.

A 1-*cocycle of* $\mathfrak{U}$ *with values in* $\mathscr{G}$ consists in giving, for any pair (U,V) of open sets of $\mathfrak{U}$, of an element $\psi_{U,V}$ of $\mathscr{G}(U \cap V)$ (for instance, in the case of $\mathrm{GL}_d(\mathscr{O}_M)$, of a holomorphic mapping $\psi_{U,V} : U \cap V \to \mathrm{GL}_d(\mathbb{C})$) satisfying the cocycle condition (3.1)(*). The set of 1-cocycles is denoted by $Z^1(\mathfrak{U},\mathscr{G})$.

Exercise. Prove that, if ψ is a 1-cocycle, then $\psi_{V,U} = (\psi_{U,V})^{-1}$.

A *coboundary* is a 1-cocycle ψ for which there exists, for any $U \in \mathfrak{U}$, an element $\eta_U \in \mathscr{G}(U)$ satisfying $\psi_{U,V} = \eta_U \cdot \eta_V^{-1}$ for all $U,V \in \mathfrak{U}$ (one checks in a straightforward way that ψ so defined satisfies the cocycle condition). We say that η is a 0-*cochain* of $\mathfrak{U}$. Let us note that, if ψ is a coboundary, then so is the 1-cocycle ψ^{-1} defined by $\psi_{U,V}^{-1}(m) = (\psi_{U,V}(m))^{-1}$ for any $m \in M$.

Two cocycles ψ and ψ' of $\mathfrak{U}$ with values in $\mathscr{G}$ are *equivalent*[3] if there exists a coboundary η such that $\psi'_{U,V} = \eta_U \cdot \psi_{U,V} \cdot \eta_V^{-1}$ for any pair (U,V) of open sets of $\mathfrak{U}$.

---

[3] Check that this defines an equivalence relation.

The first *set of nonabelian cohomology* $H^1(\mathfrak{U}, \mathscr{G})$ is the quotient set of the set of cocycles by this equivalence relation. It is a set equipped with a distinguished element: the class of the "identity" cocycle. Nevertheless, it has no other structure in general, due to the possible noncommutativity of the sheaf $\mathscr{G}$.

5.1 Exercise (Some properties of $H^1(\mathfrak{U}, \mathscr{G})$).

(1) When $d = 1$, prove that $\mathrm{GL}_1(\mathcal{O}) = \mathcal{O}^*$ (the sheaf of holomorphic functions which do not vanish) and that $H^1(\mathfrak{U}, \mathcal{O}^*)$ is an abelian group (with respect to multiplication).

(2) A *refinement* $\mathfrak{V}$ of $\mathfrak{U}$ is an open covering $(V_j)_{j \in J}$ finer than $\mathfrak{U}$, i.e., such that any V_j is contained in some U_i. Then there exists at least one mapping $\alpha : J \to I$ such that, for any $j \in J$, we have $V_j \subset U_{\alpha(j)}$. Show that, if $\psi = (\psi_{U_i, U_{i'}})$, also denoted by $(\psi_{i,i'})$, is a 1-cocycle of $\mathfrak{U}$ with values in $\mathscr{G}$, then ψ^α defined by $\psi^\alpha_{j,j'} = \psi_{\alpha(j),\alpha(j')}|_{V_j \cap V_{j'}}$ is a 1-cocycle of $\mathfrak{V}$.

(3) Prove that, if $\alpha, \beta : J \to I$ are two such mappings, the cocycles ψ^α and ψ^β are equivalent *via* the 0-cochain η defined by $\eta_j = \psi_{\beta(j),\alpha(j)}$.

(4) Conclude that there exists a refinement mapping independent of the choice of α
$$\rho : H^1(\mathfrak{U}, \mathscr{G}) \longrightarrow H^1(\mathfrak{V}, \mathscr{G}).$$

(5) Let $\psi, \psi' \in Z^1(\mathfrak{U}, \mathscr{G})$ be two 1-cocycles whose refinements ψ^α and ψ'^α are equivalent *via* some 0-cochain of $\mathfrak{V}$. Show that ψ and ψ' are equivalent *via* some 0-cochain of $\mathfrak{U}$.

(6) Conclude that the refinement mapping ρ is *injective*.

(7) Denote by $H^1(M, \mathscr{G})$ the limit of the inductive system $H^1(\mathfrak{U}, \mathscr{G})$ constructed in such a way, indexed by all the open coverings of M. Prove that the natural mappings
$$H^1(\mathfrak{U}, \mathscr{G}) \longrightarrow H^1(M, \mathscr{G})$$

are *injective* and deduce that, if one identifies the set $H^1(\mathfrak{U}, \mathscr{G})$ to its image in $H^1(M, \mathscr{G})$, the latter set is nothing but the *union* of the $H^1(\mathfrak{U}, \mathscr{G})$.

5.2 Example (Pullback and cohomology).

(1) Let $\widetilde{\mathscr{F}_i}$ be an increasing sequence of presheaves on a topological space X and let us set $\widetilde{\mathscr{F}} = \bigcup_i \widetilde{\mathscr{F}_i}$. If $\mathscr{F}_i$ denotes the sheaf associated to $\widetilde{\mathscr{F}_i}$ and $\mathscr{F}$ that associated to $\mathscr{F}$, then the sequence of sheaves $\mathscr{F}_i$ is increasing and $\mathscr{F} = \bigcup_i \mathscr{F}_i$ (in other words, for any $x \in X$, $\mathscr{F}_x = \bigcup_i \mathscr{F}_{i,x}$). Indeed, let us pick $x \in X$. It is a matter of showing that
$$\varinjlim_{V \ni x} \bigcup_i \widetilde{\mathscr{F}_i}(V) = \bigcup_i \varinjlim_{V \ni x} \widetilde{\mathscr{F}_i}(V).$$

The inclusion $\supset$ is clear. If $s \in \widetilde{\mathscr{F}_x}$, then s consists of a compatible family $s_W \in \widetilde{\mathscr{F}}(W)$ for any W contained in a sufficiently small open set V. There exists thus j such that $s_V \in \widetilde{\mathscr{F}}_j(V)$. As a consequence, for any $W \subset V$, $s_W = \rho_{V,W}(s_V) \in \widetilde{\mathscr{F}}_j(W)$, and thus $s \in \varinjlim_{V \ni x} \widetilde{\mathscr{F}}_j(V)$.

(2) Let X and S be two topological spaces, X being locally compact, and let $\mathscr{H}$ be a sheaf on $S \times X$. Then, both presheaves

$$\widetilde{\mathscr{F}}_1 : U \longmapsto H^1(S \times U, \mathscr{H}) \quad \text{and} \quad \widetilde{\mathscr{F}}_2 : U \longmapsto \bigcup_{\mathscr{X}} H^1(\mathscr{X}_{|S \times U}, \mathscr{H})$$

(where the union is taken over all open coverings $\mathscr{X}$ of $S \times X$) have the same associated sheaf. Indeed, $\widetilde{\mathscr{F}}_2(U) \subset \widetilde{\mathscr{F}}_1(U)$ is clear. If s is a section of $\mathscr{F}_{1,x}$, then s comes from $s_{\mathscr{U}} \in H^1(\mathscr{U}, \mathscr{H})$ for some open covering $\mathscr{U}$ of some open set $S \times U$ with U small enough containing x. If V is relatively compact in U and contains x, there exists $\mathscr{X}$ such that $\mathscr{X}_{|S \times V} = \mathscr{U}_{|S \times V}$. As a consequence, for any $W \subset V$, $s_W \in H^1(\mathscr{X}_{|S \times W}, \mathscr{H})$, and thus $s \in \mathscr{F}_{2,x}$.

(3) Let $\mathscr{G}$ be a sheaf on S and let $p_1^{-1}\mathscr{G}$ be its pullback by the projection $p_1 : S \times X \to S$. Let us assume that X is connected. Then there is a bijection

$$\Gamma(S, \mathscr{G}) \xrightarrow{\sim} \Gamma(S \times X, p_1^{-1}\mathscr{G})$$

(one can use that the étale cover $q : \widetilde{p_1^{-1}\mathscr{G}} \to S \times X$ is the product $p \times \mathrm{Id} : \widetilde{\mathscr{G}} \times X \to S \times X$; see §15.a for the notion of étale cover).

(4) Let us now assume that S is compact and X is locally compact and locally connected (for instance, an open set in $\mathbb{C}^n$). Then, for any $x^o \in X$, the germ at x^o of the sheaf $\mathscr{F}$ associated to the presheaf $U \mapsto H^1(S \times U, p_1^{-1}\mathscr{G})$ is equal to $H^1(S, \mathscr{G})$.

In order to prove this statement, one first shows that, for any open covering $\mathscr{X}$ of $S \times X$, there exists a *finite* open covering $\mathscr{S}$ of S, a refinement $\mathscr{V}$ of $\mathscr{X}$ and an open neighbourhood V of x^o in X such that $\mathscr{V}_{|S \times V} = \mathscr{S} \times V$. Indeed, by compactness of S, if K is a compact neighbourhood of x^o, there exists a refinement $\mathscr{X}'$ of $\mathscr{X}$ which is finite when restricted to $S \times K$ and such that each open set in $\mathscr{X}'$ which cuts $S \times K$ takes the form $S_j \times V_i$. Let us denote by V' the intersection of the V_i's which contain x^o, and let us choose as V an open neighbourhood of x^o relatively compact in V'. One now refines $\mathscr{X}'$ by taking the $S_j \times V''$s if i is such that V_i cuts $S \times \overline{V}$, and by taking the open sets in $\mathscr{X}'$ which do not cut $S \times \overline{V}$.

According to (2), the sheaf $\mathscr{F}$ is the sheaf associated to the presheaf $\bigcup_{\mathscr{X}} \widetilde{\mathscr{F}}_{\mathscr{X}}$ and thus to the presheaf $\bigcup_{\mathscr{V}} \widetilde{\mathscr{F}}_{\mathscr{V}}$, with $\widetilde{\mathscr{F}}_{\mathscr{V}}(U) = H^1(\mathscr{V}_{|S \times U}, p_1^{-1}\mathscr{G})$. On the other hand, by considering the Čech complexes and according to (3), $\widetilde{\mathscr{F}}_{\mathscr{V}}(W) = H^1(\mathscr{S}, \mathscr{G})$ for any $W \subset V$, and the restriction morphisms are reduced to identity. In particular, the sheaf $\mathscr{F}_{\mathscr{V}}$ associated to $\widetilde{\mathscr{F}}_{\mathscr{V}}$ is constant with fibre $H^1(\mathscr{S}, \mathscr{G})$ when restricted to V. Applying (1) to the increasing sequence $\widetilde{\mathscr{F}}_{\mathscr{V}}$ gives the assertion.

(5) With the same assumptions (S compact and X open in $\mathbb{C}^n$), we moreover assume that there exists a *finite* open covering $\mathscr{S}$ such that $H^1(\mathscr{S}, \mathscr{G}) =$

$H^1(S, \mathscr{G})$ (we will say that such an open covering is *good*; any refinement of a good open covering is still good). Then the sheaf $\mathscr{F}$ associated to the presheaf $U \mapsto H^1(S \times U, p_1^{-1}\mathscr{G})$ is locally constant, with germ $H^1(S, \mathscr{G})$. One can indeed adapt the previous reasoning with any compact set K instead of x^o. For any $\mathscr{X}$, one can find a refinement $\mathscr{V}$ such that the open sets in $\mathscr{V}$ which cut $S \times K$ take the form $S_j \times V_i$, where the finite family of V_i covers K and the family S_j is finite, covers S and is a good covering relative to $\mathscr{G}$. Any x in K has then an open neighbourhood V (intersection of the V_i's which contain x) such that $\mathscr{F}_{\mathscr{V}|V}$ is the constant sheaf with germ $H^1(\mathscr{S}, \mathscr{G}) = H^1(S, \mathscr{G})$. We get the assertion by applying (1).

5.3 Proposition (The cocycle of a vector bundle).

(1) *Let E be a vector bundle on M. If $\mathfrak{U}$ is a covering of M trivializing for E and ψ is the corresponding cocycle, the image of the class $[\psi] \in H^1(\mathfrak{U}, \mathrm{GL}_d(\mathscr{O}_M))$ in $H^1(M, \mathrm{GL}_d(\mathscr{O}_M))$ only depends on E (and not on the choice of $\mathfrak{U}$ and of ψ), and two bundles are isomorphic if and only if they define the same element.*
(2) *Conversely, any element of $H^1(M, \mathrm{GL}_d(\mathscr{O}_M))$ can be obtained in this way from a holomorphic bundle of rank d on M, unique up to isomorphism.*

Proof. Let us first show that two cocycles ψ, ψ' defined by two bundles E, E' for the same trivializing covering $\mathfrak{U}$ are equivalent if and only if the bundles are isomorphic. Assume that there exists some 0-cochain η such that, for all $U, V \in \mathfrak{U}$, $\psi'_{U,V} = \eta_U \psi_{U,V} \eta_V^{-1}$. In particular, η defines, for any $U \in \mathfrak{U}$, an isomorphism $U \times \mathbb{C}^d \to U \times \mathbb{C}^d$. By restricting to U, one defines an isomorphism σ_U by asking that the diagram below commutes:

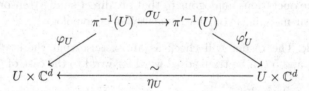

On $U \cap V$ we have

$$\varphi'_V \varphi'^{-1}_U = \psi'_{U,V} = \eta_U \psi_{U,V} \eta_V^{-1} = \eta_U \varphi_V \varphi_U^{-1} \eta_V^{-1}$$

and hence σ_U and σ_V coincide. These isomorphisms can then be glued together to produce an isomorphism $E \xrightarrow{\sim} E'$.

Conversely, if an isomorphism $E \xrightarrow{\sim} E'$ is given, one constructs, with the same formulas, a 0-cochain η which makes ψ and ψ' equivalent.

The reader will check as an exercise the independence with respect to the covering.

Let us now show the second point. Any cohomology class in the set $H^1(M, \mathrm{GL}_d(\mathscr{O}_M))$ also belongs to $H^1(\mathfrak{U}, \mathrm{GL}_d(\mathscr{O}_M))$ for a suitable covering $\mathfrak{U}$

("injectivity" in Exercise 5.1). Let us choose a cocycle ψ which represents it. We then construct a bundle E as the quotient space of the disjoint union

$$\coprod_{U \in \mathfrak{U}} (U \times \mathbb{C}^d)$$

by the equivalence relation which identifies $(m, v) \in U \times \mathbb{C}^d$ with $(n, w) \in V \times \mathbb{C}^d$ if $m = n \in U \cap V$ and $w = \psi_{U,V}(v)$. One can check that the set obtained in this way comes equipped with a projection to M (induced by the first projection on each term $U \times \mathbb{C}^d$) which makes it a holomorphic bundle with ψ as an associated cocycle. $\qquad\qquad\square$

5.4 Remark. The isomorphism class of the bundles corresponding to the element Id of the set $H^1(M, \mathrm{GL}_d(\mathscr{O}_M))$ is that of the trivial bundle

$$\pi : M \times \mathbb{C}^d \longrightarrow M.$$

5.5 Exercise (Operations on cocycles).

- Prove that a holomorphic mapping $f : M \to N$ gives rise to mappings

$$f^* : H^1(\mathfrak{U}, \mathrm{GL}_d(\mathscr{O}_N)) \longrightarrow H^1(f^{-1}\mathfrak{U}, \mathrm{GL}_d(\mathscr{O}_M))$$

(here, $\mathfrak{U}$ is a covering of N and $f^{-1}\mathfrak{U}$ is the covering of M consisting of the $f^{-1}(U)$, for $U \in \mathfrak{U}$), and

$$f^* : H^1(N, \mathrm{GL}_d(\mathscr{O}_N)) \longrightarrow H^1(M, \mathrm{GL}_d(\mathscr{O}_M))$$

passing to the limit.
- Define the operations analogous to that of "direct sum", "tensor product" and "linear mappings" on the 1-nonabelian cohomology.

5.6 Remark. The reader will check as an exercise that the statements of Sections 3, 4 and 5 can be transposed word for word to the case of C^∞ vector bundles.

6 Čech cohomology

Let $\mathscr{F}$ be a sheaf of *abelian* groups (additive notation), for instance a sheaf of $\mathscr{O}_M$-modules. The construction of 1-cocycles can be generalized, as well as that of coboundaries, and enables one to construct the *Čech cohomology* groups with coefficients in the sheaf $\mathscr{F}$.

Let then $\mathfrak{U} = (U_i)_{i \in I}$ be an open covering of M. The group $C^k(\mathfrak{U}, \mathscr{F})$ of k-cochains is the product of the groups $\Gamma(U_{i_0} \cap \cdots \cap U_{i_k}, \mathscr{F})$ taken over all the families of elements of $\mathfrak{U}$ indexed by an ordered subset of I having cardinal $k + 1$. A k-cochain can thus be written as $\sigma = (\sigma_J)_{J \prec I, \#J=k+1}$ (here, $J \prec I$ means that J is an *ordered* subset) with $\sigma_J \in \Gamma(\bigcap_{i \in J} U_i, \mathscr{F})$.

The *coboundary* operator $\delta_k : C^k(\mathfrak{U}, \mathscr{F}) \to C^{k+1}(\mathfrak{U}, \mathscr{F})$ is defined by the formula below: for $K \prec I$ having cardinal $k + 2$, we set $U_K = \bigcap_{j \in K} U_j$ and

$$(\delta_k \sigma)_K = \sum_{\ell=0}^{k+1} (-1)^\ell \sigma_{K \smallsetminus \{i_\ell\} | U_K}.$$

One checks that $\delta_{k+1} \circ \delta_k : C^k(\mathfrak{U}, \mathscr{F}) \to C^{k+2}(\mathfrak{U}, \mathscr{F})$ is the zero mapping. We set

$$H^k(\mathfrak{U}, \mathscr{F}) := \operatorname{Ker} \delta_k / \operatorname{Im} \delta_{k-1}.$$

When $\mathfrak{V}$ is a refinement of $\mathfrak{U}$, we have a natural mapping $H^k(\mathfrak{V}, \mathscr{F}) \to H^k(\mathfrak{U}, \mathscr{F})$.

The group $H^k(M, \mathscr{F})$ is defined as the limit of the inductive system above, indexed by all the open coverings of M.

One can readily verify that $H^0(\mathfrak{U}, \mathscr{F}) = \Gamma(M, \mathscr{F})$ for any open covering $\mathfrak{U}$: this is another way to express the property of unique gluing of sections for a sheaf.

The reader should take care that, for $k \geqslant 2$, the refinement mappings are not necessarily injective. Nevertheless, we have:

6.1 Leray Theorem. *If $\mathfrak{U}$ is an acyclic covering for the sheaf $\mathscr{F}$, that is, if $H^k(U_J, \mathscr{F}) = 0$ for any $k \geqslant 1$ and any $J \subset I$ having a finite cardinal, then the natural mapping $H^k(\mathfrak{U}, \mathscr{F}) \to H^k(M, \mathscr{F})$ is an isomorphism for any $k \geqslant 0$.*

Proof. See for instance [GR65, p. 189], [GH78, p. 40] or [KS90, p. 125]. □

It is much more convenient to use the formalism of homological algebra, in particular the existence of a canonical flabby resolution, to define sheaf cohomology (see [God64, p. 164] or [KS90, Chap. 2]). The isomorphism theorem between the cohomology groups and the corresponding Čech cohomology groups (see for instance [God64, p. 228]) enables one to identify both kinds of cohomology.

For the computations, the following is very useful:

6.2 Theorem (The long exact sequence). *A short exact sequence of sheaves*

$$0 \longrightarrow \mathscr{F}' \longrightarrow \mathscr{F} \longrightarrow \mathscr{F}'' \longrightarrow 0$$

gives rise to a long exact sequence in cohomology

$$0 \to H^0(M, \mathscr{F}') \to H^0(M, \mathscr{F}) \to H^0(M, \mathscr{F}'') \to H^1(M, \mathscr{F}') \to \cdots$$
$$\cdots \to H^k(M, \mathscr{F}') \to H^k(M, \mathscr{F}) \to H^k(M, \mathscr{F}'') \to H^{k+1}(M, \mathscr{F}') \to \cdots$$

Proof. See for instance [GH78, p. 40] or [God64, p. 224]. □

The existence of C^∞ partitions of unity adapted to a locally finite open covering enables one to obtain, for the sheaf of C^∞ functions on a manifold:

6.3 Theorem (The $\mathscr{C}^\infty$ sheaf has no cohomology). *For any $k \geqslant 1$, the cohomology spaces $H^k(M, \mathscr{C}_M^\infty)$ are zero.*

Proof. See for instance [GH78, p. 42]. □

7 Line bundles

According to §5, the isomorphism classes of holomorphic bundles of rank one on an analytic manifold M are in one-to-one correspondence with the elements of $H^1(M, \mathrm{GL}_1(\mathscr{O}_M))$. As the group $\mathrm{GL}_1(\mathscr{O}_M) = \mathscr{O}_M^*$ (the sections on an open set are the holomorphic functions nowhere vanishing on this open set) is abelian, this set is a (abelian) group. The product corresponds to the tensor product of line bundles (or of locally free sheaves of rank one). Any such sheaf $\mathscr{L}$ admits then an inverse element for the tensor product: if ψ is a 1-cocycle associated to this line bundle, $\psi_{U,V}$ is a holomorphic function nowhere vanishing on $U \cap V$, and the function $1/\psi_{U,V}$ gives the 1-cocycle for the inverse bundle.

7.1 Exercise. Show that the inverse element of the sheaf $\mathscr{L}$ is isomorphic to its dual sheaf, namely $\mathscr{H}om_{\mathscr{O}_M}(\mathscr{L}, \mathscr{O}_M)$.

7.a The exponential as a sheaf morphism

The mapping $z \mapsto \exp(2i\pi z)$ enables one to define, by composition, a *surjective* homomorphism of sheaves

$$(7.2) \qquad \begin{aligned} \mathscr{O}_M &\longrightarrow \mathscr{O}_M^* \\ f &\longmapsto \exp(2i\pi f) \end{aligned}$$

Indeed, let us recall that $\Gamma(U, \mathscr{O}_M^*)$ is the set of holomorphic functions on U which vanish nowhere on U and the germ $\mathscr{O}_{M,m}^*$ is the set of germs at m of holomorphic functions which do not vanish at m. Saying that the morphism above is onto is equivalent to saying that, for any m, the morphism $\mathscr{O}_{M,m} \to \mathscr{O}_{M,m}^*$ is so. Therefore, let us pick $m^o \in M$ and $g \in \mathscr{O}_{M,m^o}^*$. There exists a neighbourhood U of m^o on which g is defined and the image of which is contained in $\mathbb{C} \smallsetminus \Delta$, where Δ is the real half-line starting from 0 and passing by $ig(m^o)$. There exists on this domain a function "logarithm" and one defines $f : U \to \mathbb{C}$ by $f(m) = \dfrac{\log g(m)}{2i\pi}$. The germ of f at m^o is a preimage of g.   □

7.b Exact sequence of the exponential and Chern class

If U is an open set of M, a holomorphic function $f : U \to \mathbb{C}$ satisfies $\exp 2i\pi f \equiv 1$ if and only if f is locally constant with integral values on U, in other words, if and only if f is a section of the constant sheaf $\mathbb{Z}_M$ on U. We deduce that there is an exact sequence of sheaves of abelian groups (where one uses additive notation for $\mathbb{Z}_M$ and $\mathscr{O}_M$, and multiplicative notation for $\mathscr{O}_M^*$):

$$0 \longrightarrow \mathbb{Z}_M \longrightarrow \mathscr{O}_M \xrightarrow{\ \exp 2i\pi \cdot\ } \mathscr{O}_M^* \longrightarrow 1.$$

Theorem 6.2 shows the existence of a homomorphism of abelian groups

$$H^1(M, \mathscr{O}_M^*) \longrightarrow H^2(M, \mathbb{Z}_M).$$

If L is a holomorphic line bundle on M and if $[\psi] \in H^1(M, \mathscr{O}_M^*)$ denotes the class of a cocycle associated to L, we denote by $c_1(L)$ the image of $[\psi]$ in $H^2(M, \mathbb{Z}_M)$: this is the *Chern class of the bundle L*.

Two isomorphic line bundles have the same Chern class, but the converse can be wrong (however we shall see in §I.2 that it is true when M is the Riemann sphere $\mathbb{P}^1(\mathbb{C})$).

7.3 Exercise (The Chern class is functorial).

(1) Let L and L' be two line bundles on M. Prove that $c_1(L \otimes L') = c_1(L) + c_1(L')$.
(2) Show that, if $f : M \to N$ is a holomorphic mapping and L is a line bundle on N, we have $c_1(f^*L) = f^*c_1(L)$.

7.4 Exercise (Chern class of a C^∞ bundle).

(1) Show that we have an exact sequence

$$0 \longrightarrow \mathbb{Z}_M \longrightarrow \mathscr{C}_M^\infty \xrightarrow{\;\exp 2i\pi\cdot\;} \mathscr{C}_M^{\infty*} \longrightarrow 0.$$

(2) Prove that a C^∞ line bundle on M is determined (up to isomorphism) by its Chern class.
(3) Show that the Chern class of a holomorphic line bundle is equal to that of the underlying C^∞ bundle.

8 Meromorphic bundles, lattices

Let Z be a smooth hypersurface in a complex analytic manifold M (see §2). If U is an open set of M, the intersection $Z \cap U$ is still a smooth hypersurface of U. A holomorphic function on $U \smallsetminus Z \cap U$ is said to be *meromorphic along* $Z \cap U$ if, for any chart V of M contained in U, in which $Z \cap V$ is defined by the vanishing of some coordinate, z_1 for instance, there exists an integer m such that $z_1^m f(z_1, \ldots, z_n)$ is locally bounded in the neighbourhood of any point of $Z \cap V$, that is, can be extended as a holomorphic function on V.

One defines in this way a presheaf on M, which is a sheaf indeed, denoted by $\mathscr{O}_M(*Z)$. It contains $\mathscr{O}_M$ as a subsheaf. It also contains the subsheaf $\mathscr{O}_M(kZ)$ ($k \in \mathbb{Z}$), locally written in a chart as above as the subsheaf of meromorphic functions f such that $z_1^k f$ is holomorphic (i.e., f has a pole of order at most k if $k > 0$ and f has a zero of order at least $-k$, if $k \leqslant 0$, along Z).

8.1 Lemma. *The subsheaves $\mathscr{O}_M(kZ)$ are locally free of rank one on $\mathscr{O}_M$.* $\square$

8.2 Remark. If Z has many connected components $Z_1, \ldots, Z_p$, one can, in an analogous way, define the subsheaves $\mathscr{O}_M(k_1 Z_1 + \cdots + k_p Z_p)$, which are also locally free of rank one. If $\mathscr{E}$ is a bundle, one sets $\mathscr{E}(kZ) = \mathscr{E} \otimes_{\mathscr{O}_M} \mathscr{O}_M(kZ)$ and one defines similarly $\mathscr{E}(k_1 Z_1 + \cdots + k_p Z_p)$.

8.3 Definition (Meromorphic bundles, lattices). A *meromorphic bundle on M with poles along Z* is a locally free sheaf of $\mathscr{O}_M(*Z)$-modules of finite rank. A *lattice* of this meromorphic bundle is a locally free $\mathscr{O}_M$-submodule of this meromorphic bundle, which has the same rank.

In particular, a lattice $\mathscr{E}$ of a meromorphic bundle $\mathscr{M}$ coincides with $\mathscr{M}$ when restricted to $M \smallsetminus Z$. Moreover, we have

$$\mathscr{M} = \mathscr{O}_M(*Z) \underset{\mathscr{O}_M}{\otimes} \mathscr{E}.$$

A meromorphic bundle $\mathscr{M}$ can contain nonisomorphic lattices. It can also contain *no* lattice at all (see for instance [Mal94, Mal96] where a criterion for the existence of lattices in a meromorphic bundle is also given). Nevertheless, we have:

8.4 Proposition (The lattices do exist). *Let M be a Riemann surface and let $Z \subset M$ be a discrete set of points. Then any meromorphic bundle on M with poles at the points of Z contains at least a lattice.*

Proof. One constructs such a lattice by a gluing procedure. One chooses a covering of M by open sets $U = M \smallsetminus Z$ and U_m ($m \in Z$) in such a way that $U_m \cap U_n = \varnothing$ if $m \neq n$ and that, for any $m \in Z$, there is a trivialization of the restriction of $\mathscr{M}$ to U_m. One chooses, for any $m \in Z$, a lattice $\mathscr{E}_{(m)}$ of $\mathscr{M}_{|U_m}$ (it is enough to choose a lattice of $\mathscr{O}_{U_m}(*m)^d$ and to carry it with the trivializing isomorphism) and one takes the bundle $\mathscr{M}_{|U}$ itself as lattice on U. As $\mathscr{M}_{|U}$ and $\mathscr{E}_{(m)}$ coincide on $U \cap U_m = U_m \smallsetminus \{m\}$, one can define a lattice of $\mathscr{M}$ by gluing the local lattices defined in this way. $\qquad\square$

On the other hand, any vector bundle E gives rise to a meromorphic bundle for which it is a lattice, namely $\mathscr{M} = \mathscr{O}_M(*Z) \otimes_{\mathscr{O}_M} \mathscr{E}$.

One shows as in Proposition 5.3 that there is a one-to-one correspondence between the set of isomorphism classes of meromorphic bundles of rank d and the set $H^1(M, \mathrm{GL}_d(\mathscr{O}_M(*Z)))$. Therefore, a meromorphic bundle can be defined, up to isomorphism, by a cocycle made with matrices having meromorphic entries with poles along Z.

The natural mapping

$$H^1(M, \mathrm{GL}_d(\mathscr{O}_M)) \longrightarrow H^1(M, \mathrm{GL}_d(\mathscr{O}_M(*Z)))$$

is in general neither injective nor surjective. Proposition 8.4 above shows that, when M is a Riemann surface, it is surjective.

9 Examples of holomorphic and meromorphic bundles

9.a The cotangent bundle and the sheaves of holomorphic forms

Let U be an open set of M. A *holomorphic vector field on U* is by definition a $\mathbb{C}$-linear endomorphism of the sheaf $\mathscr{O}_U$ which is a *derivation*, that is, which satisfies the "Leibniz rule":

$$\xi(fg) = g\xi(f) + f\xi(g)$$

for any open set $V \subset U$ and all $f, g \in \mathscr{O}(V)$. We define in this way a presheaf on M, which is in fact a sheaf (the reader should check this), that we denote by Θ_M: this is the sheaf of holomorphic vector fields on M. The dual sheaf $\mathscr{H}om_{\mathscr{O}_M}(\Theta_M, \mathscr{O}_M)$ is denoted by Ω_M^1: this is the sheaf of *holomorphic differential 1-forms*.

9.1 Proposition (The bundles Θ_M and Ω_M^1). *The sheaves Θ_M and Ω_M^1 are locally free sheaves of rank n of $\mathscr{O}_M$-modules.*

Proof. As the statement is of a local nature, it is enough to check it for open sets of $\mathbb{C}^n$. Moreover, if it is true for Θ_M, it is also true for Ω_M^1. One verifies then that on an open set of $\mathbb{C}^n$ with coordinates $z_1, \ldots, z_n$, the sheaf Θ is isomorphic to $\mathscr{O}^n$ having as a basis the derivations $\partial/\partial z_1, \ldots, \partial/\partial z_n$. The dual basis of 1-forms is denoted by $dz_1, \ldots, dz_n$. $\qquad\square$

Any holomorphic function on an open set U of M gives rise to a section df of Ω_M^1 on U, defined by the property that, for any open set[4] $V \subset U$ and any holomorphic vector field ξ on V,

$$df_{|V}(\xi) = \xi(f_{|V}).$$

The bundle corresponding to the sheaf Θ_M is the *tangent bundle $TM \to M$* and that corresponding to Ω_M^1 is the *cotangent bundle $T^*M \to M$*.

In a chart of M with coordinates $z_1, \ldots, z_n$, a holomorphic 1-form can thus be written in a unique way as

$$\omega = \sum_{j=1}^n \varphi_j(z_1, \ldots, z_n) dz_j,$$

where the φ_j are holomorphic functions and, in particular,

$$df = \sum_{j=1}^n \frac{\partial f}{\partial z_j}\, dz_j.$$

Let V be an analytic submanifold of M and let $i_V : V \hookrightarrow M$ denote the inclusion. The restriction to V of the sheaf Θ_M (in the sense given in §4.4)

---

[4] Why can't we only use the open set U?

is locally free of rank dim M on V. This sheaf contains the subsheaf Θ_V of vector fields tangent to V, i.e., there exists a canonical inclusion

$$(9.2) \qquad \Theta_V \longrightarrow i_V^* \Theta_M = \mathscr{O}_V \underset{i_V^{-1}\mathscr{O}_M}{\otimes} i_V^{-1}\Theta_M.$$

In a dual way, there exists a canonical surjection

$$(9.3) \qquad i_V^* \Omega_M^1 = \mathscr{O}_V \underset{i_V^{-1}\mathscr{O}_M}{\otimes} i_V^{-1}\Omega_M^1 \longrightarrow \Omega_V^1.$$

If the manifold V is locally defined by the equations $z_{p+1} = \cdots = z_n = 0$, so that $(z_1, \ldots, z_p)$ is a system of local coordinates on V, the sheaf Ω_V^1 admits as local sections the combinations $\sum_{i=1}^{p} \varphi_i(z_1, \ldots, z_p)dz_i$ where the φ_i are holomorphic, while the sheaf $i_V^* \Omega_M^1$ admits as local sections the combinations $\sum_{j=1}^{n} \varphi_j(z_1, \ldots, z_p)dz_j$, where the φ_j are holomorphic. The surjection (9.3) consists in forgetting the $n - p$ last terms dz_j.

More generally, if $f : M' \to M$ is a holomorphic mapping between two manifolds, the associated *tangent* morphism Tf is a morphism of $\mathscr{O}_{M'}$-modules

$$(9.4) \qquad Tf : \Theta_{M'} \longrightarrow f^*\Theta_M$$

and the *cotangent* morphism is a morphism

$$(9.5) \qquad T^*f : f^* \Omega_M^1 \longrightarrow \Omega_{M'}^1.$$

We let the reader define them with coordinates.

One can apply the natural operations on vector bundles to these bundles. In particular, the *exterior product* $\Omega_M^1 \wedge \Omega_M^1$ is the sheaf of holomorphic 2-forms, denoted by Ω_M^2 or, equivalently, that of alternate bilinear forms on Θ_M. There is a $\mathbb{C}$-linear homomorphism

$$d : \Omega_M^1 \longrightarrow \Omega_M^2.$$

In a local chart as above, any 2-form η can be written in a unique way

$$\eta = \sum_{1 \leqslant i < j \leqslant n} \psi_{ij}(z_1, \ldots, z_n)dz_i \wedge dz_j,$$

where ψ_j are holomorphic functions and, in particular,

$$d\omega = \sum_{1 \leqslant i < j \leqslant n} \left[\frac{\partial \varphi_j}{\partial z_i} - \frac{\partial \varphi_i}{\partial z_j} \right] dz_i \wedge dz_j.$$

9.6 Exercise (Properties of the differential).

(1) Check that, for any germ f of holomorphic function, $d(df) = 0$ and that (cf. [GHL87, p. 45]), for any 1-form ω and any pair (ξ, η) of vector fields, $d\omega(\xi, \eta) = \xi(\omega(\eta)) - \eta(\omega(\xi)) - \omega([\xi, \eta])$.

(2) Show that, if $f : M' \to M$ is a holomorphic mapping, the cotangent mapping T^*f is compatible with the differential, i.e., $d(\varphi \circ f) = T^*f(d\varphi)$ for any germ of holomorphic function φ on M.

9.7 Lemma (Holomorphic Poincaré Lemma). *The sequence*

$$0 \longrightarrow \mathbb{C} \longhookrightarrow \mathscr{O}_M \xrightarrow{\ d\ } \Omega_M^1 \xrightarrow{\ d\ } \Omega_M^2$$

is exact, that is, for any $m \in M$,

(1) *if f is a germ of holomorphic function at m such that $df = 0$, then f is the germ of a constant function;*
(2) *if ω is a germ at m of a holomorphic 1-form such that $d\omega = 0$, then there exists a germ at m of holomorphic function f such that $\omega = df$.*

Proof. Easy by considering the Taylor expansion of f or of the φ_i.   □

9.8 Exercise (A case where the closed 1-forms are exact). Let M be a complex analytic manifold which satisfies $H^1(M, \mathbb{C}_M) = 0$, where $\mathbb{C}_M$ denotes the constant sheaf. Show that any closed holomorphic 1-form ω, i.e., such that $d\omega = 0$, is exact, i.e., of the form $\omega = df$ with f holomorphic on M (show that the short sequence of sheaves

$$0 \longrightarrow \mathbb{C}_M \longrightarrow \mathscr{O}_M \xrightarrow{\ d\ } \mathscr{L}_M^1 \longrightarrow 0,$$

where $\mathscr{L}_M^1$ denotes the subsheaf of Ω_M^1 of closed 1-forms, is exact, and then consider the associated long exact sequence in cohomology).

9.9 Exercise (The de Rham complex is exact). Define, for any $k \geqslant 1$, the sheaf of holomorphic k-forms Ω_M^k as $\wedge^k \Omega_M^1$, extend definition of the differential $d : \Omega_M^k \to \Omega_M^{k+1}$ and show that the sequence

$$0 \longrightarrow \mathbb{C}_M \longhookrightarrow \mathscr{O}_M \xrightarrow{\ d\ } \Omega_M^1 \xrightarrow{\ d\ } \cdots \xrightarrow{\ d\ } \Omega_M^k \xrightarrow{\ d\ } \cdots$$

is exact.

9.b Meromorphic and logarithmic differential forms

We assume as in §8 that Z is a smooth hypersurface of M. Let

$$\Omega_M^k(*Z) := \mathscr{O}_M(*Z) \underset{\mathscr{O}_M}{\otimes} \Omega_M^k$$

be the sheaf of meromorphic differential k-forms. This is a meromorphic bundle. The differential d defined in §9.a can be extended as a $\mathbb{C}$-linear homomorphism $d : \Omega_M^k(*Z) \to \Omega_M^{k+1}(*Z)$, that is, the meromorphy property along Z is preserved by differentiation (this is obvious by a local computation in the charts).

The meromorphic bundle $\Omega_M^1(*Z)$ contains Ω_M^1 as a lattice. It also contains another lattice which is often more useful, namely the lattice of *logarithmic differential 1-forms along Z*. We denote it by $\Omega_M^1\langle\log Z\rangle$. Let us pick $z^o \in Z$ and let U be a chart of M equipped with a system of local coordinates $z_1, \ldots, z_n$ such that z^o is the origin of coordinates and Z is defined by $z_1 = 0$. We say that a meromorphic form $\omega = \varphi_1 dz_1 + \cdots + \varphi_n dz_n$ is *logarithmic along Z* if $\varphi_2, \ldots, \varphi_n$ are holomorphic and φ_1 has a pole of order at most 1 along Z; in other words, the form can be written as

$$(9.10) \qquad \omega = \psi_1 \frac{dz_1}{z_1} + \sum_{i=2}^{n} \psi_i dz_i,$$

where $\psi_1, \ldots, \psi_n$ are holomorphic on U.

9.11 Example (Logarithmic and nonlogarithmic forms). The form dz_2/z_1 is not logarithmic along $\{z_1 = 0\}$ while dz_2 and dz_1/z_1 are logarithmic. Nevertheless, on a Riemann surface, there is no difference between $\Omega_M^1\langle\log Z\rangle$ and $\Omega_M^1(Z) = \Omega_M^1(1 \cdot Z)$.

9.12 Proposition (The residue).

(1) *The notion of logarithmic form does not depend on the choice of local coordinates adapted to Z.*
(2) *There exists a homomorphism, called residue and denoted by*

$$\mathrm{res} : \Omega_X^1\langle\log Z\rangle \longrightarrow \mathscr{O}_Z,$$

which associates to the form $\omega = \psi_1 dz_1/z_1 + \sum_{i=2}^{n} \psi_i dz_i$ *the function* $\psi_1(0, z_2, \ldots, z_n)$.

Proof. Let $(z_1, \ldots, z_n)$ and $(\tilde{z}_1, \ldots, \tilde{z}_n)$ be two systems of local coordinates adapted to Z. The coordinate change can be written as

$$\tilde{z}_i = f_i(z_1, \ldots, z_n), \quad i = 1, \ldots, n$$

and the set of zeroes of f_1 is equal to $\{z_1 = 0\}$, hence $f_1 = z_1 g_1(z_1, \ldots, z_n)$ with g_1 holomorphic. Moreover, the Jacobian matrix $((\partial f_i/\partial z_j)_{i,j})$ is invertible at the origin, which implies that $g_1(0, \ldots, 0) \neq 0$ and that the submatrix $((\partial f_i/\partial z_j)_{i,j\neq 1})$ is also invertible. If ω is logarithmic in the coordinate system $(\tilde{z}_1, \ldots, \tilde{z}_n)$, one can write

$$\omega = \tilde{\psi}_1 \frac{d\tilde{z}_1}{\tilde{z}_1} + \sum_{i=2}^{n} \tilde{\psi}_i d\tilde{z}_i$$

where the $\tilde{\psi}_i(\tilde{z}_1, \ldots, \tilde{z}_n)$ are holomorphic. Therefore,

$$\omega = \tilde{\psi}_1 \circ f \cdot \left(\frac{dz_1}{z_1} + \frac{dg_1}{g_1} \right) + \sum_{i=2}^{n} \tilde{\psi}_i \circ f \cdot df_i.$$

One can see on this expression that the form ω is also logarithmic in the coordinate system $(z_1, \ldots, z_n)$ in a neighbourhood of the origin as the function $1/g_1$ is holomorphic there. Moreover, in these coordinates, the residue can be written as

$$\widetilde{\psi}_1 \circ f_{|z_1=0} = \widetilde{\psi}_1(0, f_2(0, z_2, \ldots, z_n), \ldots, f_n(0, z_2, \ldots, z_n)),$$

in other words, it is the residue of ω in the system $(\widetilde{z}_1, \ldots, \widetilde{z}_n)$ written in the coordinates $z_2, \ldots, z_n$ of Z. $\square$

One can then define in an obvious way the sheaf $\Omega^1_M \langle \log Z \rangle$, the local sections of which are the logarithmic forms along Z, and the reader will show as an exercise:

9.13 Proposition (Freeness of logarithmic forms). *The sheaf of logarithmic 1-forms along Z is a locally free sheaf of rank n of $\mathscr{O}_M$-modules. It is a lattice of $\Omega^1_M(*Z)$.* $\square$

One can also define a new notion of *order of the pole along Z*: a meromorphic 1-form ω with poles along Z has *order $r \geqslant 0$* if in any local chart where Z is defined by $z_1 = 0$, the form $z_1^r \omega$ is logarithmic. We define in this way a lattice $\Omega^1_M \langle (r+1) \log Z \rangle$ of $\Omega^1_M(*Z)$. If Z has many connected components $Z_1, \ldots, Z_p$, one can also define the sheaves of differential forms

$$\Omega^1_M \langle (r_1 + 1) \log Z_1 + \cdots + (r_p + 1) \log Z_p \rangle.$$

When M is a Riemann surface, these sheaves are identical to the sheaves

$$\Omega^1_M((r+1)Z) \quad \text{and} \quad \Omega^1_M((r_1+1)Z_1 + \cdots + (r_p+1)Z_p)$$

introduced in Remark 8.2.

9.14 Exercise (Order of a meromorphic form).

(1) Compute, for $r \geqslant 0$, the order of the forms dz_1/z_1^r and dz_2/z_1^r.
(2) Let f be a meromorphic function on an open set U of M, with poles along $Z \cap U$. Show that the logarithmic differential $\omega = df/f$ is a section of $\Omega^1_U \langle \log Z \rangle$.

9.15 Remark (The residue and the "residue"). If ω is a meromorphic form with pole of order r along a hypersurface Z having $z_1 = 0$ as its local equation, one can write

$$\omega = \sum_{k=0}^{r} z_1^{-k} \omega_k$$

where the ω_k are logarithmic. Let us set, in local coordinates,

$$\omega_k = \psi_1^{(k)} \frac{dz_1}{z_1} + \sum_{i=2}^{n} \psi_i^{(k)} dz_i.$$

The decomposition of ω above is not unique, but it becomes so if one more-over asks that, for any $k \geqslant 1$, the functions $\psi_i^{(k)}$ $(i = 1, \ldots, n)$ depend only on $z_2, \ldots, z_n$. The dependence of the coefficients on z_1 only appears in the logarithmic term ω_0.

Assume now that M is the product of a disc with Z. Then

- on the one hand, one can define a homomorphism

$$\text{res} : \Omega_M^1\langle(r+1)\log Z\rangle \longrightarrow \mathscr{O}_Z$$

by integrating ω along a circle centered at the origin in D: this is also the residue of ω_0;
- on the other hand, the residue of $z_1^r\omega$, that is, that of ω_r, depends on the choice of coordinates up to multiplication by an invertible holomorphic function. In particular, if one *fixes* a local coordinate z_1 on D, one can define in this way a "residue" homomorphism

$$\text{"res"} : \Omega_M^1\langle(r+1)\log Z\rangle \longrightarrow \mathscr{O}_Z.$$

9.16 Remarks (cf. [Del70, §3]).

(1) As usual, we set, for any $k \geqslant 1$,

$$\Omega_M^k\langle\log Z\rangle = \wedge^k \Omega_M^1\langle\log Z\rangle.$$

It can be useful to have an intrinsic definition (i.e., without coordinates) of these sheaves. A local section ω of $\Omega_M^k(*Z)$ is a section of $\Omega_M^k\langle\log Z\rangle$ if and only if ω and $d\omega$ have at most simple poles along Z (see [Del70, Prop. 3.2]).

(2) The notion of a logarithmic form can be extended in a natural way to the case where Z is a union (not necessarily disjoint) of smooth hypersurfaces $Z_1, \ldots, Z_p$ which meet transversally. In the neighbourhood of an intersec-tion point of exactly m hypersurfaces, say $Z_1, \ldots, Z_m$, there exist local coordinates $z_1, \ldots, z_n$ such that Z_i is defined by $z_i = 0$ $(i = 1, \ldots, m)$ and a logarithmic form can be written as

$$\omega = \sum_{i=1}^m \psi_i\frac{dz_i}{z_i} + \sum_{j=m+1}^n \psi_j dz_j$$

where the ψ_j are holomorphic. All the properties of the sheaf $\Omega_M^1\langle\log Z\rangle$ seen above can be extended to this situation.

(3) The definition given in (1) above remains meaningful even if the hyper-surface Z has singularities. When these singularities are *normal crossings* (as in (2) above), nothing really new happens. For more complicated sin-gularities, the sheaf of logarithmic forms is harder to analyze. When this sheaf is locally free (of rank $\dim M$), one says that the hypersurface is a *free divisor*. We will see a remarkable example in §VII.1.10.

10 Affine varieties, analytization, algebraic differential forms

We will now take up in an algebraic framework some of the notions introduced above. This will enable us to work with objects having polynomial—and not only analytic—coefficients. The affine setting will suffice. For more details (in particular the projective setting), the reader can refer for instance to [Har80] or [Per95].

10.a Affine varieties

An *affine algebraic set* is by definition the zero set, in the affine space $\mathbb{A}^n$ with coordinates $u_1, \ldots, u_n$, of a family of polynomials in the u_i. One can assume that this family is finite, as the ring of polynomials is Noetherian. The *Zariski topology* on $\mathbb{A}^n$ is the topology for which the open sets are the complements of affine algebraic subsets. Such a set is *irreducible* if it is not the union of two affine algebraic proper subsets. One equips it with the Zariski topology, induced by that of $\mathbb{A}^n$.

10.1 Example (Elementary affine open sets). Fix $f \in \mathbb{C}[u_1, \ldots, u_n]$. The "elementary" open set $U = \{f \neq 0\} \subset \mathbb{A}^n$ is an affine algebraic set:

$$U = \{1 - u_0 f(u_1, \ldots, u_n) = 0\} \subset \mathbb{A}^{n+1}.$$

In particular, the complex torus $(\mathbb{C}^*)^n = \{u_1 \cdots u_n \neq 0\}$ is an affine algebraic set. On the other hand, the open set $\mathbb{A}^n \smallsetminus \{0\}$ is not affine as soon as $n \geqslant 2$.

Let U be a Zariski open set of an affine algebraic set $Z \subset \mathbb{A}^n$. A function $f : U \to \mathbb{C}$ is said to be *regular* if, for any point $z^o \in U$, there exist two polynomials $g, h \in \mathbb{C}[u_1, \ldots, u_n]$ such that $h(z^o) \neq 0$ and $f \equiv g/h$ on the open set $V = U \smallsetminus \{h = 0\}$. We define in this way a sheaf $U \mapsto \mathscr{O}_Z(U)$ on Z equipped with its Zariski topology: this is the sheaf of *regular functions* on the set Z, which is called an *affine variety*.

Let $I_Z \subset \mathbb{C}[u_1, \ldots, u_n]$ be the ideal of polynomials which vanish on Z. Then (cf. [Har80, Chap. 1, Th. 3.2]),

$$\Gamma(Z, \mathscr{O}_Z) = \mathbb{C}[u_1, \ldots, u_n]/I_Z.$$

The locally free sheaves of finite type (resp. coherent) of $\mathscr{O}_Z$-modules correspond to the modules which are free of finite type (resp. of finite type) over the ring $\mathbb{C}[u_1, \ldots, u_n]/I_Z$ by the functors (cf. [Ser55]).

$$\mathscr{F} \longmapsto \Gamma(Z, \mathscr{F}) = F, \qquad F \longmapsto \mathscr{O}_Z \underset{\Gamma(Z, \mathscr{O}_Z)}{\otimes} F = \mathscr{F}.$$

10.b Analytization

Let $Z \subset \mathbb{A}^n$ be an affine variety and let Z' be the same subset, equipped with the topology induced by the usual topology of $\mathbb{C}^n$. Then the identity mapping $\mathrm{Id} : Z' \to Z$ is continuous (as a Zariski open set is also an open set for the usual topology). A function f defined on an open set U' of Z' is said to be *holomorphic* if, for any $z^o \in U'$, there exist an open neighbourhood V' of z^o in $\mathbb{C}^n$ and a holomorphic function $g : V' \to \mathbb{C}$ such that $g_{|V' \cap U'} = f_{|U'}$. The space Z', equipped with the sheaf of holomorphic functions, is denoted by Z^{an}.

If $\mathscr{F}$ is a locally free (or coherent) sheaf of $\mathcal{O}_Z$-modules and if $\mathscr{F}' = \mathrm{Id}^{-1}\mathscr{F}$ denotes the sheaf that it induces on Z', one sets $\mathscr{F}^{\mathrm{an}} = \mathcal{O}_{Z^{\mathrm{an}}} \otimes_{\mathrm{Id}^{-1}\mathcal{O}_Z} \mathscr{F}'$: this is the *analytization* of the sheaf $\mathscr{F}$. It is a locally free (or coherent) sheaf of $\mathcal{O}_{Z^{\mathrm{an}}}$-modules. The essential property of this functor $\mathscr{F} \mapsto \mathscr{F}^{\mathrm{an}}$ comes from the following result (cf. for instance [AM69] for the notion of flatness):

10.2 Theorem (Faithful flatness, [Ser56]). *For any $z^o \in Z$, the ring $\mathcal{O}_{Z^{\mathrm{an}},z^o}$ of germs of holomorphic functions on Z^{an} at z^o is faithfully flat over the ring $\mathcal{O}_{Z,z^o}$ of germs of regular functions on Z at z^o.* $\square$

A consequence of faithful flatness (see for instance [Mat80] for this notion) is the following statement, maybe more "concrete":

10.3 Corollary. *Let $\varphi : \mathscr{F} \to \mathscr{F}'$ be a homomorphism of sheaves of $\mathcal{O}_Z$-modules. If $\varphi^{\mathrm{an}} : \mathscr{F}^{\mathrm{an}} \to \mathscr{F}'^{\mathrm{an}}$ is an isomorphism, then so is φ.* $\square$

When Z is irreducible, one defines the *dimension* of Z as the *Krull dimension* of the integral ring $\Gamma(Z, \mathcal{O}_Z)$, that is, the maximal length of increasing sequences of prime ideals.

10.c Nonsingular affine varieties

The sheaf $\mathrm{Der}\,\mathcal{O}_Z$ of derivations of the sheaf $\mathcal{O}_Z$ can be defined as in §9.a and the sheaf of differential algebraic 1-forms Ω^1_Z is by definition the sheaf $\mathscr{H}om_{\mathcal{O}_Z}(\mathrm{Der}\,\mathcal{O}_Z, \mathcal{O}_Z)$.

The following result collects the definitions that we will have to use.

10.4 Theorem (The regular points, cf. for instance [Per95, Chap. 5]). *Let $Z \subset \mathbb{A}^n$ be an irreducible affine variety. The following properties are equivalent:*

(1) *If $f_1, \ldots, f_p \in \mathbb{C}[u_1, \ldots, u_n]$ generate the ideal I_Z, then the Jacobian matrix $((\partial f_i/\partial u_j)_{ij})$ has constant rank on Z, equal to $n - \dim Z$.*
(2) *The set Z^{an} is a complex analytic submanifold of $\mathbb{C}^n$.*
(3) *The sheaf Ω^1_Z is locally free of rank $\dim Z$.*
(4) *There exists a finite covering of Z by Zariski open sets U_i and, on each U_i, regular functions $x_1^{(i)}, \ldots, x_{\dim Z}^{(i)}$ which form an étale chart.* $\square$

If these properties are satisfied, we say that Z is a *nonsingular* affine variety. On a nonsingular affine variety, the sheaves of forms $\Omega_Z^k := \wedge^k \Omega_Z^1$ are thus also locally free and $(\Omega_Z^k)^{\mathrm{an}} = \Omega_{Z^{\mathrm{an}}}^k$ (use étale charts).

10.5 Exercise (The hyperelliptic curve). Let $a_1, \ldots, a_d$ be complex numbers. Show that the subset of $\mathbb{C}^2$ having equation

$$y^2 = \prod_{j=1}^{d} (x - a_j)$$

is an affine variety of dimension one, and that it is nonsingular if and only if the a_j are pairwise distinct.

11 Holomorphic connections on a vector bundle

It is not in general possible to define an action by derivation of vector fields on any vector bundle, as for the trivial bundle $\mathscr{O}_M$ or its powers. Moreover, there exist various ways to make vector fields act as derivation on the trivial bundles. The notion of holomorphic connection (also called *covariant derivative*) will enable us to take these phenomena into account.

11.1 Definition (of a connection). A *holomorphic connection* ∇ on a holomorphic vector bundle $\pi : E \to M$ is a $\mathbb{C}$-linear homomorphism of sheaves

$$\nabla : \mathscr{E} \longrightarrow \Omega_M^1 \otimes_{\mathscr{O}_M} \mathscr{E}$$

satisfying the *Leibniz rule*: for any open set U of M, any section $s \in \Gamma(U, \mathscr{E})$ and any holomorphic function $f \in \mathscr{O}(U)$,

$$\nabla(f \cdot s) = f \nabla(s) + df \otimes s \in \Gamma(U, \Omega_M^1 \otimes_{\mathscr{O}_M} \mathscr{E}).$$

Any holomorphic vector field ξ on an open set U of M induces (by definition of Ω_M^1) a $\mathscr{O}_U$-linear homomorphism of contraction (also called *interior product*) $\iota_\xi : \Omega_U^1 \to \mathscr{O}_U$. We deduce from it a homomorphism of contraction $\iota_\xi : \Omega_U^1 \otimes \mathscr{E}_{|U} \to \mathscr{E}_{|U}$ for any $\mathscr{O}_U$-module $\mathscr{E}_U$.

The connection ∇ enables then to define a *derivation* along ξ on the sheaf $\mathscr{E}_{|U}$: for any open set $V \subset U$ and any section $s \in \Gamma(V, \mathscr{E})$, one defines $\nabla_\xi(s)$ as the result of the contraction of $\nabla s \in \Gamma(V, \Omega_M^1 \otimes \mathscr{E})$ by the field ξ. We have

$$\nabla_\xi(f \cdot s) = f \nabla_\xi(s) + \xi(f) \cdot s.$$

11.2 Proposition (The affine space of connections). *Two holomorphic connections ∇ and ∇' on a bundle $\mathscr{E}$ differ by a $\mathscr{O}_M$-linear homomorphism*

$$\Omega : \mathscr{E} \longrightarrow \Omega_M^1 \otimes_{\mathscr{O}_M} \mathscr{E}.$$

Conversely, adding such a homomorphism Ω to a connection ∇ gives a new connection ∇'.

Proof. If one applies the Leibniz rule to $\nabla - \nabla'$, the terms which are independent of the connection cancel, hence the $\mathscr{O}_M$-linearity. $\qquad\square$

11.3 Example (Connections on the trivial bundle). The trivial bundle of rank one is equipped with the connection $d : \mathscr{O}_M \to \Omega^1_M$. In an analogous way, the trivial bundle of rank δ is equipped with the direct sum connection $d : \mathscr{O}^\delta_M \to \left(\Omega^1_M\right)^\delta$. Any other connection ∇ on the trivial bundle of rank δ can be written as $\nabla = d + \Omega$ where Ω is a global section of the sheaf $M_\delta(\Omega^1_M)$ of $\delta \times \delta$ matrices with entries in Ω^1_M.

11.a Local expression of the connection

Let U be a trivializing open set for E and let $e = (e_1, \ldots, e_d)$ be a basis of $\Gamma(U, \mathscr{E})$; any holomorphic section s of E on U can thus be written in a unique way as $\sum_j s_j e_j$ with $s_j \in \mathscr{O}(U)$. There exists then a $d \times d$ matrix $\Omega = (\omega_{ij})$ of holomorphic 1-forms such that

$$(11.4) \qquad \nabla(e_i) = \sum_j \omega_{ji} \otimes e_j.$$

We can write this equality with matrices[5]:

$$\begin{pmatrix} \nabla(e_1) \\ \vdots \\ \nabla(e_d) \end{pmatrix} = {}^t\Omega \otimes \begin{pmatrix} e_1 \\ \vdots \\ e_d \end{pmatrix}$$

We say that Ω is the *connection matrix* ∇ in the basis e. If $s = \sum_i s_i e_i$ is a holomorphic section of E on U, one can write, using matrices[6],

$$\nabla(s) = \nabla \left((s_1, \ldots, s_d) \cdot \begin{pmatrix} e_1 \\ \vdots \\ e_d \end{pmatrix} \right)$$

$$= (ds_1, \ldots, ds_d) \otimes \begin{pmatrix} e_1 \\ \vdots \\ e_d \end{pmatrix} + (s_1, \ldots, s_d) \cdot \nabla \begin{pmatrix} e_1 \\ \vdots \\ e_d \end{pmatrix}$$

$$= \left((ds_1, \ldots, ds_d) + (s_1, \ldots, s_d) \cdot {}^t\Omega \right) \otimes \begin{pmatrix} e_1 \\ \vdots \\ e_d \end{pmatrix}$$

---

[5] The choice that we make to put the differential form "on the left" does not simplify the matrix expression and forces us to write the basis as a column vector, hence the need to transpose; nevertheless this choice is more natural, because the vector fields act on the left on the bundle with connection.

[6] Here also, we transpose of the usual way or writing.

Let $\varepsilon = (\varepsilon_1, \ldots, \varepsilon_d)$ be another basis of sections of E on U and let $P \in \Gamma(U, \mathrm{GL}_d(\mathscr{O}_M))$ be the base change from e to ε:

$$(\varepsilon_1, \ldots, \varepsilon_d) = (e_1, \ldots, e_d) \cdot P.$$

Then the matrix Ω' of ∇ in the basis ε can be obtained from Ω by the formula

$$(11.5) \qquad \Omega' = P^{-1}\Omega P + P^{-1}dP.$$

11.6 Exercise. Write explicitly the formula for the base change in a local chart of M with coordinates $z_1, \ldots, z_n$ in which the sheaf $\mathscr{E}$ has basis e: set $P = (P_{ij})_{1 \leqslant i,j \leqslant d}$ and $\Omega = (\omega_{ij})_{1 \leqslant i,j \leqslant d}$ with $\omega_{ij} = \sum_{\ell=1}^{n} \varphi_{ij}^{(\ell)} dz_\ell$.

11.7 Example. The connection matrix of d on the trivial bundle $\mathscr{O}_M^\delta$ is zero in the canonical basis.

11.b Operations on bundles with connection

Let (E, ∇) and (E', ∇') be two holomorphic bundles equipped with holomorphic connections. One can equip in a natural way the bundles $E \oplus E'$, $\mathrm{L}(E, E')$ and $E \otimes E'$ with holomorphic connections that we denote by ∇''. If s and s' are two holomorphic local sections of E and E' respectively, we set for any vector field ξ,

- $\nabla_\xi''(s \oplus s') := \nabla_\xi(s) \oplus \nabla_\xi(s')$, in other words $\nabla'' = \nabla \oplus \nabla'$;
- $\nabla_\xi''(s \otimes s') := (\nabla_\xi(s) \otimes s') + (s \otimes \nabla_\xi'(s'))$, in other words $\nabla'' = (\nabla \otimes \mathrm{Id}) + (\mathrm{Id} \otimes \nabla')$;
- if $\varphi : E \to E'$ is a homomorphism, $\nabla_\xi''(\varphi)$ is the local section of $\mathrm{L}(E, E')$ defined by

$$\nabla_\xi''(\varphi)(s) = \nabla_\xi'\left(\varphi(s)\right) - \varphi\left(\nabla_\xi(s)\right).$$

11.8 Exercise (The connection on the bundle of endomorphisms).

(1) Define the natural connection on the bundle E^* dual to the bundle E. Show that both natural connections on $E^* \otimes E'$ and $\mathrm{L}(E, E')$ coincide.
(2) Compute in a local basis of $\mathscr{E}$ and $\mathscr{E}'$ the connection matrices of ∇'' from Ω and Ω'.
(3) Check in particular that, if Ω is the matrix of ∇ in the basis e, that of ∇'' on $\mathrm{End}(E) = \mathrm{L}(E, E)$ in the basis deduced from e is

$$\mathrm{ad}\,\Omega : A \longmapsto [\Omega, A].$$

11.c Pullback of bundles with connection

Let (E, ∇) be a holomorphic bundle with connection on an analytic manifold M and let $f : M' \to M$ be a holomorphic mapping from an analytic manifold M' to M. We know how to define the pullback $E' = f^*E$ of the bundle E on the manifold M' (cf. §§3.3 and 4.4). The sheaf $\mathscr{E}'$ of holomorphic sections of E' is given by the formula

$$(11.9) \qquad \mathscr{E}' = \mathscr{O}_{M'} \underset{f^{-1}\mathscr{O}_M}{\otimes} f^{-1}\mathscr{E}.$$

We use the notation $E' = f^*E$ and $\mathscr{E}' = f^*\mathscr{E}$.

When E is the cotangent bundle T^*M, the *cotangent mapping* T^*f is a morphism of bundles

$$T^*f : f^*T^*M \longrightarrow T^*M'$$

defining a morphism of sheaves of $\mathscr{O}_{M'}$-modules (cf. §9.a)

$$T^*f : f^*\Omega^1_M \longrightarrow \Omega^1_{M'}.$$

Let (E, ∇) be a holomorphic bundle with connection. Let us define the *pullback*

$$\nabla' : \mathscr{E}' \longrightarrow \Omega^1_{M'} \underset{\mathscr{O}_{M'}}{\otimes} \mathscr{E}'$$

of the connection ∇. The problem comes from the non $\mathscr{O}_M$-linearity of ∇. Its pullback cannot be defined by simply using the pullback functor for vector bundles, as given by Formula (11.9).

11.10 Example (Pullback by a projection). Assume that M' is the product $M \times M''$ and that f is the first projection. The pullback ∇' of the connection ∇ has a simple definition in this case: indeed, in such a case, any vector field ξ' on an open set of M' can be written in a unique way as the sum of a "horizontal" field ξ (i.e., section of $f^*\Theta_M$) and of a "vertical" field ξ'' (i.e., section of $p^*\Theta_{M''}$ if p denotes the second projection); by definition, the action of $\nabla'_{\xi''}$ on the sections of $\mathscr{E}'$ like $1 \otimes e$, e a section of $\mathscr{E}$, is *zero*; on the other hand, if ξ is the sum of fields like $\varphi \otimes \eta$, where η is a field on an open set of M and φ a function on an open set of M', then $\nabla'_\xi(1 \otimes e)$ is determined by the formula $\nabla'_{\varphi \otimes \eta}(1 \otimes e) = \varphi \otimes \nabla_\eta(e)$.

We will start with an *a priori* nonintrinsic definition, in local coordinates, when M' is a complex analytic submanifold of M and f denotes the inclusion mapping (this case is called *restriction* of the bundle with connection).

Assume that M is an open set of $\mathbb{C}^n$ with coordinates $z_1, \ldots, z_n$ and that M' is the trace on this open set of the subspace with equations $z_1 = \cdots = z_p = 0$. Assume also that E is trivialized and equipped with a basis e of holomorphic sections on M. We can then write the connection matrix in this basis as

$$\Omega = \sum_{i=1}^n \Omega^{(i)}(z_1, \ldots, z_n)\, dz_i.$$

The matrix of ∇' on the bundle E', in the basis e' restriction of e to M', is then given by the formula

$$\Omega' = \sum_{i=p+1}^n \Omega^{(i)}(0, \ldots, 0, z_{p+1}, \ldots, z_n)\, dz_i.$$

In an intrinsic way and for f arbitrary, one uses the composition of the "sheaf-theoretic" restriction

$$f^{-1}\nabla : f^{-1}\mathscr{E} \longrightarrow f^{-1}\Omega^1_M \underset{f^{-1}\mathscr{O}_M}{\otimes} f^{-1}\mathscr{E}$$

(i.e., one only considers the points of M') with the operation $\mathscr{O}_{M'}\otimes_{f^{-1}\mathscr{O}_M}$ (i.e., "one makes $z_1 = \cdots = z_p = 0$") to reach

$$f^*\Omega^1_M \underset{\mathscr{O}_{M'}}{\otimes} f^*\mathscr{E} = f^*\Omega^1_M \underset{\mathscr{O}_{M'}}{\otimes} \mathscr{E}',$$

then one composes with the morphism induced by the cotangent mapping

$$f^*\Omega^1_M \underset{\mathscr{O}_{M'}}{\otimes} \mathscr{E}' \xrightarrow{\ T^*f \otimes \mathrm{Id}\ } \Omega^1_{M'} \underset{\mathscr{O}_{M'}}{\otimes} \mathscr{E}'$$

(i.e., one forgets $dz_1,\ldots,dz_p$). We have now obtained a $\mathbb{C}$-linear homomorphism

$$f^{-1}\mathscr{E} \longrightarrow \Omega^1_{M'} \underset{\mathscr{O}_{M'}}{\otimes} \mathscr{E}'$$

which satisfies the Leibniz rule with respect to the multiplication by any function on an open set of M' of the form $\varphi \circ f$, where φ is holomorphic on an open set of M.

We then define the pullback connection $f^*\nabla$ as a connection

$$\mathscr{E}' \longrightarrow \Omega^1_{M'} \underset{\mathscr{O}_{M'}}{\otimes} \mathscr{E}'$$

by extending the previous homomorphism to $\mathscr{E}' = \mathscr{O}_{M'} \otimes_{f^{-1}\mathscr{O}_M} f^{-1}\mathscr{E}$ in such a way that the Leibniz rule is satisfied. We let the reader check that such an extension is well defined and is unique.

We will denote by $f^+(E, \nabla)$ the pullback bundle f^*E equipped with the connection ∇' constructed in this way.

11.11 Exercise (The connection matrix of the pullback). Let U be an open set of M trivializing for $\mathscr{E}$, let $\boldsymbol{e} = (e_1,\ldots,e_d)$ be a basis of $\mathscr{E}_{|U}$ and $\Omega = (\omega_{ij})$ the connection matrix of ∇ in this basis. Let $(e'_1,\ldots,e'_d) = (1\otimes e_1,\ldots,1\otimes e_d)$ be the basis of the pullback $\mathscr{E}'_{|f^{-1}(U)}$ obtained from $\boldsymbol{e}$. Show that the entries of the matrix of ∇' in this basis are the 1-forms $T^*f(\omega_{ij})$.

11.12 Exercise. Use the decomposition $f : M' \to M$ as $p \circ i$, where $i : M \hookrightarrow M \times M'$ is the inclusion of the graph of f defined by $m \mapsto (m, f(m))$, and p is the second projection, to compute the pullback of a connection, by reducing the problem to the case of the inclusion of a submanifold and to that of a projection.

12 Holomorphic integrable connections and Higgs fields

12.a Holomorphic integrable connections

One can, in a natural way, extend the action of ∇ on the sections of $\mathscr{E} = \mathscr{O}_M \otimes \mathscr{E}$ as an action on the sections of $\Omega^1_M \otimes \mathscr{E}$ (with values in $\Omega^2_M \otimes \mathscr{E}$):

$$\nabla(\omega \otimes s) = d\omega \otimes s - \omega \wedge \nabla(s).$$

12.1 Exercise (The curvature is $\mathscr{O}$-linear).

(1) Show (cf. Exercise 9.6-(1)) that, if σ is a section of $\Omega^1_M \otimes \mathscr{E}$,

$$\nabla\sigma(\xi, \eta) = \nabla_\xi(\sigma(\eta)) - \nabla_\eta(\sigma(\xi)) - \sigma([\xi, \eta]).$$

(2) Prove that the $\mathbb{C}$-linear homomorphism

$$R_\nabla := \nabla \circ \nabla : \mathscr{E} \longrightarrow \Omega^2_M \otimes_{\mathscr{O}_M} \mathscr{E}$$

is in fact $\mathscr{O}_M$-linear. Compute $R_\nabla(s)(\xi, \eta)$.

The homomorphism R_∇ is the *curvature* of the connection ∇.

12.2 Definition (of integrability for a connection). The connection

$$\nabla : \mathscr{E} \longrightarrow \Omega^1_M \otimes_{\mathscr{O}_M} \mathscr{E}$$

is said to be *integrable*, or also *flat*, if its curvature R_∇ vanishes identically.

12.3 Remark. Exercise 9.6 shows that the trivial bundle $\mathscr{O}_M$ (or its powers), equipped with the connection d, is flat.

An easy computation gives:

12.4 Proposition (The structure equations). *The connection ∇ is flat if and only if, in any local basis e of $\mathscr{E}$, the connection matrix Ω satisfies*

$$d\Omega + \Omega \wedge \Omega = 0. \qquad \square$$

We define $\Omega \wedge \Omega$ as the usual matrix product, in which the product of entries is the exterior product of 1-forms (hence is not commutative).

12.5 Remark (In dimension one, any connection is integrable). When $\dim M = 1$, the sheaf Ω^2_M is zero and therefore the integrability condition is fulfilled by any connection. Moreover, the matrix Ω in any local basis simultaneously satisfies both equations $d\Omega = 0$ and $\Omega \wedge \Omega = 0$.

12.6 Exercise (Explicit form of the structure equations).

(1) if one writes in local coordinates $\Omega = \sum_{\ell=1}^{n} \Omega^{(\ell)} dz_\ell$ where $\Omega^{(\ell)} = (\omega_{ij}^{(\ell)})$ is a matrix of holomorphic functions, prove that the integrability condition can be written as

$$\forall k, \ell \in \{1, \ldots, n\}, \quad \frac{\partial \Omega^{(\ell)}}{\partial z_k} - \frac{\partial \Omega^{(k)}}{\partial z_\ell} = [\Omega^{(\ell)}, \Omega^{(k)}].$$

(2) Check that, if the integrability condition is fulfilled by the matrix Ω, it is also fulfilled by the matrix Ω' obtained by Formula (11.5).

A ∇-*horizontal holomorphic section* of E on an open set U of M is a section $s \in \Gamma(U, \mathscr{E})$ satisfying $\nabla(s) = 0$, in other words a section on U of the sheaf[7]

$$\mathrm{Ker}[\nabla : \mathscr{E} \longrightarrow \Omega_M^1 \otimes_{\mathscr{O}_M} \mathscr{E}].$$

If the components of s in some local basis e of $\mathscr{E}$ are $s_1, \ldots, s_d$, the equation $\nabla(s) = 0$ can be written as

$$(12.7) \qquad \begin{pmatrix} ds_1 \\ \vdots \\ ds_d \end{pmatrix} + \Omega \begin{pmatrix} s_1 \\ \vdots \\ s_d \end{pmatrix} = 0$$

or also, using the notation of Exercise 12.6,

$$\forall k \in \{1, \ldots, n\}, \quad \frac{\partial}{\partial z_k} \begin{pmatrix} s_1 \\ \vdots \\ s_d \end{pmatrix} = -\Omega^{(k)} \begin{pmatrix} s_1 \\ \vdots \\ s_d \end{pmatrix}.$$

In other words, s is a solution of a *linear homogeneous differential system with holomorphic coefficients*.

12.8 Theorem (Cauchy-Kowalevski theorem). *Let $\nabla : \mathscr{E} \to \Omega_M^1 \otimes_{\mathscr{O}_M} \mathscr{E}$ be a holomorphic integrable connection on some bundle E of rank d.*

(1) *The sheaf $E^\nabla := \mathrm{Ker}\,\nabla$ is a locally constant sheaf of $\mathbb{C}$-vector spaces of rank d, i.e., is locally isomorphic to the constant sheaf $\mathbb{C}_M^d$.*
(2) *The sheaf $\mathscr{O}_M \otimes_{\mathbb{C}_M} E^\nabla$ is a locally free sheaf of $\mathscr{O}_M$-modules, the connection on this sheaf defined by $\nabla(f \otimes s) = df \otimes s$ is flat and $(\mathscr{O}_M \otimes_{\mathbb{C}_M} E^\nabla)^\nabla = E^\nabla$.*
(3) *The natural homomorphism $\mathscr{O}_M \otimes_{\mathbb{C}_M} E^\nabla \to \mathscr{E}$ is an isomorphism of bundles with connection.*

---

[7] Take care: as ∇ is not $\mathscr{O}_M$-linear, this sheaf is not a sheaf of $\mathscr{O}_M$-modules; in other words, it is not stable by multiplication by the sections of $\mathscr{O}_M$; it is only a sheaf of $\mathbb{C}$-vector spaces.

12.9 Remarks.

(1) This statement is equivalent to the local existence of a basis of E consisting of ∇-horizontal sections.
(2) In general there is no *global* isomorphism between the sheaves $\mathrm{Ker}\,\nabla$ and $\mathbb{C}_M^d$. Such is the case however if M is *simply connected* (see §15 below).

Proof.

(1) As the statement is of a local nature, one can work in some local chart of M in which the bundle E is trivialized. It is a matter of showing that, on a polydisc Δ of $\mathbb{C}^n$ centered at 0, the solutions of the differential system (12.7) form a vector space of dimension d on $\mathbb{C}$. We begin with fixing the variables $z_2, \ldots, z_n$ and then solve the one-variable system depending on parameters $z_2, \ldots, z_d$:

$$(12.10) \qquad \frac{\partial}{\partial z_1} \begin{pmatrix} s_1 \\ \vdots \\ s_d \end{pmatrix} = -\Omega^{(1)} \begin{pmatrix} s_1 \\ \vdots \\ s_d \end{pmatrix}.$$

For any holomorphic vector $\sigma(z_2, \ldots, z_n)$ on the given polydisc, there exists a unique solution $s(z_1, \ldots, z_n)$, which is holomorphic in z_1, of the system (12.10) satisfying $s(0, z_2, \ldots, z_n) = \sigma(z_2, \ldots, z_n)$ (see [Car95, Chap. VII] or [BG91, Chap. 5, §15]). This solution is holomorphic with respect to all the variables. By induction on n, we can apply the theorem to (12.7) in which we make $z_1 = 0$ and we forget the derivation with respect to z_1. Hence, for any vector v of $\mathbb{C}^d$, there exists a unique holomorphic solution $\sigma_v(z_2, \ldots, z_n)$ of the previous system, such that $\sigma_v(0, \ldots, 0) = v$. Let s_v be the solution of (12.10) with σ_v as initial condition. Then the integrability condition on the connection implies that s_v is the unique holomorphic solution of the system (12.7) such that $s_v(0, \ldots, 0) = v$.
We thus have proved that the mapping which associates to any section $s \in \Gamma(\Delta, E^\nabla)$ its value at 0 (or, in an analogous way, at a given point $m \in \Delta$) is an isomorphism of $\mathbb{C}$-vector spaces $\Gamma(\Delta, E^\nabla) \xrightarrow{\sim} \mathbb{C}^d$.
(2) Let us only make precise the construction of ∇ (the remaining part is quite obvious). We identify the sheaf $\Omega_M^1 \otimes_{\mathcal{O}_M} (\mathcal{O}_M \otimes_{\mathbb{C}_M} E^\nabla)$ with $\Omega_M^1 \otimes_{\mathbb{C}} E^\nabla$ by "putting the holomorphic functions as coefficients of 1-forms". The connection ∇ is then defined by the fact that, for any $m \in M$, any germ s of section of E^∇ at m and any germ of holomorphic function $f \in \mathcal{O}_{M,m}$, there is the relation $\nabla(f \otimes s) = df \otimes s$.
(3) That the natural homomorphism $\mathcal{O}_M \otimes_{\mathbb{C}_M} E^\nabla \to \mathcal{E}$, defined at the level of germs by $f \otimes s \mapsto fs$, is an isomorphism means that, at any $m \in M$, the germs of horizontal sections generate the germ $\mathcal{E}_m$. In the local situation of (1), it is a matter of checking that the values at the point $m \in \Delta$ of horizontal sections generate the vector space $\mathbb{C}^d$ over $\mathbb{C}$ (because then they generate the fibre $\mathcal{E}_m = \mathcal{O}_{M,m}^d$ over $\mathcal{O}_M$, after Nakayama's lemma).

But this is exactly what gives the proof, as any vector $v \in \mathbb{C}^d$ is the value at m of a section of E^∇ on Δ. □

12.11 Exercise (Operations on horizontal sections).

(1) If the bundles with connection (E, ∇) and (E', ∇') are integrable, prove that so are the bundles $E \oplus E'$, E^*, $\mathrm{L}(E, E')$ and $E \otimes E'$ equipped with their natural connection. Moreover, show that

$$(E \oplus E')^{\nabla''} = E^\nabla \oplus E'^{\nabla'}$$

$$E^{*\nabla''} = \mathscr{H}om_{\mathbb{C}_M}(E^\nabla, \mathbb{C}_M)$$

$$\mathrm{L}(E, E')^{\nabla''} = \mathscr{H}om_{\mathbb{C}_M}(E^\nabla, E'^{\nabla'})$$

$$(E \otimes E')^{\nabla''} = E^\nabla \otimes_{\mathbb{C}_M} E'^{\nabla'}.$$

(2) Prove that $\mathrm{L}(E, E')^{\nabla''}$ can be identified with the subsheaf of the sheaf $\mathscr{H}om_{\mathscr{O}_M}(\mathscr{E}, \mathscr{E}')$ consisting of the homomorphisms which commute to the action of ∇ and ∇' (we also say that they are morphisms of bundles with connection), or also which send E^∇ in $E'^{\nabla'}$.

12.12 Exercise. With the notation of §11.c, show that, if ∇ is integrable, then so is its pullback by f on the manifold M'.

12.13 Corollary (Analytic extension). *The homomorphisms of bundles with connection satisfy the principle of analytic extension: if $V \subset U$ is the inclusion of connected open sets of M which induces an isomorphism between the fundamental groups of V and U, and if $\varphi : \mathscr{E}_{|V} \to \mathscr{E}'_{|V}$ is a homomorphism of bundles with connection, then φ can be extended in a unique way as a homomorphism of bundles with connection $\mathscr{E}_{|U} \to \mathscr{E}'_{|U}$.*

Proof. This is a direct consequence of the results on locally constant sheaves explained in §15. □

12.b Higgs fields

12.14 Definition. A *Higgs field*[8] on a holomorphic bundle E is a $\mathscr{O}_M$-linear homomorphism

$$\Phi : \mathscr{E} \longrightarrow \Omega^1_M \underset{\mathscr{O}_M}{\otimes} \mathscr{E}$$

which fulfills the "integrability" condition $\Phi \wedge \Phi = 0$. We also say that (E, Φ) is a *Higgs bundle*.

---

[8] The relation with the physicist Higgs and the boson having the same name would need some explanation. We have here an example of appropriation by mathematicians of concepts coming from physics. Originally, a Higgs field is related to a connection on a principal bundle, and is used to perturb Yang-Mills equations (see for instance [AH88] or [JT80]). The terminology that we use has been introduced by C. Simpson [Sim92], after N. Hitchin; this terminology is reasonable when there also exists a "harmonic metric". Here, we keep the terminology, in spite of the absence of a harmonic metric.

Let us be more explicit with regard to this condition: the homomorphism Φ enables us to define a homomorphism

$$^{(1)}\Phi : \Omega^1_M \underset{\mathscr{O}_M}{\otimes} \mathscr{E} \longrightarrow \Omega^2_M \underset{\mathscr{O}_M}{\otimes} \mathscr{E}$$

by setting $^{(1)}\Phi(\omega \otimes e) = \omega \wedge \Phi(e)$ (the reader will make precise the meaning of $\wedge$). Then $-\Phi \wedge \Phi$ is by definition the composition $^{(1)}\Phi \circ \Phi$.

Let e be a local basis of $\mathscr{E}$ and let us still denote by Φ the matrix of Φ in this basis, defined with the same convention (11.4) as for connections. The integrability condition can then be written as $\Phi \wedge \Phi = 0$, now in the matrix sense that we also used for connections (in Proposition 12.4).

If ξ is a holomorphic vector field on an open set U of M, we denote by $\Phi_\xi : \mathscr{E}_{|U} \to \mathscr{E}_{|U}$ the $\mathscr{O}_U$-linear endomorphism obtained by contraction of Φ with ξ. One can also regard Φ as a $\mathscr{O}_M$-linear homomorphism $\Theta_M \otimes \mathscr{E} \to \mathscr{E}$. From this point of view, we have $\Phi_\xi(\bullet) = \Phi(\xi, \bullet)$.

In a chart of M equipped with local coordinates $z_1, \ldots, z_n$, one can write

$$\Phi = \sum_{i=1}^{n} \Phi^{(k)} \, dz_k$$

where the $\Phi^{(k)} = \Phi_{\partial/\partial z_k}$ are endomorphisms of the bundle E (in the given chart). The "integrability" condition is equivalent to

$$\forall k, \ell \in \{1, \ldots, n\}, \qquad [\Phi^{(k)}, \Phi^{(\ell)}] = 0.$$

One does not have an analogue of the Cauchy-Kowalevski theorem for Higgs fields. If for instance E is the trivial bundle of rank one on M, giving a Higgs field on E is equivalent to giving a holomorphic 1-form on M. Where this form does not vanish, the kernel of Φ reduces to $\{0\}$.

12.15 Remark (Operations on Higgs fields). The formulas describing the operations $(\oplus, \otimes, \text{L})$ on bundles equipped with a Higgs fields are analogous to that for connections (Exercise 14.9 will justify this choice).

- If $f : M' \to M$ is a holomorphic mapping and $(\mathscr{E}, \Phi)$ a Higgs bundle on M, then the pullback Higgs bundle $f^+(\mathscr{E}, \Phi)$ is defined as in §11.c. In a local basis of $\mathscr{E}$ (cf. Exercise 11.11), the matrix of $f^+\Phi$ is obtained by applying T^*f to the entries of the matrix of Φ.
- For $\mathscr{E}'' = \mathscr{E} \oplus \mathscr{E}'$ we set, for any field ξ, $\Phi''_\xi = \Phi_\xi \oplus \Phi'_\xi$.
- For $\mathscr{E}'' = \mathscr{E} \otimes \mathscr{E}'$ we set, for any field ξ, $\Phi''_\xi = (\Phi_\xi \otimes \text{Id}) + (\text{Id} \otimes \Phi'_\xi)$.
- For $\mathscr{E}'' = \mathscr{H}om_{\mathscr{O}_M}(\mathscr{E}, \mathscr{E}')$ and φ a local section of $\mathscr{E}''$ we set, for any field ξ, $\Phi''_\xi(\varphi)(e) = \Phi'_\xi(\varphi(e)) - \varphi(\Phi_\xi(e))$.

The reader should check the integrability of the fields constructed in this way.

13 Geometry of the tangent bundle

When the bundle E that one considers is the tangent bundle TM, the objects that we introduced above (connections and Higgs fields) get supplementary symmetry properties. The integrability condition is then related to the existence of coordinates of a particular type on any simply connected open set of M or on the universal covering space of M.

13.a Some points of holomorphic differential geometry

The reader can refer for instance to [GHL87, Chap. 1] for more details on what follows. Recall that a holomorphic vector field on an open set U of M is a derivation of $\mathcal{O}_U$. Any holomorphic vector field admits locally a holomorphic flow parametrized by a complex time: this is an immediate consequence of Cauchy's theorem for holomorphic differential equations (see for instance [Car95, Chap. VII, Th. 1 and 2]).

The *Lie bracket* of two vector fields ξ, η on U is the vector field denoted by $[\xi, \eta]$ defined by

$$[\xi, \eta](\varphi) = \xi(\eta(\varphi)) - \eta(\xi(\varphi))$$

for any holomorphic function φ on an open set $V \subset U$. For ξ fixed, the operator $\eta \mapsto [\xi, \eta]$ is only $\mathbb{C}$-linear relatively to η and satisfies the Leibniz rule when one considers Θ_M as a $\mathcal{O}_M$-module. The is the *Lie derivative* associated to ξ. We will denote

$$\mathscr{L}_\xi(\eta) = [\xi, \eta].$$

We let the reader check that, if $f : M' \to M$ is a holomorphic mapping between manifolds and if ξ', η' are vector fields on an open set U' of M', then, in $f^*\Theta_M$ (cf. (9.4)),

$$Tf([\xi', \eta']) = [Tf(\xi), Tf(\eta)].$$

The action of the Lie derivative can be extended in a natural way to differential forms. It is a derivation of the graded algebra $\Omega_M = \bigoplus_p \Omega_M^p$, of degree 0 with respect to this gradation (i.e., the degree of the form $\mathscr{L}_\xi \omega$ is equal to that of ω).

The derivation d has degree one. The interior product ι_ξ by the vector field ξ, defined by the property that, for any p-form ω and any $p - 1$ vector fields $\xi_2, \ldots, \xi_p$,

$$(\iota_\xi \omega)(\xi_2, \ldots, \xi_p) = \omega(\xi, \xi_2, \ldots, \xi_p),$$

has degree -1. These three operators are related by the formulas

$$\mathscr{L}_\xi = \iota_\xi \circ d + d \circ \iota_\xi \quad \text{and} \quad \iota_{[\xi, \eta]} = \mathscr{L}_\xi \iota_\eta - \iota_\eta \mathscr{L}_\xi.$$

13.b Holomorphic foliations

A *holomorphic foliation* of dimension p on a complex analytic manifold M is a subbundle $\mathscr{F}$ (as defined in Exercise 4.5) of rank p of the tangent sheaf Θ_M, which is stable by the Lie bracket, that is, such that, for any $m \in M$ and all $\xi, \eta \in \mathscr{F}_m$, we have $[\xi, \eta] \in \mathscr{F}_m$.

An *integral submanifold* of the foliation $\mathscr{F}$ is a (not necessarily closed) connected analytic submanifold V of M of dimension p such that the tangent map Ti_V to the inclusion i_V (cf. §9.a) induces an isomorphism

$$\Theta_V \xrightarrow{\sim} i_V^* \mathscr{F} \subset i_V^* \Theta_M.$$

A *leaf* of the foliation $\mathscr{F}$ is a *maximal* integral submanifold. Two distinct leaves are disjoint.

The holomorphic variant of the Frobenius Integrability Theorem shows in particular the existence of a partition of the manifold M into leaves:

13.1 Theorem (Local normal form of a foliation). *Let $\mathscr{F}$ be a holomorphic foliation of dimension p of M. There exists then, in the neighbourhood of any point of M, a coordinate system $(z_1, \ldots, z_n)$ centered at this point such that, in this neighbourhood, the sheaf $\mathscr{F}$ is generated by the fields $\partial_{z_1}, \ldots, \partial_{z_p}$.*

In the neighbourhood considered in the theorem, the leaves are the level submanifolds of the functions $z_{p+1}, \ldots, z_n$.

Proof. One can refer for instance to [HH81, Chap. II, §2.3] for the proof of the Frobenius theorem in the real C^∞ framework. The adaptation of this proof to the holomorphic framework is straightforward.   □

Let us consider the exact sequence of sheaves of $\mathcal{O}_M$-modules

$$0 \longrightarrow \mathscr{F} \longrightarrow \Theta_M \longrightarrow \mathscr{Q} \longrightarrow 0.$$

By assumption, the sheaf $\mathscr{Q}$ is locally free. Dually, we thus get an exact sequence of locally free $\mathcal{O}_M$-modules

$$0 \longrightarrow \mathscr{J} \longrightarrow \Omega_M^1 \longrightarrow \mathscr{F}^\vee \longrightarrow 0$$

where $\mathscr{F}^\vee = \mathscr{H}om_{\mathcal{O}_M}(\mathscr{F}, \mathcal{O}_M)$ and $\mathscr{J} = \mathscr{H}om_{\mathcal{O}_M}(\mathscr{Q}, \mathcal{O}_M)$. It is equivalent to giving the subbundle $\mathscr{F}$ of Θ_M or the subbundle $\mathscr{J}$ of Ω_M^1.

13.2 Lemma (Frobenius condition). *The subbundle $\mathscr{F}$ is stable by the Lie bracket if and only if $\mathscr{J}$ fulfills one of the equivalent conditions:*

(1) $d\mathscr{J} \subset \mathscr{J} \wedge \Omega_M^1$.
(2) *For any $m \in M$ and all germs $\xi, \eta \in \mathscr{F}_m$, the germ $\xi \wedge \eta$ is killed by all elements of $d\mathscr{J}$.*

Proof. See for instance [HH81, Chap. II, §2.4].   □

13.3 Remark. Let $f : M' \to M$ be a holomorphic mapping and let $\mathscr{J}$ be a locally free subsheaf of Ω^1_M satisfying the integrability condition $d\mathscr{J} \subset \mathscr{J} \wedge \Omega^1_M$. Then the subsheaf $T^*f(\mathscr{J}) \subset \Omega^1_{M'}$ is possibly not locally free, but nevertheless satisfies the integrability condition: indeed, if ω is a germ of 1-form on M, then $T^*f(d\omega) = d(T^*f(\omega))$; therefore,

$$d(T^*f(\mathscr{J})) = T^*f(d\mathscr{J}) \subset T^*f(\mathscr{J} \wedge \Omega^1_M)$$
$$= T^*f(\mathscr{J}) \wedge T^*f(\Omega^1_M) \subset T^*f(\mathscr{J}) \wedge \Omega^1_{M'}.$$

13.c Torsion free connections

The *torsion* of a connection ∇ on the tangent bundle of a manifold M is the $\mathscr{O}_M$-bilinear operator

$$\Theta_M \underset{\mathscr{O}_M}{\otimes} \Theta_M \xrightarrow{\mathscr{T}_\nabla} \Theta_M$$
$$(\xi, \eta) \longmapsto \nabla_\xi \eta - \nabla_\eta \xi - [\xi, \eta].$$

The assertion of $\mathscr{O}_M$-bilinearity is not free, so we will check for instance the $\mathscr{O}_M$-linearity in η: by definition, the term $\nabla_\eta \xi$ is $\mathscr{O}_M$-linear in η; on the other hand, $\eta \mapsto \nabla_\xi \eta$ and $\eta \mapsto [\xi, \eta]$ are two derivations along ξ, hence their difference is $\mathscr{O}_M$-linear.

A *flat coordinate system* on an open set U of M is a family $t_1, \ldots, t_n$ of functions on U with complex values, such that in any point of U the differentials are independent (hence for any $m \in M$, there exist constants $c_i(m)$ such that the functions $t_i + c_i(m)$ form a local coordinate system at m) and that the vector fields $\partial/\partial t_i$ defined by

$$\frac{\partial}{\partial t_i}(dt_j) = \begin{cases} 0 & \text{if } i \neq j \\ 1 & \text{if } i = j \end{cases}$$

are ∇-horizontal. The map $(t_1, \ldots, t_n) : U \to \mathbb{C}^n$ is thus an *étale chart*, as defined in Remark 2.4-(2).

The existence of such a system implies that the connection ∇ is flat (as there locally exists a basis of ∇-horizontal vector fields) and that it has no torsion, i.e., has identically vanishing torsion (as, in the basis (∂_{t_i}), we have $\nabla_{\partial_{t_i}} \partial_{t_j} = 0$ by horizontality and $[\partial_{t_i}, \partial_{t_j}] = 0$ for all i, j).

If $(t_1, \ldots, t_n)$ is a flat coordinate system on U, then for any $(c_1, \ldots, c_n) \in \mathbb{C}^n$ the family $(t_1 + c_1, \ldots, t_n + c_n)$ is another one. We say that it is obtained by *translation* from the first one.

13.4 Theorem (Flat coordinates). *Let M be a 1-connected complex analytic manifold and ∇ a flat torsion free connection on the tangent bundle TM. Fix $m^o \in M$ and let $(\xi^o_1, \ldots, \xi^o_n)$ be a basis of the space $T_{m^o}M$.*

(1) *There exists then a flat coordinate system $(t_1, \ldots, t_n)$ on M such that we have $\partial_{t_i}(m^o) = \xi_i^o$ for any $i = 1, \ldots, n$.*
(2) *The systems which satisfy these conditions are those which can be deduced from the first one by translation.*
(3) *Two flat coordinate systems on M can be deduced one from the other by an affine coordinate change.*

13.5 Remark. With these conditions, we say that the manifold M is equipped with an *affine structure*.

13.6 Example. Let U be a connected open set of $\mathbb{C}^n$ and let $\rho : \tilde{U} \to U$ be a covering. The tangent bundle of U is canonically trivialized. So is the tangent bundle of $\tilde{U}$, which thus has a flat torsion free connection. Then the components $\rho_1, \ldots, \rho_n$ of ρ in the canonical coordinates of $\mathbb{C}^n$ form a flat coordinate system on $\tilde{U}$.

Proof (of Theorem 13.4). Let us choose n independent vectors $\xi_1^o, \ldots, \xi_n^o$ in the fibre $T_{m^o}M$. As the connection ∇ is flat, we can apply the Cauchy-Kowalevski theorem 12.8 and Lemma 15.9: as M is 1-connected, there exists a unique family of ∇-*horizontal* vector fields $(\xi_1, \ldots, \xi_n)$ whose restriction to $T_{m^o}M$ is the given family. As the connection is torsion free and as these fields are horizontal, their Lie brackets vanish.

Let us consider the dual basis $\omega_1, \ldots, \omega_n$. The 1-forms ω_i are *closed*: indeed, for all $i, j, k = 1, \ldots, n$,

$$
\begin{aligned}
d\omega_i(\xi_j, \xi_k) &= \iota_{\xi_k}\left(\iota_{\xi_j} d\omega_i\right) \\
&= \iota_{\xi_k}\left(\mathscr{L}_{\xi_j}\omega_i\right) && \text{because } d(\iota_{\xi_j}\omega_i) = 0 \\
&= \mathscr{L}_{\xi_j}\left(\iota_{\xi_k}\omega_i\right) && \text{because } [\xi_j, \xi_k] = 0 \\
&= 0 && \text{because } \iota_{\xi_k}\omega_i = \delta_{ki}.
\end{aligned}
$$

As M is 1-connected, the space $H^1(M, \mathbb{C})$ is zero, hence the forms ω_i are exact (see Exercise 9.8). The functions t_i such that $\omega_i = dt_i$ induce a mapping having everywhere maximal rank, as the ω_i are everywhere independent. They form thus a flat étale chart.

The remaining part of the proof is left to the reader. $\square$

The torsion free connection associated to a "metric". The result which follows is classic in Riemannian geometry (see for instance [GHL87, Th. 2.51]). It can be adapted without any trouble to the holomorphic framework.

13.7 Theorem (The "metric" connection). *Let g be a symmetric nondegenerate $\mathscr{O}_M$-bilinear form g on the sheaf Θ_M of vector fields on M. There exists then a unique torsion free connection ∇ such that, for any triple (ξ, η, ζ) of germs of vector fields,*

$$
\mathscr{L}_\zeta\left(g(\xi, \eta)\right) = g\left(\nabla_\zeta\xi, \eta\right) + g\left(\xi, \nabla_\zeta\eta\right).
$$
$\square$

We say that the connection ∇ given by the theorem is the *Levi-Civita connection* associated to the "metric". One can formulate the relation above by extending the action of the connection ∇ to sections of the tensor bundles (tensor products, possibly symmetric or skew-symmetric, of copies of Θ_M and Ω_M^1). The relation expresses that the metric tensor is ∇-horizontal.

If the connection ∇ is flat, we say that the "metric" is flat.

13.8 Remark. The notion of *flat function* has then a meaning: a holomorphic function $f : U \to \mathbb{C}$ on an open set U of M is said to be *flat* if the field ξ_f, deduced from the 1-form df by the isomorphism $\Omega_M^1 \xrightarrow{\sim} \Theta_M$ induced by the "metric", is ∇-horizontal.

13.d Symmetric Higgs fields

Let Φ be a Higgs field on the tangent bundle TM of the manifold M. Let us recall that this is a $\mathscr{O}_M$-linear homomorphism

$$\Phi : \Theta_M \longrightarrow \Omega_M^1 \otimes \Theta_M$$

which satisfies the "integrability" condition $\Phi \wedge \Phi = 0$. By using the natural pairing of 1-forms with vector fields, one can regard it as a bilinear homomorphism

$$\Theta_M \otimes \Theta_M \longrightarrow \Theta_M$$
$$\xi, \eta \longmapsto \Phi_\xi(\eta).$$

Therefore, the Higgs field Φ defines a product[9] $\star$ "fibre by fibre" on the tangent bundle by the formula
$$\xi \star \eta = -\Phi_\xi(\eta)$$

(the sign $-$ is introduced for convenience with a later use). The analogue, for Higgs fields, of the absence of torsion for connections is the property of *symmetry*: we say that the Higgs field Φ is symmetric if the associated product $\star$ is *commutative*.

13.9 Proposition (Higgs field and product). *The field Φ is a symmetric Higgs field if and only if the product $\star$ is associative and commutative.*

Proof. The problem is local on M, so we can work in a system of local coordinates $(z_1, \ldots, z_n)$. We can write $\Phi = \sum_{i=1}^n dz_i \otimes \Phi^{(i)}$, where the $\Phi^{(i)} = \Phi_{\partial/\partial z_i}$ define, for any m of the given chart, an endomorphism of $\Theta_{M,m}$. The condition $\Phi \wedge \Phi = 0$ is then equivalent to:

$$\forall i, k, \quad \Phi^{(i)}\Phi^{(k)} = \Phi^{(k)}\Phi^{(i)}.$$

---

[9] One should not confuse this supplementary structure on Θ_M with the natural structure of Lie algebra defined by the bracket of vector fields.

By definition, $\partial_{z_i} \star \partial_{z_j} = -\Phi^{(i)}(\partial_{z_j})$. We conclude that, when Φ is a symmetric Higgs fields, the following holds for all i, j, k:

$$\partial_{z_i} \star (\partial_{z_j} \star \partial_{z_k}) = \partial_{z_i} \star (\partial_{z_k} \star \partial_{z_j}) \qquad \text{(symmetry)}$$
$$= \Phi^{(i)}\left(\Phi^{(k)}(\partial_{z_j})\right)$$
$$= \Phi^{(i)} \circ \Phi^{(k)}(\partial_{z_j}),$$

and similarly

$$(\partial_{z_i} \star \partial_{z_j}) \star \partial_{z_k} = \partial_{z_k} \star (\partial_{z_i} \star \partial_{z_j})$$
$$= \Phi^{(k)} \circ \Phi^{(i)}(\partial_{z_j}),$$

hence the associativity. The converse is proved in the same way.   □

When, moreover, there exists on M a *unit field*, that is, a section $e \in \Gamma(M, \Theta_M)$ such that
$$\xi \star e = e \star \xi = \xi$$
for any germ ξ of holomorphic vector field[10], any tangent space $T_m M$ is equipped with the structure of an associative and commutative algebra with unit, a structure which varies in a holomorphic way with m.

Let $(z_1, \ldots, z_n)$ be a system of local coordinates on an open set U of M. Any polynomial
$$\sum_{|\alpha| \leqslant d} a_\alpha(z_1, \ldots, z_n) \eta_1^{\alpha_1} \cdots \eta_n^{\alpha_n}$$
in the variables $\eta_1, \ldots, \eta_n$ with holomorphic coefficients in $z_1, \ldots, z_n$ defines a holomorphic vector field on U, namely,
$$\sum_{|\alpha| \leqslant d} a_\alpha(z_1, \ldots, z_n) e \star \underbrace{\partial_{z_1} \star \cdots \star \partial_{z_1}}_{\alpha_1 \text{ times}} \star \cdots \star \underbrace{\partial_{z_n} \star \cdots \star \partial_{z_n}}_{\alpha_n \text{ times}}$$

(the field e is useful for the degree 0 term). The set of polynomials such that the associated vector field is zero is an ideal of the ring $\mathscr{O}_M(U)[\eta_1, \ldots, \eta_n]$; the zero locus of this family of polynomials is a subset of $U \times \mathbb{C}^n$. One can check that, if Φ is symmetric and admits a unit field on M, this set is well defined as a subset L_Φ of the total space of the *cotangent bundle* T^*M (of which the open set $U \times \mathbb{C}^n$ considered above is a chart).

13.10 Definition (Canonical coordinates). A system $(x_1, \ldots, x_n)$ of coordinates on an open set of M is said to be *canonical* if the vector fields ∂_{x_i} satisfy
$$\partial_{x_i} \star \partial_{x_j} = \delta_{ij} \partial_{x_i} = \begin{cases} \partial_{x_i} & \text{if } i = j \\ 0 & \text{if } i \neq j. \end{cases}$$

---

[10] Such a field e is then unique.

On such an open set, the field $e = \sum_{i=1}^{n} \partial_{x_i}$ is the unit field. In such a system, we will also have to consider the field $\mathfrak{E} = \sum_{i=1}^{n} x_i \partial_{x_i}$, called the *Euler field*. In the basis $(\partial_{x_1}, \ldots, \partial_{x_n})$, the matrix of the endomorphism $R_0(\bullet) = \mathfrak{E} \star \bullet$ of multiplication by $\mathfrak{E}$ is the diagonal matrix $\mathrm{diag}(x_1, \ldots, x_n)$.

13.11 Remark (On the existence of canonical coordinates). Unlike the case of flat torsion free connections, one does not have, without any supplementary assumption, an analogue of Theorem 13.4 which would give the local existence of canonical coordinates for any symmetric Higgs field on Θ_M. Note that, if such a system exists, then, for any vector field ξ, the operator ℓ_ξ of multiplication by ξ on the fibres of the tangent bundle of M is semisimple. Moreover, the canonical coordinates are the eigenfunctions of the operator $\ell_\mathfrak{E}$ for some field $\mathfrak{E}$.

Conversely, let us assume that, for any ξ, the operator ℓ_ξ is semisimple. There exists then locally[11] a basis of fields $(\eta_i)_{i=1,\ldots,n}$ satisfying $\eta_i \star \eta_j = \delta_{ij} \eta_i$, where δ_{ij} denotes the Kronecker symbol: indeed, there exists locally a basis $(\xi_i)_{i=1,\ldots,n}$ of vector fields which simultaneously diagonalize the operators ℓ_ξ (as they commute); we deduce that there exist functions $\lambda_i(z)$ such that $\xi_i \star \xi_j = \delta_{ij} \lambda_i \xi_i$; it is a matter of showing that the λ_i do not vanish (we will set then $\eta_i = \xi_i/\lambda_i$); this follows from the existence of a unit field e as, if we set $e = \sum_i a_i \xi_i$,

$$\xi_i = \xi_i \star e = \xi_i \star \left(\sum_j a_j \xi_j \right) = a_j \xi_i \star \xi_i = a_i \lambda_i \xi_i.$$

The reader will note[12] that this argument does not give however the local existence of canonical coordinates, because it does not show that the brackets $[\eta_i, \eta_j]$ vanish.

13.12 Exercise (The manifold L_Φ is Lagrangian). Show that, in canonical coordinates, the ideal of polynomials considered above is generated by the polynomials

$$\eta_i \eta_j \quad (i \neq j), \qquad \eta_i(\eta_i - 1), \qquad 1 - \sum_i \eta_i.$$

Deduce that the set L_Φ is the subset of $U \times \mathbb{C}^n$ consisting of the points $(x_1, \ldots, x_n, \eta_1, \ldots, \eta_n)$ where all the η_i except one are zero, the nonzero one being equal to 1.

---

[11] If one assumes the existence of a unit field.

[12] In order to obtain the existence of canonical coordinates, it is necessary and sufficient to ask for a supplementary compatibility condition between the product $\star$ and the Lie derivatives $\mathscr{L}_\xi$. This leads to the notion of *weak Frobenius manifold*, introduced in [HM99]. This condition is equivalent to the geometric condition "L_Φ is a Lagrangian manifold in T^*M" (cf. [Her02]). This condition is realized for Frobenius manifolds that we will meet in Chapter VII (cf. [Aud98a] and §VII.1.8).

Compatibility with a "metric". Given, as in §13.c, a symmetric nondegenerate bilinear form g on TM, that we also call a "metric", we will say that the Higgs field Φ, or the product $\star$, is *compatible* with the metric g if, for any triple (ξ_1, ξ_2, ξ_3) of vector fields on M,

$$g(\xi_1 \star \xi_2, \xi_3) = g(\xi_1, \xi_2 \star \xi_3).$$

Unlike the case of connections, a "metric" does not naturally define by itself a symmetric Higgs field compatible with it.

If a unit field exists, one can consider the differential 1-form e^* defined by adjunction. As a linear form on the tangent bundle, it is expressed by

$$e^*(\xi) = g(e, \xi).$$

We then have $g(\xi_1, \xi_2) = e^*(\xi_1 \star \xi_2)$.

It can be useful to emphasize the trilinear form

$$c(\xi_1, \xi_2, \xi_3) = g(\xi_1 \star \xi_2, \xi_3) = e^*((\xi_1 \star \xi_2) \star \xi_3).$$

13.13 Exercise (commutative Frobenius algebras). Let V be a finite dimensional $\mathbb{C}$-vector space equipped with a linear form ℓ, with a nondegenerate bilinear form b and with a trilinear form t. By considering the isomorphism $B : V \xrightarrow{\sim} V^*$ defined by b, the trilinear form t defines a product on V: one asks that, for all $v_1, v_2 \in V$, both linear forms $B(v_1 \star v_2)$ and $c(v_1, v_2, \bullet)$ coincide. Give conditions on ℓ, b, t expressing that $B^{-1}(\ell)$ is the unit and that the product is commutative and associative.

One then says that $(V, \star, e, b)$ or, in an equivalent way (V, ℓ, b, t), is a *Frobenius algebra*[13].

14 Meromorphic connections

Let $\mathcal{M}$ be a meromorphic bundle with poles along an analytic submanifold Z, i.e., a locally free $\mathcal{O}_M(*Z)$-module of rank d. A connection on $\mathcal{M}$ is defined as in the holomorphic case, namely as a $\mathbb{C}$-linear homomorphism $\nabla : \mathcal{M} \to \Omega^1_M \otimes_{\mathcal{O}_M} \mathcal{M}$ satisfying the Leibniz rule.

Notice that, in a local basis of $\mathcal{M}$ over $\mathcal{O}_M(*Z)$, the matrix Ω of the connection has entries in $\mathcal{O}_M(*Z) \otimes_{\mathcal{O}_M} \Omega^1_M = \Omega^1_M(*Z)$.

The considerations of §11 can be readily extended to the meromorphic case. Note that the matrix P of §11.a now belongs to $\Gamma(U, \mathrm{GL}_d(\mathcal{O}_M(*Z)))$. The sheaf of horizontal sections $\mathcal{M}^\nabla_{|M \smallsetminus Z}$ is then a locally constant sheaf of finite

---

[13] The example originally considered by Frobenius when working on characters of finite groups is the (noncommutative in general) algebra of such a group on a field (see for instance [Lit40, Chap. IV]). That such an algebra is "Frobenius" as defined above is shown for instance in [CR62, Th. 62.1], where one will also find (Th. 61.3) other characterizations of Frobenius algebras.

dimensional vector spaces which corresponds, after Theorem 15.8, to a linear representation of the fundamental group $\pi_1(M \smallsetminus Z, o)$: this is the *monodromy representation* attached to the meromorphic bundle with connection $(\mathcal{M}, \nabla)$.

14.1 Exercise (Torsion of a connection by a logarithmic differential form). Let $f \in \Gamma(M, \mathscr{O}_M(*Z))$ be a meromorphic function on M, with poles along Z, and let $\omega = df/f$ be its logarithmic differential.

(1) Prove that the meromorphic bundle with connection $(\mathscr{O}_M(*Z), d+\omega)$ has trivial monodromy, i.e., that the sheaf of its horizontal sections on $M \smallsetminus Z$ is isomorphic to the constant sheaf $\mathbb{C}_{M \smallsetminus Z}$ (write $d + \omega = f^{-1} \circ d \circ f$).

(2) Deduce that, if $(\mathcal{M}, \nabla)$ is a meromorphic bundle with connection on M with poles along Z, the sheaves of horizontal sections of $(\mathcal{M}, \nabla)_{|M \smallsetminus Z}$ and $(\mathcal{M}, \nabla + \omega \operatorname{Id})_{|M \smallsetminus Z}$ are isomorphic (use Exercise 12.11).

We will say that a connection on a meromorphic bundle is *integrable* (or *flat*) if its restriction to $M \smallsetminus Z$ is an integrable connection on the holomorphic bundle $\mathcal{M}_{|M \smallsetminus Z}$.

If $\mathscr{E}$ is a lattice of a meromorphic bundle with connection $(\mathcal{M}, \nabla)$, it may happen that $\nabla(\mathscr{E})$ is not contained in $\Omega^1_M \otimes_{\mathscr{O}_M} \mathscr{E}$. However,

$$\nabla(\mathscr{E}) \subset \nabla(\mathcal{M}) \subset \Omega^1_M \underset{\mathscr{O}_M}{\otimes} \mathcal{M} = \Omega^1_M(*Z) \underset{\mathscr{O}_M}{\otimes} \mathscr{E}.$$

Therefore, ∇ defines a *meromorphic connection*, which is not necessarily holomorphic, on the bundle $\mathscr{E}$.

We will say that $\mathscr{E}$ is a *logarithmic lattice* of the meromorphic bundle with connection $(\mathcal{M}, \nabla)$ if

$$\nabla(\mathscr{E}) \subset \Omega^1_M \langle \log Z \rangle \underset{\mathscr{O}_M}{\otimes} \mathscr{E}$$

and more generally that it is a *lattice of order* $\leqslant r$ if

$$\nabla(\mathscr{E}) \subset \Omega^1_M \langle (r+1) \log Z \rangle \underset{\mathscr{O}_M}{\otimes} \mathscr{E}$$

and of order r if moreover it does not have order $\leqslant r-1$ (logarithmic $=$ order 0). Therefore, $\mathscr{E}$ is a logarithmic lattice if, in any local basis of $\mathscr{E}$, the connection matrix of ∇ has forms with logarithmic poles along Z as entries.

14.2 Exercise (Behaviour of the order by operations on lattices). Prove that the order of $\mathscr{E} \oplus \mathscr{E}'$ (resp. of $\mathscr{E} \otimes \mathscr{E}'$, resp. of $L(\mathscr{E}, \mathscr{E}')$) is equal to the maximum of the orders of $\mathscr{E}$ and $\mathscr{E}'$.

We will say that a meromorphic bundle with *flat* connection $(\mathcal{M}, \nabla)$ has *regular singularity along Z* if, in the neighbourhood U of any point $z^o \in Z$, there exists a *logarithmic lattice* $\mathscr{E}_U$ of $\mathcal{M}_{|U}$. More generally, we will call *Poincaré rank* of $(\mathcal{M}, \nabla)$ along Z at z^o the minimal order along Z of a lattice in the neighbourhood of z^o.

The condition of regular singularity is *local* on Z and one does not ask for the existence of a logarithmic lattice $\mathscr{E}$ globally on M. Nevertheless, one can show the existence of such a lattice (see [Del70]). We will analyze with more details the structure of connections of one variable, with regular or irregular singularity, in Chapter II.

14.3 Remark (Regular singularity and logarithmic pole). One should clearly distinguish between the notion of regular singularity, which concerns *meromorphic* bundles with a flat connection, and that of logarithmic pole, which concerns *holomorphic* bundles with a flat meromorphic connection. In fact, a meromorphic bundle of rank $\geqslant 2$ with a regular singular connection can contain lattices of arbitrarily large order (see Exercise II.4.4).

14.a Restriction of a meromorphic connection

Let M be the product $D \times M'$, where M' is a complex analytic manifold, D is a disc centered at the origin in $\mathbb{C}$. We will identify M' with the submanifold $\{0\} \times M'$ of M and we will denote by $i : M' \hookrightarrow M$ the inclusion. We will assume below that the hypersurface Z is equal to $D \times Z'$, where Z' is a smooth hypersurface of M'.

Let $(\mathscr{M}, \nabla)$ be a meromorphic bundle with connection on M with poles along Z. The meromorphic bundle $\mathscr{M}$ can be restricted as a meromorphic bundle $\mathscr{M}'$ on M' by the formula $\mathscr{M}' = \mathscr{O}_{M'} \otimes_{i^{-1}\mathscr{O}_M} i^{-1}\mathscr{M}$, analogous to Equation (11.9). It is important, in order that this be meaningful, that the polar set Z of $\mathscr{M}$ intersects *properly*, i.e., along a hypersurface, the submanifold M'.

One can then, exactly as in §11.c, define the restriction

$$\nabla' : \mathscr{M}' \longrightarrow \Omega^1_{M'} \underset{\mathscr{O}_{M'}}{\otimes} \mathscr{M}'$$

of the meromorphic connection ∇. We will say that the meromorphic bundle with connection $(\mathscr{M}', \nabla')$ is the *restriction* (or pullback by the inclusion) to M' of the meromorphic bundle with connection $(\mathscr{M}, \nabla)$.

Notice that, if one restricts $(\mathscr{M}, \nabla)$ to $M \smallsetminus Z$ to obtain a holomorphic bundle with connection, the restricted bundle that one obtains from the latter with the definition of §11.c is nothing but $(\mathscr{M}', \nabla')_{|M' \smallsetminus Z'}$.

If ∇ is integrable, then so is ∇' (cf. Exercise 12.12).

Notice also that, if $\mathscr{E}$ is a lattice of $\mathscr{M}$, its restriction $i^*\mathscr{E}$ is a lattice of $\mathscr{M}'$.

14.b Restriction and residue of connections with logarithmic poles

Let us keep the notation of §14.a but let us now assume that the hypersurface Z is equal to $\{0\} \times M'$; one could, for the logarithmic case which follows, consider a more general situation (where the "normal bundle" of Z in M is trivial), but we will not have to consider such a situation.

Given a holomorphic bundle with meromorphic connection $(\mathscr{E}, \nabla)$ (and not only a meromorphic bundle with connection), assume that $(\mathscr{E}, \nabla)$ is logarithmic along Z. One can define a "restriction" of $(\mathscr{E}, \nabla)$ to Z; this is a holomorphic bundle of rank d with holomorphic connection $\bigtriangledown$ on Z: as a bundle, it is the restriction $E_{|Z}$ of E to Z (i.e., the pullback by the inclusion mapping $i_Z : Z \hookrightarrow M$). Let us be more explicit concerning the connection matrix of the "restriction" $\bigtriangledown$ in a local basis of E and in local coordinates $z_1, \ldots, z_n$ with $Z = \{z_1 = 0\}$: if

$$\Omega = \Omega^{(1)} \frac{dz_1}{z_1} + \sum_{i \geqslant 2} \Omega^{(i)} \, dz_i$$

is the connection matrix, where the $\Omega^{(i)}$ are matrices with holomorphic entries, the desired connection $\bigtriangledown$ has matrix

$$\sum_{i \geqslant 2} \Omega^{(i)}(0, z_2, \ldots, z_n) \, dz_i$$

in the corresponding basis of $E_{|Z}$. One checks that this is independent of the choices and well defines a holomorphic connection on $E_{|Z}$.

The logarithmic connection ∇ also equips the bundle $E_{|Z}$ with an endomorphism: this is the *residue* $\mathrm{Res}_Z \nabla$ of the connection along Z. With the local choices above, it has matrix $\Omega^{(1)}(0, z_2, \ldots, z_n)$. In order to check that this definition is really intrinsic, one considers the composed homomorphism

$$(14.4) \qquad i_Z^{-1} \mathscr{E} \xrightarrow{\ \nabla\ } i_Z^{-1} \Omega_M^1 \langle \log Z \rangle \underset{i_Z^{-1} \mathscr{O}_M}{\otimes} i_Z^{-1} \mathscr{E} \xrightarrow{\ \mathrm{res} \otimes \mathrm{Id}\ } i_Z^* \mathscr{E}.$$

Then, if σ is a local section of $\mathscr{E}$ which vanishes on Z, its image by (14.4) is zero and therefore (14.4) passes to the quotient and defines $\mathrm{Res}_Z \nabla : i_Z^* \mathscr{E} \to i_Z^* \mathscr{E}$.

14.5 Exercise (Restriction, residue and operations).

(1) Prove that the construction of $\bigtriangledown$ from ∇ is compatible with the operations on bundles with connection.

(2) Determine the behaviour of the residue by the operations $\oplus$, $\otimes$ and L on the bundles with logarithmic connection. Check in particular that, if $(\mathscr{O}_M, d)$ denotes the trivial bundle of rank one equipped with the trivial connection, the dual bundle $\mathscr{H}om_{\mathscr{O}_M}(\mathscr{E}, \mathscr{O}_M)$ has residue $- {}^t\mathrm{Res}_Z \nabla$.

14.6 Exercise (Restriction, residue and integrability). Assume that the connection ∇ is integrable.

(1) Prove that so is its "restriction" $\bigtriangledown$.

(2) Show that the residue $\mathrm{Res}_Z \nabla$, regarded as an endomorphism of the bundle $E_{|Z}$, is compatible with the connection $\bigtriangledown$, i.e., when regarded as a section of the bundle $\mathrm{Hom}(E_{|Z}, E_{|Z})$, is a horizontal section with respect to the natural flat connection constructed from $\bigtriangledown$.

14.c Connections of order one

When the connection has order one along Z, one can write locally its matrix as

$$\Omega = z_1^{-1}\left[\Omega^{(1)}\frac{dz_1}{z_1} + \sum_{i\geqslant 2}\Omega^{(i)}\,dz_i\right]$$

where $\Omega^{(i)}$ are holomorphic. One cannot define the restriction to Z of a flat connection by the procedure above, as $\sum_{i\geqslant 2}\Omega^{(i)}(0, z_2,\ldots, z_n)\,dz_i$ is a form which does not necessarily satisfy the integrability condition. However, if *one fixes* a coordinate z_1 on D and if one considers a system of local coordinates $z_2,\ldots, z_n$ on Z, one can equip the bundle $E_{|Z}$ with a "residue" endomorphism R_0 (cf. Remark 9.15) and with a 1-form Φ with values in the endomorphisms of $E_{|Z}$: in a local basis, R_0 has matrix $\Omega^{(1)}(0, z_2,\ldots, z_n)$ and Φ is given by $\sum_{i\geqslant 2}\Omega^{(i)}(0, z_2,\ldots, z_n)\,dz_i$.

14.7 Exercise ("Residue" and integrability). Prove that the integrability of ∇ implies that the following relations are satisfied by R_0 and Φ (they are the analogues of the integrability of the connection $\triangledown$ and of the horizontality with respect to $\triangledown$ of the residue endomorphism, in the logarithmic case)

$$\Phi\wedge\Phi = 0,\quad \Phi_\xi\circ R_0 = R_0\circ\Phi_\xi\quad \text{for any vector field } \xi \text{ on } M.$$

In particular, Φ is a Higgs field on $E_{|Z}$ and $(E_{|Z},\Phi)$ is a *Higgs bundle*. These objects depend on the choice of coordinate on D by a multiplicative constant.

14.8 Remark. However, it is not in general possible to define, for connections with pole of order $\geqslant 1$ along Z, a residue endomorphism whose matrix in a local basis is formed by the residues of the coefficients of Ω. Indeed, if the rank of the bundle is $\geqslant 2$, such a matrix does not have a suitable behaviour under a holomorphic base change.

14.9 Exercise (Higgs field, "residue" and operations).

(1) Prove that the construction of Φ from ∇ is compatible with the operations on Higgs bundles.
(2) Determine the behaviour of the "residue" R_0 by the operations $\oplus$, $\otimes$, L.

15 Locally constant sheaves

15.a Locally constant sheaves of sets

Let $\mathscr{F}$ be a sheaf of sets on a topological space X. The disjoint union $\widetilde{\mathscr{F}} := \coprod_{x\in X}\mathscr{F}_x$ of the germs of $\mathscr{F}$ at the points of X is equipped with a natural topology, for which a basis of open sets is given by the sets

$$U_s = \coprod_{x\in U}\{s_x\} \subset \coprod_{x\in X}\mathscr{F}_x,$$

where U is an open set of X and $s \in \Gamma(U, \mathscr{F})$. For this topology, the natural projection $p : \widetilde{\mathscr{F}} \to X$ is continuous and the set $\Gamma(U, \mathscr{F})$ is identified with that of continuous sections $\sigma : U \to \widetilde{\mathscr{F}}$ of the projection p, i.e., the continuous mappings which satisfy $p \circ \sigma = \mathrm{Id}_U$. Moreover, the projection p is a local homeomorphism and the topology induced on each fibre $p^{-1}(x) = \mathscr{F}_x$ is the discrete topology (cf. [God64]). We say that $\widetilde{\mathscr{F}}$, equipped with the projection p, is the *étale cover* associated to the sheaf $\mathscr{F}$.

15.1 Exercise (Locally constant sheaves and coverings). Assume that X is connected. Prove that the sheaf $\mathscr{F}$ is constant (cf. §2.6) if and only if the étale cover $p : \widetilde{\mathscr{F}} \to X$ is homeomorphic to the projection $p_2 : \mathscr{F}_x \times X \to X$, for any $x \in X$. Deduce that the following properties are equivalent:

(1) the sheaf $\mathscr{F}$ is locally isomorphic to a constant sheaf;
(2) the projection $p : \widetilde{\mathscr{F}} \to X$ is a covering map.

When one of both equivalent properties above is satisfied, we say that the sheaf $\mathscr{F}$ is *locally constant*. One thus sees that the category of locally constant sheaves of sets is isomorphic to that of coverings (check the behaviour of morphisms). From the theory of coverings (see for instance [God71]), we deduce:

15.2 Corollary. *Any locally constant sheaf on a 1-connected space is constant.* □

15.3 Exercise (Locally constant sheaves and coverings, continuation).

(1) Prove that, if $f : X' \to X$ is a continuous mapping and if $\mathscr{F}$ is a locally constant sheaf on X, the pullback sheaf $f^{-1}\mathscr{F}$ is locally constant.
(2) Let $\pi : X \to X'$ be a finite covering and let $\mathscr{F}$ be a locally constant sheaf on X. Show that the direct image sheaf $\pi_*\mathscr{F}$ is locally constant.

15.b Locally constant sheaves of finite dimensional $\mathbb{C}$-vector spaces

We will now be more precise concerning the correspondence

$$\textit{locally constant sheaves} \longleftrightarrow \textit{coverings}$$

for sheaves of finite dimensional $\mathbb{C}$-vector spaces. Such a sheaf is locally constant if there exist an open covering $\mathfrak{U}$ of X and, for any open set U of $\mathfrak{U}$, an isomorphism of $\mathscr{F}_{|U}$ with the constant sheaf $\mathbb{C}^d_U$. In the following, all the sheaves will be sheaves of finite dimensional $\mathbb{C}$-vector spaces and we will simply write "locally constant". One also finds in the literature the term *local system*[14] instead of locally constant sheaf.

[14] This in fact abbreviates "local system of coefficients" when one considers cohomology with coefficients in a locally constant sheaf.

15.4 Exercise. In Exercise 15.3, compute the rank of the sheaves $f^{-1}\mathscr{F}$ and $\pi_*\mathscr{F}$ in term of that of $\mathscr{F}$.

We thus have defined a subcategory of the category (cf. §17) of sheaves on X, namely that of locally constant sheaves of finite rank. We say that this subcategory is *full*, as the set of homomorphisms of locally constant sheaves is equal to that of all homomorphisms of sheaves.

If o is a given point of X (also called the *base point*), we have a "restriction" functor from this category to that of finite dimensional vector spaces: one associates to $\mathscr{F}$ its germ $\mathscr{F}_o$ and to $\varphi : \mathscr{F} \to \mathscr{G}$ its germ $\varphi_o : \mathscr{F}_o \to \mathscr{G}_o$.

If $\pi : \widetilde{X} \to X$ is a fixed universal covering space of X, we have a "global multivalued sections" functor from this category to that of finite dimensional vector spaces: one associates to $\mathscr{F}$ the space $\Gamma(\widetilde{X}, \pi^{-1}\mathscr{F})$ and to $\varphi : \mathscr{F} \to \mathscr{G}$ the induced morphism $\Gamma(\widetilde{X}, \pi^{-1}\mathscr{F}) \to \Gamma(\widetilde{X}, \pi^{-1}\mathscr{G})$.

15.5 Lemma (Locally constant sheaves on the square are constant). *Let us assume that X is equal to the interval $[0,1]$ or to the square $[0,1] \times [0,1]$ and let us denote by 0 the origin of X. Let $\mathscr{F}, \mathscr{G}$ be locally constant sheaves on X.*

(1) *If $s_0 \in \mathscr{F}_0$, there exists a unique section s of $\mathscr{F}$ on X such that $s(0) = s_0$.*
(2) *Let $\varphi_0 : \mathscr{F}_0 \to \mathscr{G}_0$ be a homomorphism. There exists a unique homomorphism $\varphi : \mathscr{F} \to \mathscr{G}$ whose restriction to $\mathscr{F}_0$ is φ_0.*

Proof. For the first point, adapt the proof of [God71, Lemmes 1.1 and 1.2, p. 129]. The second point follows from the first one, if we remark that $\mathscr{H}om_{\mathbb{C}_X}(\mathscr{F}, \mathscr{G})$ is also a locally constant sheaf.   □

15.6 Exercise.

(1) Prove that, in the lemma above, if φ_0 is an isomorphism, then so is φ (consider the determinant of φ).
(2) Show that, for X as in the lemma, any locally constant sheaf $\mathscr{F}$ is isomorphic to the constant sheaf $\pi^{-1}\mathscr{F}_0$, if $\pi : X \to \{0\}$ is the constant map, and that such an isomorphism is unique if one moreover asks that its germ at 0 is equal to identity.
(3) Prove that the lemma above can be interpreted by saying that, on the space $X = [0,1]$ or $X = [0,1]^2$ (with 0 as base point), the restriction functor is an equivalence (cf. §17) between the category of locally constant sheaves on X and that of finite dimensional $\mathbb{C}$-vector spaces.

15.c Linear representations of groups (abstract)

Let Π be a group and let F be a $\mathbb{C}$-vector space (here, $\mathbb{C}$ could denote any field). A *linear representation* of Π in F (cf. for instance [Kir76, §7]) is a group homomorphism $\rho : \Pi \to \mathrm{Aut}(F)$, in other words a left linear action of Π on F.

If Π acts on F_1 and F_2, then Π acts on $\mathrm{Hom}_{\mathbb{C}}(F_1, F_2)$ by

$$\gamma \in \Pi \longmapsto \rho(\gamma) : \left[\varphi \longmapsto \rho_2(\gamma) \circ \varphi \circ \rho_1(\gamma)^{-1}\right].$$

A *homomorphism* of the representation ρ_1 to the representation ρ_2 is a homomorphism $\varphi \in \mathrm{Hom}_{\mathbb{C}}(F_1, F_2)$ such that

$$\forall \gamma \in \Pi, \quad \rho(\gamma)(\varphi) = \varphi.$$

We denote this set by

$$\mathrm{Hom}_{\Pi}(\rho_1, \rho_2) = \mathrm{Hom}_{\mathbb{C}}(F_1, F_2)^{\Pi} \subset \mathrm{Hom}_{\mathbb{C}}(F_1, F_2).$$

In other words, $\varphi : F_1 \to F_2$ is a homomorphism of representations (one also says *intertwining operator*) if and only if, for any $\gamma \in \Pi$, the following diagram commutes:

$$
\begin{array}{ccc}
F_1 & \xrightarrow{\;\varphi\;} & F_2 \\
\rho_1(\gamma) \downarrow \wr & & \wr \downarrow \rho_2(\gamma) \\
F_1 & \xrightarrow{\;\varphi\;} & F_2
\end{array}
$$

In the language of the theory of categories (cf. §17), we have defined the category of linear representations of the group Π, whose objects are the linear representations and the morphisms are the homomorphisms of representations. It contains the subcategory of finite dimensional representations, which is *full*. It is equipped with a "forget" functor to the category of vector spaces.

In what follows, we will take for Π the *fundamental group* $\pi_1(X, o)$ of the space X with base point o (cf. for instance [God71]). We will denote by $[\gamma] \in \pi_1(X, o)$ the homotopy class of a continuous loop $\gamma : [0, 1] \to X$ based at o, i.e., such that $\gamma(0) = \gamma(1) = o$. It will be convenient to use a convention opposite to that of [God71] for the product. In other words, we will denote by $[\gamma] \cdot [\gamma']$ the class of the loop obtained by following first γ', then γ (in accordance with the composition of mappings, as in [Del70]).

15.d An equivalence of categories

Let $\mathscr{F}$ be a locally constant sheaf on X. If γ is a continuous loop based at o, the sheaf $\gamma^{-1}\mathscr{F}$ is locally constant on $[0, 1]$ (Exercise 15.3). The first part of Lemma 15.5 shows the existence of a map

$$T_\gamma : \mathscr{F}_o = (\gamma^{-1}\mathscr{F})_0 \longrightarrow (\gamma^{-1}\mathscr{F})_1 = \mathscr{F}_o.$$

It associates to any $s_0 \in \mathscr{F}_o$ the value $s(1) \in (\gamma^{-1}\mathscr{F})_1 = \mathscr{F}_o$ of the unique section s of $\gamma^{-1}\mathscr{F}$ such that $s(0) = s_0$. The uniqueness property implies that this mapping T_γ is *linear*.

The second part of the same lemma also shows that this mapping only depends on the homotopy class of γ and we denote it by $T_{[\gamma]}$, or also by $T_{[\gamma]}^{\mathscr{F}}$ to recall the dependence with respect to $\mathscr{F}$.

15.7 Exercise (The monodromy representation is functorial).

(1) Prove that $[\gamma] \mapsto T_{[\gamma]}$ is a linear representation of $\pi_1(X, o)$ in $\mathscr{F}_o$.
(2) Show that $\mathscr{F} \mapsto T^{\mathscr{F}}$ is a functor from the category of locally constant sheaves of finite rank to that of finite dimensional linear representations of the group $\pi_1(X, o)$.

15.8 Theorem (The monodromy representation is an equivalence).
The functor $\mathscr{F} \mapsto T^{\mathscr{F}}$ is an equivalence between the category of locally constant sheaves of finite rank on X and that of finite dimensional representations of $\pi_1(X, o)$.

Proof. Let us apply the criterion given in §17. For the full faithfulness, it is a matter of verifying that any element $\varphi_o \in \mathrm{Hom}_{\mathbb{C}}(\mathscr{F}_o, \mathscr{G}_o)^{\pi_1}$ can be lifted in a unique way as a homomorphism $\varphi : \mathscr{F} \to \mathscr{G}$. This is a consequence of the following lemma, applied to the locally constant sheaf $\mathscr{H}om_{\mathbb{C}_X}(\mathscr{F}, \mathscr{G})$.

15.9 Lemma (An existence criterion for a global section). *Let $\mathscr{F}$ be a locally constant sheaf on X and pick $s_o \in \mathscr{F}_o$. There exists a section $s \in \Gamma(X, \mathscr{F})$ such that $s(o) = s_o$ if and only if $T_{[\gamma]}(s_o) = s_o$ for any $[\gamma] \in \pi_1(X, o)$. If such a section s exists, it is unique.*

Proof. Fix $x \in X$ and let $\eta : [0, 1] \to X$ be a continuous path from o to x (recall that X is assumed to be path-connected). Lemma 15.5 applied to the sheaf $\eta^{-1}\mathscr{F}$ enables us to define $s_x \in \mathscr{F}_x$ from s_o. Moreover, the condition $T_{[\gamma]}(s_o) = s_o$ for any $[\gamma] \in \pi_1(X, o)$ implies that s_x does not depend on the chosen path η. We then see that $x \mapsto s(x) := s_x$ is a section of $\mathscr{F}$.

The set of points where the germs of two sections coincide is open in X. As $\mathscr{F}$ is locally constant, it is also closed, hence the uniqueness assertion. $\square$

Let us now sketch the proof of the essential surjectivity. Let $(\widetilde{X}, \widetilde{o})$ be a universal covering space of (X, o). Recall (cf. [God71, p. 134]) that $\pi_1(X, o)$ acts on $\widetilde{X}$: if $\widetilde{\gamma}$ is the lift of a loop γ based at $\widetilde{o}$, there exists a unique automorphism $h_{[\gamma]}$ of the covering $\widetilde{X}$ such that $h_{[\gamma]}(\widetilde{o}) = \widetilde{\gamma}(1)$. One then denotes by $\widetilde{F}$ the quotient space of $\widetilde{X} \times \mathscr{F}_o$ by the relation $(\widetilde{x}, s_o) \sim (h_{[\gamma]^{-1}}(\widetilde{x}), \rho_{[\gamma]}(s_o))$ ($[\gamma] \in \pi_1(X, o)$). One checks that the natural projection to X makes $\widetilde{F}$ a covering: this is the étale cover of the desired sheaf $\mathscr{F}$. In order to check that the representation associated to $\mathscr{F}$ is equal to ρ, one takes the pullback of $\widetilde{X} \times \mathscr{F}_o$ by $\widetilde{\gamma}$: the section which takes value s_o at $\widetilde{o}$ also takes value s_o at $h_{[\gamma]}(\widetilde{o})$; one identifies then the points $(h_{[\gamma]}(\widetilde{o}), s_o)$ and $(\widetilde{o}, \rho_{[\gamma]}(s_o))$. $\square$

15.10 Corollary (Analytic extension). *Let $V \subset U$ be two connected open sets of X, containing the base point, such that the canonical homomorphism $\pi_1(V, o) \to \pi_1(U, o)$ is an isomorphism, and let $\mathscr{G}$ be a locally constant sheaf on V.*

(1) *There exists a locally constant sheaf $\mathscr{F}$ on U and an isomorphism $\mathscr{F}_{|V} \xrightarrow{\sim} \mathscr{G}$.*

(2) *The restriction mapping* $\Gamma(U, \mathscr{F}) \to \Gamma(V, \mathscr{F})$ *is an isomorphism.*

(3) *If we have two locally constant sheaves* $\mathscr{F}$ *and* $\mathscr{F}'$ *on* U, *any homomorphism* $\psi : \mathscr{F}_{|V} \to \mathscr{F}'_{|V}$ *can be extended in a unique way as a homomorphism* $\mathscr{F} \to \mathscr{F}'$. □

15.11 Exercise (Another equivalence of categories). Let $\pi : \widetilde{X} \to X$ be a fixed universal covering space of X.

(1) Let $\mathscr{F}$ be a locally constant sheaf of rank d on X. Prove that $\Gamma(\widetilde{X}, \pi^{-1}\mathscr{F})$ is a finite dimensional vector space, equipped with an action of the group G of automorphisms of the covering π. We denote by $\rho^{\mathscr{F}}$ the corresponding representation of G.

(2) Prove that the functor $\mathscr{F} \mapsto \rho^{\mathscr{F}}$ is an equivalence of categories.

(3) Compare this functor with that defined in Exercise 15.7.

15.12 Exercise (Examples of locally constant sheaves).

(1) Describe all locally constant sheaves of rank d on $X = \mathbb{C}^*$.

(2) Describe all locally constant sheaves of rank one on the space $X = \mathbb{P}^1 \smallsetminus \{0, 1, \infty\} = \mathbb{C}^* \smallsetminus \{1\}$.

(3) Prove that two linear representations $\rho, \rho' : \pi_1(X, o) \to \mathrm{GL}_d(\mathbb{C})$ define *isomorphic* locally constant sheaves if and only if they are conjugate, i.e., there exists an invertible matrix C such that, for any $[\gamma] \in \pi_1$ we have $\rho'([\gamma]) = C^{-1}\rho([\gamma])C$.

16 Integrable deformations and isomonodromic deformations

Let us start with an analytic family of complex analytic manifolds whose topology (*a fortiori* the fundamental group) does not change. All these manifolds are thus pairwise homeomorphic. Given a family of linear differential systems on this family of manifolds, that is, a family of vector bundles, each one equipped with a holomorphic flat connection depending analytically on the parameters, we produce the locally constant sheaf of horizontal sections of this connection on each manifold in the family. As the notion of locally constant sheaf is topological, one can regard all these sheaves as sheaves on the same topological manifold. When these sheaves are all isomorphic to the same constant sheaf, we say that the family is *isomonodromic*.

Let us be more precise. We thus have an analytic mapping $\pi : M \to X$ *everywhere of maximal rank* between two connected complex analytic manifolds. We make the assumption that π is a locally trivial topological fibration, that is, for any $x^o \in X$, there exists a neighbourhood V of x^o in X and a homeomorphism h making the following diagram commute:

$$V \times \pi^{-1}(x^o) \xrightarrow[\sim]{h} \pi^{-1}(V)$$

$$p_1 \searrow \quad \swarrow \pi$$
$$V$$

and whose restriction to $\pi^{-1}(x^o)$ is the identity. Note that, as π is everywhere of maximal rank, the fibres of π are analytic submanifolds of M.

16.1 Examples. It is often necessary to consider more general situations than the projection π of a product $M \times X$ on X.

(1) When π is a *proper* mapping, a theorem by Ehresmann (cf. [Ehr47, Ehr51], as well as [Wol64] for a proof) says that π is a locally trivial C^∞ fibration (i.e., one can choose a C^∞ diffeomorphism for h in the neighbourhood of any x^o). It is in general impossible to choose h holomorphic.
(2) When π is not proper, the *First Isotopy Lemma* of R. Thom (cf. for instance [Dim92, Chap. 1] and the references given therein) often enables one to show that π is a locally trivial topological fibration over an open set dense of X. Such is the case for instance when there exists an analytic manifold $\overline{M}$ containing M as a dense open subset, equipped with a *proper* holomorphic mapping $\overline{\pi} : \overline{M} \to X$ extending π. For example, a n-variable polynomial defines a fibration $\pi : \mathbb{C}^n \smallsetminus \pi^{-1}(\Sigma) \to \mathbb{C} \smallsetminus \Sigma$, where Σ is a finite set of points, called the "generalized critical values" of the polynomial π.
(3) Let $X = X_d \subset \mathbb{C}^d$ be the open set consisting of the points with pairwise distinct coordinates and set $M = X_{d+1}$. Then the projection $\pi : X_{d+1} \to X_d$ which forgets the last coordinate is a locally trivial C^∞ fibration.

The sheaf $\Theta_{M/X}$ of vector fields tangent to the fibres of π is by definition the kernel of the tangent map $T\pi : \Theta_M \to \pi^*\Theta_X$. This is a locally free sheaf, as π has maximal rank. One constructs by duality the sheaf $\Omega^1_{M/X}$ of relative differential 1-forms.

In any system of local coordinates $(z_1, \ldots, z_p, x_1, \ldots, x_q)$ where π is written as $\pi(z, x) = x$, any relative 1-form can be written as $\sum_{i=1}^p \varphi_i(z, x) \, dz_i$.

The sheaf $\Omega^1_{M/X}$ is naturally a quotient of the sheaf Ω^1_M. It is equipped with a relative differential $d_{M/X} : \Omega^1_{M/X} \to \Omega^2_{M/X}$. Similarly, there is a relative differential $d_{M/X} : \mathscr{O}_M \to \Omega^1_{M/X}$ such that the identity $d_{M/X} \circ d_{M/X} = 0$ holds: in local coordinates (z, x) as above,

$$d_{M/X} f = \sum_{i=1}^p \frac{\partial f}{\partial z_i} \, dz_i.$$

Let E be a bundle on M. A *relative connection* $\nabla_{M/X}$ on E is a homomorphism $\mathscr{E} \to \Omega^1_{M/X} \otimes_{\mathscr{O}_M} \mathscr{E}$ which is linear over the sheaf of rings $\pi^{-1}\mathscr{O}_X$

and which satisfies the Leibniz rule

$$\nabla_{M/X}(f \cdot s) = f\nabla_{M/X}s + d_{M/X}f \otimes s.$$

The relative connection is said to be *integrable* (or *flat*) if the relative curvature $\nabla_{M/X} \circ \nabla_{M/X}$ vanishes. One can readily express these notions on the connection matrix in a local basis of E and in local coordinates adapted to π as above. One also verifies that, when X is reduced to a point, one recovers the notions introduced in §§11 and 12.

16.2 Example (The relative connection associated to a connection). If $\nabla : \mathscr{E} \to \Omega^1_M \otimes \mathscr{E}$ is a connection on the bundle E, the composition of ∇ with the projection $\Omega^1_M \otimes \mathscr{E} \to \Omega^1_{M/X} \otimes \mathscr{E}$ is a relative connection $\nabla_{M/X}$. If ∇ is integrable, then so is $\nabla_{M/X}$. In a local basis of E and in local coordinates adapted to π, the matrix of $\nabla_{M/X}$ is obtained by forgetting the terms containing dz_j $(j = 1, \ldots, q)$ in that of ∇.

16.3 Exercise (The Cauchy-Kowalevski theorem with parameters). Let $\nabla_{M/X}$ be an *integrable* relative connection on some bundle E on M. Prove the analogue of Theorem 12.8 where one changes the sheaf $\mathbb{C}_M$ with $\pi^{-1}\mathscr{O}_X$ and the notion of a locally constant sheaf of $\mathbb{C}$-vector spaces with that of a locally constant sheaf of locally free $\pi^{-1}\mathscr{O}_X$-modules (hint: see [Del70, Th. 2.23]).

16.4 Definition (of an isomonodromic family). An integrable relative connection $\nabla_{M/X}$ on a bundle E is an *isomonodromic* family if the isomorphism class of the locally constant sheaf (of $\mathbb{C}$-vector spaces) $\operatorname{Ker} \nabla_{M/X|\pi^{-1}(x)}$ does not depend on the point $x \in X$.

Let us make it clear here that the restriction is taken in the analytic sense, that is, recalling (cf. previous exercise) that $\operatorname{Ker} \nabla_{M/X}$ is a $\pi^{-1}\mathscr{O}_X$-module,

$$\operatorname{Ker} \nabla_{M/X|\pi^{-1}(x)} = \mathscr{O}_{\pi^{-1}(x)} \otimes_{\pi^{-1}\mathscr{O}_X} \operatorname{Ker} \nabla_{M/X}.$$

Let us now explain with more detail the expression "does not depend on $x \in X$". For any $x^o \in X$, let V be a neighbourhood of x^o and let h be a homeomorphism $V \times \pi^{-1}(x^o) \xrightarrow{\sim} \pi^{-1}(V)$ as above. We deduce, for any $x \in V$, a homeomorphism $h_x : \pi^{-1}(x^o) \to \pi^{-1}(x)$. It enables us to pull back the sheaf $\operatorname{Ker} \nabla_{M/X|\pi^{-1}(x)}$ on $\pi^{-1}(x^o)$. We get in this way a locally constant sheaf $\mathscr{F}_x$ on $\pi^{-1}(x^o)$. The isomonodromy property means that, for x in a sufficiently small neighbourhood of x^o, the sheaves $\mathscr{F}_x$ are pairwise isomorphic.

16.5 Exercise. Prove that the local isomonodromy property above does not depend on the choice of the homeomorphism h (it is a matter of verifying that, if $\mathscr{F}$ is a locally constant sheaf of $\mathbb{C}$-vector spaces on a manifold N and if $g : N \to N$ is a homeomorphism homotopic to identity, then the sheaves $\mathscr{F}$ and $g^{-1}\mathscr{F}$ are isomorphic).

This exercise gives a nonambiguous meaning to the local notion of isomonodromy and therefore enables one to define it globally.

16.6 Proposition (Integrability implies isomonodromy). *Let* $\nabla : \mathscr{E} \to \Omega^1_M \otimes \mathscr{E}$ *be a flat connection on some bundle E on M. Then the associated relative connection* $\nabla_{M/X}$ *defines an isomonodromic family.*

Proof. We have already mentioned that this relative connection is flat. One then checks that
$$\operatorname{Ker} \nabla_{M/X} = \pi^{-1}\mathscr{O}_X \otimes_{\mathbb{C}} \operatorname{Ker} \nabla.$$

Lastly, if V is a sufficiently small 1-connected neighbourhood of $x^o \in X$, the fundamental group of $\pi^{-1}(V) \simeq V \times \pi^{-1}(x^o)$ is equal to that of $\pi^{-1}(x^o)$. It follows from Equivalence 15.8 that $\mathscr{F} = \operatorname{Ker} \nabla$ is isomorphic to the pullback by $h^{-1} \circ p_2$ of its restriction to $\pi^{-1}(x^o)$. We thus get isomonodromy. $\square$

It is remarkable that, with only a weak assumption on fibres of π, the converse of this statement is true, at least locally on X:

16.7 Theorem (Isomonodromy implies integrability). *Let us assume that the fundamental group of the fibres of π has finite type. Then, if $\nabla_{M/X}$ is an integrable relative connection on some bundle E on M, defining an isomonodromic family, it comes, locally on X, from an integrable connection ∇ on E.*

16.8 Remark. This assumption on the fundamental group of the fibres is satisfied for all the manifolds that we will consider in the remaining part of this book (quasiprojective varieties). In general, verifying this assumption will be easy. Our main example will be that of the complex line minus a finite number of points.

Proof (Sketch). Fix $x^o \in X$. According to the relative Cauchy-Kowalevski theorem, it is a matter of proving that, if $\mathscr{G}$ is a locally constant sheaf of locally free $\pi^{-1}\mathscr{O}_X$-modules on M, satisfying the isomonodromy property, there exists, up to changing X with a sufficiently small neighbourhood of x^o, a locally constant sheaf $\mathscr{F}$ of $\mathbb{C}$-vector spaces on M such that $\mathscr{G} = \pi^{-1}\mathscr{O}_X \otimes_{\mathbb{C}} \mathscr{F}$.

We assume that the fundamental group $\pi_1(\pi^{-1}(x^o), \star)$ is generated by a finite number of classes $\gamma_1, \ldots, \gamma_p$ of loops based at $\star$. A linear representation of rank d of this group consists in giving p matrices $T_1, \ldots, T_p \in \operatorname{GL}_d(\mathbb{C})$ satisfying the same relations as the γ_i do. The set Rep of these representations is thus the closed subset of $(\operatorname{GL}_d(\mathbb{C}))^p$ defined by the algebraic equations induced by the relations. These are equations like
$$T_{i_1}^{n_1} \cdots T_{i_r}^{n_r} - \operatorname{Id} = 0.$$

The group $\operatorname{GL}_d(\mathbb{C})$ acts on the product by
$$P \cdot (T_1, \ldots, T_p) = (PT_1P^{-1}, \ldots, PT_pP^{-1}).$$

The orbit of a representation ρ^o consists in the set of representations equivalent to ρ^o.

The assumption of the theorem implies that, on some neighbourhood V of x^o, there exists an analytic mapping $V \to \text{Rep}$, the image of which is contained in the orbit $\text{GL}_d(\mathbb{C}) \cdot \rho^o$ (isomonodromy), if ρ^o is the representation corresponding to the locally constant sheaf $\mathscr{G}_{|\pi^{-1}(x^o)}$, and which sends x^o to ρ^o. It is then a matter of showing that this mapping can be lifted, possibly after restricting V, as an analytic mapping $V \to \text{GL}_d(\mathbb{C})$ sending x^o to Id, that is, such that the following diagram commutes:

$$
\begin{array}{ccc}
 & \text{GL}_d(\mathbb{C}) & \\
 \nearrow & \downarrow P \mapsto P \cdot \rho^o & \\
 V \longrightarrow & \text{GL}_d(\mathbb{C}) \cdot \rho^o &
\end{array}
$$

This is possible because the mapping $\text{GL}_d(\mathbb{C}) \to \text{GL}_d(\mathbb{C}) \cdot \rho^o$ has everywhere maximal rank, as the group acts transitively. $\qquad\square$

16.9 Remark. It is possible to extend the domain of existence of the integrable connection ∇, by using a vanishing theorem, cf. [Bol98].

17 Appendix: the language of categories

We refer to [ML71] for more details on what follows.

A *category* $\mathscr{C}$ consists

(1) of a family of *objects* $\text{Ob}(\mathscr{C})$,
(2) for any pair (X, Y) of objects of $\mathscr{C}$, of a *set* $\text{Hom}_\mathscr{C}(X, Y)$, the elements of which are called *morphisms* of X to Y,
(3) for any triple (X, Y, Z) of objects of $\mathscr{C}$, of a mapping (called *composition*)

$$
\text{Hom}_\mathscr{C}(X, Y) \times \text{Hom}_\mathscr{C}(Y, Z) \xrightarrow{\ \circ\ } \text{Hom}_\mathscr{C}(X, Z)
$$
$$
(f, g) \longmapsto g \circ f
$$

satisfying the following properties:

(a) the composition is associative,
(b) for any object X of $\mathscr{C}$, there exists an element $\text{Id}_X \in \text{Hom}_\mathscr{C}(X, X)$ satisfying $f \circ \text{Id}_X = f$ and $\text{Id}_Y \circ f = f$, for any $f \in \text{Hom}_\mathscr{C}(X, Y)$. This element is necessarily unique.

An $f \in \text{Hom}_\mathscr{C}(X, Y)$ is an *isomorphism* if there exists $g \in \text{Hom}_\mathscr{C}(Y, X)$ with $f \circ g = \text{Id}_Y$ and $g \circ f = \text{Id}_X$. According to associativity, such an inverse is unique.

A *functor* F from a category $\mathscr{C}$ to a category $\mathscr{C}'$ consists in giving a mapping $F : \mathrm{Ob}(\mathscr{C}) \to \mathrm{Ob}(\mathscr{C})$ and, for any pair (X, Y) of objects of $\mathscr{C}$, of a mapping $F : \mathrm{Hom}(X, Y) \to \mathrm{Hom}(F(X), F(Y))$ compatible with composition and preserving the identity morphisms.

We let the reader define the notion of morphism between two functors, also called *natural transformation* of functors, and then the notion of isomorphism of functors.

If, for a given functor F, there exists an inverse functor F', that is, such that $F \circ F'$ and $F' \circ F$ are *equal* to the identity functors of each category, we say that the functor F is an isomorphism of categories.

17.1 Exercise. Check that the functor which, to any locally constant sheaf of sets associates its étale cover, is an isomorphism from the category of these sheaves to that of coverings.

More often, there exists a *quasi-inverse* functor F', i.e., such that $F \circ F'$ and $F' \circ F$ are *isomorphic* (and not necessarily equal) to the "identity" functors of the corresponding categories. Such a quasi-inverse functor is not necessarily unique. If such a quasi-inverse functor exists, we say that the categories are *equivalent*. It can be difficult to explicitly construct a quasi-inverse functor for a given functor. In order to check that a functor admits a quasi-inverse functor (and thus that it induces an equivalence of categories), one usually refers to the criterion below.

We say that a functor $F : \mathscr{C} \to \mathscr{C}'$ is *fully faithful* if, for any pair (X, Y) of objects of $\mathscr{C}$ the map $F : \mathrm{Hom}(X, Y) \to \mathrm{Hom}(F(X), F(Y))$ is a bijection; we say that it is *essentially surjective* if, for any object X' of $\mathscr{C}'$, there exists an object X of $\mathscr{C}$ such that $F(X)$ is isomorphic to X'.

The criterion says that a functor F is an *equivalence of categories* if and only if it is fully faithful and essentially surjective. Indeed, when the latter properties are satisfied, one can show that there exists a quasi-inverse functor F' (see for instance [ML71, p. 91]).

Remark. In this book, we repeatedly use the notion of equivalence of categories. This is a convenient way, although somewhat formal, to express many properties:

- The equivalence of categories enables us to manipulate in different ways the objects of a category, by working directly in an equivalent category. This is what we have done with vector bundles. In this case, neither category (bundles and locally free sheaves) is simpler than the other, but depending on the question, we have "a better feeling" with one of them.

- One of the two categories is simpler to manipulate than the other one. One can then consider that equivalence plays the role of a *classification*, up to isomorphism, of the objects of the more complicated category. The equivalence bundles $\leftrightarrow$ nonabelian 1-cohomology of §5 can be regarded—deceptively however—in this framework. On the other hand, for instance, the Riemann-Hilbert correspondence of Chapter II really falls within the scope of this framework.

- It may happen that the equivalence emphasizes hidden properties of one category which are visible in the other one. The equivalence enables then to simply exhibit these hidden properties. A known example is the equivalence between the category of *regular holonomic $\mathscr{D}$-modules* and that of *perverse sheaves*, which shows that the latter is abelian.

Lastly, when both categories possess supplementary operations (direct sum or tensor product for instance), it is in general important to verify that the functor which establishes the equivalence preserves these operations.

I

Holomorphic vector bundles on the Riemann sphere

In the following, the Riemann sphere is denoted by $\mathbb{P}^1$: this is the complex projective line equipped with its usual topology and with the structure of complex analytic manifold as described in §0.2.2-4.

1 Cohomology of $\mathbb{C}$, $\mathbb{C}^*$ and $\mathbb{P}^1$

We denote by t the coordinate on $\mathbb{C}$. Let $\mathscr{C}_{\mathbb{C}}^\infty$ be the sheaf of C^∞ functions on $\mathbb{C}$. It is a sheaf of rings, which contains the sheaf $\mathscr{O}_{\mathbb{C}}$ of holomorphic functions as a subsheaf of rings. It is then in particular a sheaf of $\mathscr{O}_{\mathbb{C}}$-modules.

The operator $\overline{\partial}_t := \dfrac{\partial}{\partial \overline{t}}$ defines a homomorphism of sheaves

$$(1.1) \qquad \mathscr{C}_{\mathbb{C}}^\infty \xrightarrow{\ \overline{\partial}_t\ } \mathscr{C}_{\mathbb{C}}^\infty$$

which is $\mathscr{O}_{\mathbb{C}}$-linear (but not $\mathscr{C}_{\mathbb{C}}^\infty$-linear). By definition, its kernel is $\mathscr{O}_{\mathbb{C}}$. The results below are fundamental.

1.2 The Poincaré Lemma for $\overline{\partial}_t$. *Let Δ be an open disc of the complex plane and let $\overline{\Delta}$ be its closure. Let f be a C^∞ function defined in some neighbourhood of $\overline{\Delta}$. There exists then a C^∞ function $g : \Delta \to \mathbb{C}$ such that $\dfrac{\partial g}{\partial \overline{t}} = f_{|\Delta}$.*

Proof. See for instance [GH78, p. 5]. □

1.3 Theorem (The Dolbeault resolution on an open set). *For any open set Ω of the complex plane, the map $\overline{\partial}_t : \mathscr{C}^\infty(\Omega) \to \mathscr{C}^\infty(\Omega)$ is onto (and the kernel is $\mathscr{O}(\Omega)$).*

Proof. See for instance [BG91, Th. 3.2.1, p. 221]. □

1.4 Corollary (The local Dolbeault resolution). *The morphism* (1.1) *is onto.*

Proof. It is of course a consequence of Theorem 1.3. It is also a consequence of the Poincaré Lemma 1.2: it is a matter of showing the surjectivity of the germ at t^o of (1.1), for any $t^o \in \mathbb{C}$; if $[f]_{t^o}$ is a C^∞ germ at t^o, there exist an open disc centered at t^o with radius $r > 0$ and a function f which is C^∞ on this disc and with germ $[f]_{t^o}$ at t^o; one takes for Δ the disc centered at t^o having radius $r/2$; Poincaré Lemma gives a C^∞ solution $g : \Delta \to \mathbb{C}$ of $\overline{\partial}_t g = f$ and the germ $[g]_{t^o}$ of g at t^o satisfies thus $\overline{\partial}_t[g]_{t^o} = [f]_{t^o}$. $\qquad\square$

1.5 Corollary (The holomorphic cohomology of an open set of $\mathbb{C}$ vanishes). *For any open set $\Omega \subset \mathbb{C}$, $H^k(\Omega, \mathscr{O}_\Omega) = 0$ for any $k \geqslant 1$.*

Proof. Corollary 1.4 shows that we have a short exact sequence of sheaves of $\mathbb{C}$-vector spaces

$$0 \longrightarrow \mathscr{O}_\Omega \longrightarrow \mathscr{C}_\Omega^\infty \longrightarrow \mathscr{C}_\Omega^\infty \longrightarrow 0.$$

Theorem 0.6.2 gives then a long exact sequence

$$\cdots \to H^k(\Omega, \mathscr{O}_\Omega) \to H^k(\Omega, \mathscr{C}_\Omega^\infty) \to H^k(\Omega, \mathscr{C}_\Omega^\infty) \to H^{k+1}(\Omega, \mathscr{O}_\Omega) \to \cdots$$

from which we deduce, according to Theorem 0.6.3, that $H^k(\Omega, \mathscr{O}_\Omega) = 0$ for $k \geqslant 2$. Theorem 1.3 expresses then the vanishing of $H^1(\Omega, \mathscr{O}_\Omega)$. $\qquad\square$

1.6 Corollary (The holomorphic cohomology of $\mathbb{P}^1$ vanishes). *The spaces $H^k(\mathbb{P}^1, \mathscr{O}_{\mathbb{P}^1})$ are zero for any $k \geqslant 1$, and $H^0(\mathbb{P}^1, \mathscr{O}_{\mathbb{P}^1}) = \mathbb{C}$.*

Proof. The second equality follows from Liouville's Theorem, as a holomorphic function on $\mathbb{P}^1$ can be restricted as a holomorphic function on $\mathbb{C} = U_0$, which is bounded by compactness of $\mathbb{P}^1$.

The covering of $\mathbb{P}^1$ by the charts U_0 and U_∞ (cf. Example 0.2.2-4) is acyclic for the sheaf $\mathscr{O}_{\mathbb{P}^1}$, after the previous corollary; Leray's Theorem 0.6.1 computes the cohomology of $\mathbb{P}^1$ with values in $\mathscr{O}$ with this covering. As it only consists of two open sets, we trivially have $H^k(\mathbb{P}^1, \mathscr{O}_{\mathbb{P}^1}) = 0$ for $k \geqslant 2$. In order to show that $H^1(\mathbb{P}^1, \mathscr{O}_{\mathbb{P}^1}) = 0$, it is enough to see that any holomorphic function f on $\mathbb{C}^*$ is the difference between the restrictions to $\mathbb{C}^*$ of a holomorphic function on U_0 and of a holomorphic function on U_∞: this is exactly what gives the Laurent expansion of f on the punctured line $\mathbb{C}^*$. $\qquad\square$

1.7 Corollary (Bundles of rank one on $\mathbb{P}^1$). *A holomorphic bundle of rank one on $\mathbb{P}^1$ is determined (up to isomorphism) by its Chern class.*

Proof. The exact sequence of the exponential (cf. §0.7.b) induces a cohomology exact sequence

$$\cdots \to H^1(\mathbb{P}^1, \mathscr{O}_{\mathbb{P}^1}) \to H^1(\mathbb{P}^1, \mathscr{O}_{\mathbb{P}^1}^*) \xrightarrow{c_1} H^2(\mathbb{P}^1, \mathbb{Z}_{\mathbb{P}^1}) \to H^2(\mathbb{P}^1, \mathscr{O}_{\mathbb{P}^1}) \to \cdots$$

and both extreme terms vanish, after the previous corollary. This shows that c_1 is an isomorphism. $\qquad\square$

2 Line bundles on $\mathbb{P}^1$

2.a The tautological line bundle $\mathscr{O}_{\mathbb{P}^1}(-1)$

Let $L \subset \mathbb{P}^1 \times \mathbb{C}^2$ be the subset consisting of pairs (m, v), where m is a line of $\mathbb{C}^2$ and v is a vector of $\mathbb{C}^2$, *such that* $v \in m$. We denote by $\pi : L \to \mathbb{P}^1$ the restriction to L of the first projection. If $\dot{L} \subset L$ denotes the open set of pairs (m, v) such that $v \neq 0$, $\dot{L}$ is identified, through the second projection, to $\mathbb{C}^2 \smallsetminus \{0\}$ and $\pi : \dot{L} \to \mathbb{P}^1$ to the *Hopf fibration* $\mathbb{C}^2 \smallsetminus \{0\} \to \mathbb{P}^1$.

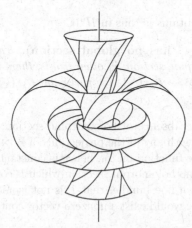

Fig. I.1. The Hopf fibration restricted to the sphere $S^3 = \mathbb{R}^3 \cup \infty$.

2.1 Proposition. *The projection π makes L a holomorphic line bundle on $\mathbb{P}^1$.*

We denote by $\mathscr{O}_{\mathbb{P}^1}(-1)$ the sheaf associated to the bundle L, called *tautological*.

Proof. We will first compute the restriction of L to the open sets U_0 and U_∞. Let us remark that L is a closed subset of $\mathbb{P}^1 \times \mathbb{C}^2$; we will equip this set with the induced topology.

Let t be the coordinate on U_0, let $m^o \in U_0$ be the point with homogeneous coordinates $(1; t^o)$ and let us set $v = (v_0, v_\infty) \in \mathbb{C}^2$. Then, $v \in m^o$ if and only if $v_\infty = t^o v_0$. We deduce a homeomorphism

$$L_{|U_0} \xrightarrow{\ \varphi_0\ } U_0 \times \mathbb{C}$$
$$(t, v) \longmapsto (t, v_0)$$

and in the same way, if t' denotes the coordinate on U_∞ and if the homogeneous coordinates of $m^o \in U_\infty$ is $(t'^o; 1)$, $v \in m^o$ if and only if $v_0 = t'^o v_\infty$, hence a homeomorphism

$$L_{|U_\infty} \xrightarrow{\ \varphi_\infty\ } U_\infty \times \mathbb{C}$$
$$(t', v) \longmapsto (t', v_\infty).$$

On $U_0 \cap U_\infty$ the change of chart is given by

$$U_0 \cap U_\infty \times \mathbb{C} \xrightarrow{\ \varphi_\infty \circ \varphi_0^{-1}\ } U_0 \cap U_\infty \times \mathbb{C}$$
$$(t, v_0) \longmapsto (t, tv_0).$$

The cocycle that one obtains is thus in $H^0(\mathbb{C}^*, \mathscr{O}^*)$. □

2.2 Corollary ($\mathscr{O}_{\mathbb{P}^1}(-1)$ has no global section). *The bundle L has no nonzero global holomorphic section; in other words, there is no non identically zero homomorphism $\mathscr{O}_{\mathbb{P}^1} \to \mathscr{O}_{\mathbb{P}^1}(-1)$ (in particular $\mathscr{O}_{\mathbb{P}^1}(-1)$ is not isomorphic to the trivial bundle).*

Proof. Let us show the absence of holomorphic section: the composition of such a section $\mathbb{P}^1 \to L$ with any of the projections of $\mathbb{P}^1 \times \mathbb{C}^2 \to \mathbb{C}$ restricts as a bounded holomorphic function on U_0, hence is constant. It remains to note that the bundle L has no holomorphic section which is constant and nonzero; in other words, as L is a line bundle, that L is not isomorphic to the trivial bundle; otherwise, there would exist a nonzero vector contained in all the lines of $\mathbb{C}^2$, which is absurd.

We have seen (cf. the remark in the proof of Proposition 0.4.1) that a homomorphism $\mathscr{O}_{\mathbb{P}^1} \to \mathscr{O}_{\mathbb{P}^1}(-1)$ is nothing but a global section of the bundle $\mathscr{O}_M(-1)$, and such a section must be zero, by the previous argument. □

2.b The line bundles $\mathscr{O}_{\mathbb{P}^1}(k)$ for $k \in \mathbb{Z}$

We define the bundle $\mathscr{O}_{\mathbb{P}^1}(1)$ as the bundle dual to $\mathscr{O}_{\mathbb{P}^1}(-1)$, that is, $\mathscr{H}om_{\mathscr{O}_{\mathbb{P}^1}}(\mathscr{O}_{\mathbb{P}^1}(-1), \mathscr{O}_{\mathbb{P}^1})$. It is also the inverse bundle of the bundle $\mathscr{O}_{\mathbb{P}^1}(-1)$ with respect to the tensor product. Is is called the *canonical bundle* on $\mathbb{P}^1$. Then, for any $k \in \mathbb{Z}$, we set

$$\mathscr{O}_{\mathbb{P}^1}(k) = \begin{cases} \mathscr{O}_{\mathbb{P}^1}(1)^{\otimes k} & \text{if } k \geqslant 1 \\ \mathscr{O}_{\mathbb{P}^1}(-1)^{\otimes |k|} & \text{if } k \leqslant -1 \\ \mathscr{O}_{\mathbb{P}^1} & \text{if } k = 0. \end{cases}$$

2.3 Exercise (First properties of the bundles $\mathscr{O}_{\mathbb{P}^1}(k)$). Prove the following assertions:

(1) The bundle $\mathscr{O}_{\mathbb{P}^1}(k)$ is trivial when restricted to the open sets U_0 and U_∞ and the corresponding cocycle is $\varphi_\infty \circ \varphi_0^{-1} : (t, v_0) \mapsto (t, t^{-k}v_0)$.
(2) For any $k \in \mathbb{Z}$ we have $\mathscr{O}_{\mathbb{P}^1}(-k) = \mathscr{H}om_{\mathscr{O}_{\mathbb{P}^1}}(\mathscr{O}_{\mathbb{P}^1}(k), \mathscr{O}_{\mathbb{P}^1})$.

(3) $\mathrm{Hom}_{\mathscr{O}_{\mathbb{P}^1}}(\mathscr{O}_{\mathbb{P}^1}(k), \mathscr{O}_{\mathbb{P}^1}(\ell)) = \mathrm{Hom}_{\mathscr{O}_{\mathbb{P}^1}}(\mathscr{O}_{\mathbb{P}^1}, \mathscr{O}_{\mathbb{P}^1}(\ell - k))$.

(4) $H^0(\mathbb{P}^1, \mathscr{O}_{\mathbb{P}^1}(k)) \neq 0 \iff k \geqslant 0$.

(5) $\mathrm{Hom}_{\mathscr{O}_{\mathbb{P}^1}}(\mathscr{O}_{\mathbb{P}^1}(k), \mathscr{O}_{\mathbb{P}^1}(\ell)) \neq 0 \iff \ell - k \geqslant 0$.

(6) If $\mathscr{O}_{\mathbb{P}^1}(k)$ is isomorphic to $\mathscr{O}_{\mathbb{P}^1}(\ell)$, then $k = \ell$.

(7) $c_1(\mathscr{O}_{\mathbb{P}^1}(k)) = k \cdot c_1(\mathscr{O}_{\mathbb{P}^1}(1))$.

2.c Line bundle associated to a divisor

We will now be more explicit concerning the notions introduced in §0.8 when $M = \mathbb{P}^1$. Let us pick $m^o \in \mathbb{P}^1$. For $k \leqslant 0$, we define $\mathscr{O}_{\mathbb{P}^1}(km^o)$ as the subsheaf of $\mathscr{O}_{\mathbb{P}^1}$ consisting of germs of functions which vanish at order at least $-k = |k|$ at m^o: the germ of $\mathscr{O}_{\mathbb{P}^1}(km^o)$ at $m \neq m^o$ is equal to $\mathscr{O}_{\mathbb{P}^1, m}$, while the germ at m^o is equal to $t^{|k|}\mathbb{C}\{t\}$ if t is a local coordinate vanishing at m^o.

One can extend this definition to $k > 0$: one first considers the sheaf $\mathscr{O}_{\mathbb{P}^1}(*m^o)$ of germs of *meromorphic functions with pole at m^o* on $\mathbb{P}^1$, whose germ at $m \neq m^o$ is $\mathscr{O}_{\mathbb{P}^1, m}$ and whose germ at m^o is the ring $\mathbb{C}\{t\}[1/t]$; one then defines $\mathscr{O}_{\mathbb{P}^1}(km^o) \subset \mathscr{O}_{\mathbb{P}^1}(*m^o)$, for $k > 0$, as the subsheaf of germs which have a pole of order *at most k* at m^o.

We clearly have an exact sequence of sheaves

$$(2.4) \qquad 0 \longrightarrow \mathscr{O}_{\mathbb{P}^1}(km^o) \longrightarrow \mathscr{O}_{\mathbb{P}^1}((k+1)m^o) \longrightarrow \mathbb{C}_{m^o} \longrightarrow 0$$

where $\mathbb{C}_{m^o}$ is the "skyscraper" sheaf on $\mathbb{P}^1$ defined by

$$\Gamma(U, \mathbb{C}_{m^o}) = \begin{cases} \mathbb{C} & \text{if } m^o \in U, \\ 0 & \text{otherwise.} \end{cases}$$

2.5 Exercise (The bundles $\mathscr{O}(k)$ and their divisors).

(1) Prove that, for any $k \in \mathbb{Z}$, the sheaf $\mathscr{O}_{\mathbb{P}^1}(km^o)$ is locally free of rank one.

(2) Show that $\mathscr{O}_{\mathbb{P}^1}(km^o) \xrightarrow{\sim} \mathscr{O}_{\mathbb{P}^1}(km)$ for any $m \in \mathbb{P}^1$.

(3) Taking $m^o = \infty$, compute a cocycle for $\mathscr{O}_{\mathbb{P}^1}(km^o)$ and deduce that $\mathscr{O}_{\mathbb{P}^1}(km^o) \simeq \mathscr{O}_{\mathbb{P}^1}(k)$.

2.d The universal quotient bundle

By its very construction, the tautological bundle is a subbundle of the trivial bundle of rank 2, i.e., $\mathscr{O}_{\mathbb{P}^1} \oplus \mathscr{O}_{\mathbb{P}^1}$.

2.6 Proposition. *The quotient sheaf $\mathscr{Q} := (\mathscr{O}_{\mathbb{P}^1} \oplus \mathscr{O}_{\mathbb{P}^1})/\mathscr{O}_{\mathbb{P}^1}(-1)$ is a locally free sheaf of rank one on $\mathbb{P}^1$, isomorphic to the canonical bundle $\mathscr{O}_{\mathbb{P}^1}(1)$.*

Proof. Let us use the notation presented at the beginning of this section. Let us pick $m^o = (1; t^o) \in U_0$. The quotient $\mathbb{C}^2/L_{m^o}$ is identified with $\mathbb{C}$ by the mapping $v \mapsto v_\infty - t^o v_0$. There is thus an exact sequence of bundles

$$0 \longrightarrow L_{|U_0} \longrightarrow U_0 \times \mathbb{C}^2 \xrightarrow{\varphi} U_0 \times \mathbb{C} \longrightarrow 0$$

with $\varphi(t, v) = (t, v_\infty - tv_0)$, which gives a trivialization of $\mathscr{Q}$ on U_0. In the same way, there is a trivialization on U_∞ by the map $\varphi'(t', v) = (t', t'v_\infty - v_0)$. The cocycle associated to $\mathscr{Q}$ is $(t, w) \mapsto (t, t^{-1}w)$: one recognizes the cocycle of $\mathscr{O}_{\mathbb{P}^1}(1)$.   □

2.e Extensions of line bundles

We have constructed an exact sequence

$$(2.7) \qquad 0 \longrightarrow \mathscr{O}_{\mathbb{P}^1}(-1) \longrightarrow \mathscr{O}_{\mathbb{P}^1} \oplus \mathscr{O}_{\mathbb{P}^1} \overset{p}{\longrightarrow} \mathscr{O}_{\mathbb{P}^1}(1) \longrightarrow 0.$$

One can ask whether this exact sequence splits, in other words if there exists a homomorphism $\mathscr{O}_{\mathbb{P}^1}(1) \to \mathscr{O}_{\mathbb{P}^1} \oplus \mathscr{O}_{\mathbb{P}^1}$ which becomes the identity after composition with p. Such a homomorphism does not exist, as any homomorphism $\mathscr{O}_{\mathbb{P}^1}(1) \to \mathscr{O}_{\mathbb{P}^1}$ is zero (cf. Corollary 2.2). More generally:

2.8 Theorem (Splitting of extensions).

(1) *If $\mathscr{E}$ is a bundle of rank 2 on $\mathbb{P}^1$, any exact sequence*

$$0 \longrightarrow \mathscr{O}_{\mathbb{P}^1}(-k) \longrightarrow \mathscr{E} \longrightarrow \mathscr{O}_{\mathbb{P}^1}(\ell) \longrightarrow 0$$

splits as soon as $k + \ell \leqslant 1$.
(2) *If $k + \ell \geqslant 2$, there exists an extension $\mathscr{E}$ of $\mathscr{O}_{\mathbb{P}^1}(\ell)$ by $\mathscr{O}_{\mathbb{P}^1}(-k)$ which does not split.*

Proof. We will say that two extensions $\mathscr{E}, \mathscr{E}'$ of $\mathscr{O}_{\mathbb{P}^1}(\ell)$ by $\mathscr{O}_{\mathbb{P}^1}(-k)$ are *isomorphic* if there exist isomorphisms making the following diagram commute:

$$
\begin{array}{ccccccccc}
0 & \longrightarrow & \mathscr{O}_{\mathbb{P}^1}(-k) & \longrightarrow & \mathscr{E} & \longrightarrow & \mathscr{O}_{\mathbb{P}^1}(\ell) & \longrightarrow & 0 \\
 & & \wr\downarrow & & \wr\downarrow & & \downarrow\wr & & \\
0 & \longrightarrow & \mathscr{O}_{\mathbb{P}^1}(-k) & \longrightarrow & \mathscr{E} & \longrightarrow & \mathscr{O}_{\mathbb{P}^1}(\ell) & \longrightarrow & 0
\end{array}
$$

(the extreme isomorphisms are then a nonzero constant times the identity). The proof of the theorem is thus reduced to both statements below.   □

2.9 Lemma. *The set of isomorphism classes of extensions of the bundle $\mathscr{O}_{\mathbb{P}^1}(\ell)$ by the bundle $\mathscr{O}_{\mathbb{P}^1}(-k)$ is in one-to-one correspondence with $H^1(\mathbb{P}^1, \mathscr{O}_{\mathbb{P}^1}(-\ell - k))$.*

2.10 Theorem (Vanishing Theorem). *We have $H^1(\mathbb{P}^1, \mathscr{O}_{\mathbb{P}^1}(-k)) = 0$ if and only if $k \leqslant 1$.*

Proof (of Lemma 2.9). We can tensor all terms by $\mathscr{O}_{\mathbb{P}^1}(-\ell)$, so that we can assume that $\ell = 0$. Consider first an extension $\mathscr{E}$:

$$0 \longrightarrow \mathscr{O}_{\mathbb{P}^1}(-k) \longrightarrow \mathscr{E} \longrightarrow \mathscr{O}_{\mathbb{P}^1} \longrightarrow 0.$$

One chooses a covering $\mathfrak{U}$ of $\mathbb{P}^1$ finer than (U_0, U_∞) and fine enough so that, on any open set U_j, the exact sequence splits (it is enough to insure that, on any U_j, one can complete a basis of $\mathscr{O}_{\mathbb{P}^1}(-k)$ in a basis of $\mathscr{E}$). Let σ_j be a section of $\mathscr{E} \to \mathscr{O}_{\mathbb{P}^1}$ on U_j. On $U_j \cap U_{j'}$, we have $\sigma_j - \sigma_{j'} \in \Gamma(U_j \cap U_{j'}, \mathscr{O}_{\mathbb{P}^1}(-k))$. The cocycle condition is clearly satisfied, hence (σ_j) defines an element of $H^1(\mathfrak{U}, \mathscr{O}_{\mathbb{P}^1}(-k))$. We have inclusions

$$H^1((U_0, U_\infty), \mathscr{O}_{\mathbb{P}^1}(-k)) \subset H^1(\mathfrak{U}, \mathscr{O}_{\mathbb{P}^1}(-k)) \subset H^1(\mathbb{P}^1, \mathscr{O}_{\mathbb{P}^1}(-k))$$

and, on the other hand, the covering (U_0, U_∞) is acyclic for $\mathscr{O}_{\mathbb{P}^1}(-k)$ because it is so for $\mathscr{O}_{\mathbb{P}^1}$ and the restrictions of both bundles to U_0 and U_∞ are isomorphic. Therefore, after Leray's Theorem 0.6.1, these inclusions are equalities.

Conversely, any cocycle $\sigma \in Z^1(\mathfrak{U}, \mathscr{O}_{\mathbb{P}^1}(-k))$ associated to a covering $\mathfrak{U}$ of $\mathbb{P}^1$ defines an extension $\mathscr{E}$ of $\mathscr{O}_{\mathbb{P}^1}$ by $\mathscr{O}_{\mathbb{P}^1}(-k)$: on any open set U_j of $\mathfrak{U}$ we set $\mathscr{E}_j = (\mathscr{O}_{\mathbb{P}^1}(-k) \oplus \mathscr{O}_{\mathbb{P}^1})_{|U_j}$; on $U_i \cap U_j$, we glue $\mathscr{E}_{i|U_i \cap U_j}$ and $\mathscr{E}_{j|U_i \cap U_j}$ by means of the matrix

$$\begin{pmatrix} \mathrm{Id} & \sigma_{ij} \\ 0 & \mathrm{Id} \end{pmatrix}$$

where the component σ_{ij} of cocycle σ is regarded as an element of

$$\Gamma(U_i \cap U_j, \mathscr{H}om(\mathscr{O}_{\mathbb{P}^1}, \mathscr{O}_{\mathbb{P}^1}(-k))).$$

An extension $\mathscr{E}$ defined by a cocycle σ splits if and only if this cocycle is a coboundary (exercise). Therefore, any element of $H^1(\mathbb{P}^1, \mathscr{O}_{\mathbb{P}^1}(-k))$ defines an isomorphism class of extension. $\qquad\square$

Proof (of Vanishing Theorem 2.10). We know that

$$H^1(\mathbb{P}^1, \mathscr{O}_{\mathbb{P}^1}(-k)) = H^1((U_0, U_\infty), \mathscr{O}_{\mathbb{P}^1}(-k)).$$

By using the natural basis of $\mathscr{O}_{\mathbb{P}^1}(-k)$ on U_0, we see that a cocycle is a holomorphic function h on $U_0 \cap U_\infty$ and that a coboundary is a function of the form $f(t) - t^{-k} g(t)$ with f holomorphic on U_0 and g holomorphic on U_∞. The existence of a Laurent expansion shows that a function h can be written as

$$h(t) = \sum_{n \geqslant 0} a_n t^n - t^{-k}\left(\sum_{n \geqslant 0} a_{-n} t^{-n}\right),$$

where both series have an infinite radius of convergence, if and only if $-k \geqslant -1$. $\qquad\square$

2.f Vector fields and differential forms on $\mathbb{P}^1$

As above, let us denote by t the coordinate in the chart U_0 and by t' the coordinate in the chart U_∞. The sheaf $\Theta_{\mathbb{P}^1}$ of vector fields on $\mathbb{P}^1$ has an everywhere nonvanishing section on U_0, namely $\partial/\partial t$, and an everywhere nonvanishing section on U_∞, namely $\partial/\partial t'$. We deduce isomorphisms

$$\Theta_{\mathbb{P}^1}|_{U_0} \xrightarrow{\varphi_0} \mathscr{O}_{U_0} \qquad\qquad \Theta_{\mathbb{P}^1}|_{U_\infty} \xrightarrow{\varphi_\infty} \mathscr{O}_{U_\infty}$$

$$f(t)\frac{\partial}{\partial t} \longmapsto f(t) \qquad\qquad g(t')\frac{\partial}{\partial t'} \longmapsto g(t')$$

and the cocycle $\varphi_\infty \circ \varphi_0^{-1} : \mathscr{O}_{\mathbb{C}^*} \to \mathscr{O}_{\mathbb{C}^*}$ sends 1 to $-1/t^2$. It easily follows:

2.11 Proposition (Computation of the tangent and cotangent bundles of $\mathbb{P}^1$). *We have isomorphisms*

$$\Theta_{\mathbb{P}^1} \simeq \mathscr{O}_{\mathbb{P}^1}(2), \qquad \Omega^1_{\mathbb{P}^1} \simeq \mathscr{O}_{\mathbb{P}^1}(-2)$$

$$\Omega^1_{\mathbb{P}^1}\left(\sum_{i=1}^{p}(r_i+1)m_i\right) \simeq \mathscr{O}_{\mathbb{P}^1}\left(-2+\sum_{i=1}^{p}(r_i+1)\right),$$

if we denote by $\Omega^1_{\mathbb{P}^1}(\sum_{i=1}^{p}(r_i+1)m_i)$ the sheaf of meromorphic 1-forms with poles at the points $m_1,\ldots,m_p$ and with order r_i at m_i (cf. §0.9.b). □

3 A finiteness theorem and some consequences

The following result, that we will not prove, enables one to analyze the bundles on $\mathbb{P}^1$. It is one of the key points in the theory of Riemann surfaces, as well as in analytic and algebraic geometry; it is in fact true in a much more general framework (that of coherent sheaves on a *compact* complex analytic manifold).

3.1 Theorem (Finiteness Theorem). *Let E be a holomorphic vector bundle on $\mathbb{P}^1$ and let $\mathscr{E}$ be the associated sheaf. Then we have $\dim H^k(\mathbb{P}^1, \mathscr{E}) < +\infty$ for any $k \geqslant 0$.*

Proof. See for instance [Rey89, Chap. IX] for the finiteness of H^0 and H^1. By using the *Dolbeault resolution*

$$0 \longrightarrow \mathscr{O}_{\mathbb{P}^1} \longrightarrow \mathscr{C}^\infty_{\mathbb{P}^1} \xrightarrow{\bar{\partial}} \mathscr{C}^{(0,1)}_{\mathbb{P}^1} \longrightarrow 0$$

where $\mathscr{C}^{(0,1)}_{\mathbb{P}^1}$ denotes the sheaf of differential 1-forms with C^∞ coefficients of type $(0,1)$ on $\mathbb{P}^1$ (see for instance [GH78]), as well as Theorems 0.6.3 and 0.6.2, one verifies the vanishing of $H^k(\mathbb{P}^1, \mathscr{E})$ for $k \geqslant 2$. □

If E is a holomorphic vector bundle on $\mathbb{P}^1$ and $\mathscr{E}$ is the associated sheaf, we will denote by $\mathscr{E}(k)$ the tensor product $\mathscr{E} \otimes_{\mathscr{O}_{\mathbb{P}^1}} \mathscr{O}_{\mathbb{P}^1}(k)$: it is a locally free sheaf over $\mathscr{O}_{\mathbb{P}^1}$ having the same rank as $\mathscr{E}$.

3.2 Exercise. Describe a cocycle of $\mathscr{E}(k)$ from a cocycle of $\mathscr{E}$.

3.3 Corollary. *For any holomorphic bundle E on $\mathbb{P}^1$, there exists an integer $k(E)$ such that $H^0(\mathbb{P}^1, \mathscr{E}(k)) = 0$ for any $k < k(E)$, and $H^0(\mathbb{P}^1, \mathscr{E}(k)) \neq 0$ for any $k \geqslant k(E)$.*

Proof. Let us pick $m^o \in \mathbb{P}^1$. Let us choose an isomorphism $\mathscr{O}_{\mathbb{P}^1}(1) \simeq \mathscr{O}_{\mathbb{P}^1}(m^o)$. Denote by $\mathbb{E}_{m^o}$ the sheaf such that $\Gamma(U, \mathbb{E}_{m^o}) = E_{m^o}$ if $m^o \in U$ and $= 0$ otherwise. For any $j \in \mathbb{Z}$, we have an exact sequence analogous to (2.4):

$$(3.4) \qquad 0 \longrightarrow \mathscr{E}(jm^o) \longrightarrow \mathscr{E}((j+1)m^o) \longrightarrow \mathbb{E}_{m^o} \longrightarrow 0.$$

Any open covering $\mathfrak{U}$ of $\mathbb{P}^1$ such that m^o is contained in only one open set is acyclic for $\mathbb{E}_{m^o}$; we deduce that $H^j(\mathbb{P}^1, \mathbb{E}_{m^o}) = 0$ for any $j \geqslant 1$. Hence, the long exact sequence associated to (3.4) becomes

$$(3.5) \quad 0 \longrightarrow H^0(\mathbb{P}^1, \mathscr{E}(jm^o)) \longrightarrow H^0(\mathbb{P}^1, \mathscr{E}((j+1)m^o)) \longrightarrow \mathbb{E}_{m^o} \longrightarrow$$
$$\longrightarrow H^1(\mathbb{P}^1, \mathscr{E}(jm^o)) \longrightarrow H^1(\mathbb{P}^1, \mathscr{E}((j+1)m^o)) \longrightarrow 0.$$

It follows that, if $H^0(\mathbb{P}^1, \mathscr{E}(km^o)) \neq 0$ for some $k \in \mathbb{Z}$, then $H^0(\mathbb{P}^1, \mathscr{E}(jm^o)) \neq 0$ for any $j \geqslant k$; in the same way, if $H^0(\mathbb{P}^1, \mathscr{E}(km^o)) = 0$ for some $k \in \mathbb{Z}$, then $H^0(\mathbb{P}^1, \mathscr{E}(jm^o)) = 0$ for any $j \leqslant k$.

Therefore, we are reduced to showing that there exists k such that $H^0(\mathbb{P}^1, \mathscr{E}(km^o)) \neq 0$ and that there exists ℓ such that $H^0(\mathbb{P}^1, \mathscr{E}(\ell m^o)) = 0$.

Let us show the first assertion: if it is false, for any $j \in \mathbb{Z}$,

$$\dim H^1(\mathbb{P}^1, \mathscr{E}(j)) = \dim \mathbb{E}_{m^o} + \dim H^1(\mathbb{P}^1, \mathscr{E}(j+1)),$$

after the exact sequence (3.5). It follows that, for any $j \geqslant 0$,

$$\dim H^1(\mathbb{P}^1, \mathscr{E}) = j \dim \mathbb{E}_{m^o} + \dim H^1(\mathbb{P}^1, \mathscr{E}(j)) \geqslant j \dim \mathbb{E}_{m^o} \geqslant j,$$

which is in contradiction with the finiteness theorem 3.1.

If the second assertion is false, there exists $k_0 \in \mathbb{Z}$ such that, for any $k \leqslant k_0$, we have

$$0 \neq H^0(\mathbb{P}^1, \mathscr{E}(km^o)) = H^0(\mathbb{P}^1, \mathscr{E}(k_0 m^o)).$$

One can choose $k_0 \leqslant 0$. One would then be able to find a *strictly increasing* sequence of integers ℓ_n ($n \in \mathbb{N}$) and a sequence of sections $s_n \in H^0(\mathbb{P}^1, \mathscr{E}(-\ell_n m^o))$ such that, for any $n \in \mathbb{N}$, the section s_n vanishes at order ℓ_n *exactly* at m^o. It is clear that such a family is free, which is in contradiction with the finiteness statement of Theorem 3.1. $\qquad\square$

4 Structure of vector bundles on $\mathbb{P}^1$

4.a Birkhoff-Grothendieck theorem

Let us begin with the classification of line bundles on $\mathbb{P}^1$, the first step in the inductive procedure of Theorem 4.4.

4.1 Theorem (Classification of line bundles on $\mathbb{P}^1$). *Any line bundle on $\mathbb{P}^1$ is isomorphic to $\mathcal{O}_{\mathbb{P}^1}(k)$ for one and only one $k \in \mathbb{Z}$.*

Proof. Let $\mathcal{L}$ be a locally free sheaf of rank one on $\mathbb{P}^1$. It follows from Corollary 3.3 that there exists a unique $k_0 \in \mathbb{Z}$ such that $H^0(\mathbb{P}^1, \mathcal{L}(k_0)) \neq 0$ and $H^0(\mathbb{P}^1, \mathcal{L}(k_0 - 1)) = 0$. There exists thus a non identically zero section of $\mathcal{L}(k_0)$ and such a section does not vanish anywhere in $\mathbb{P}^1$, otherwise it would be a section of $\mathcal{L}(k_0 - 1)$. The homomorphism $\mathcal{O}_{\mathbb{P}^1} \to \mathcal{L}(k_0)$ defined by this section is thus an isomorphism. □

We deduce from Corollary 1.7:

4.2 Corollary. *We have $H^2(\mathbb{P}^1, \mathbb{Z}_{\mathbb{P}^1}) = \mathbb{Z} \cdot c_1(\mathcal{O}_{\mathbb{P}^1}(1))$.* □

4.3 Definition (Quasitrivial bundles). We will say that bundle E of rank d on $\mathbb{P}^1$ is *quasitrivial* (of weight k) if it is isomorphic to $\mathcal{O}_{\mathbb{P}^1}(k)^d$ for some $k \in \mathbb{Z}$.

Therefore, any line bundle on $\mathbb{P}^1$ is quasitrivial.

4.4 Theorem (Birkhoff-Grothendieck theorem). *Let E be a vector bundle of rank d on $\mathbb{P}^1$. There exists then a unique sequence of integers $a_1 \geqslant \cdots \geqslant a_d$ such that we have an isomorphism*

$$\mathcal{E} \simeq \mathcal{O}_{\mathbb{P}^1}(a_1) \oplus \cdots \oplus \mathcal{O}_{\mathbb{P}^1}(a_d).$$

We say that the sequence $a_1 \geqslant \cdots \geqslant a_d$ is the *type* of the bundle E. This theorem gives a multiplicative matrix variant of the Laurent expansion:

4.5 Corollary (Matrix Laurent expansion). *Let A be an invertible matrix $d \times d$ with holomorphic entries on an annulus*

$$\mathscr{C} = \{t \in \mathbb{C} \mid 0 \leqslant r < |t| < R \leqslant +\infty\}.$$

There exist then two invertible holomorphic matrices P and Q respectively on $|t| < R$ and $|t| > r$ such that, on $\mathscr{C}$, the matrix PAQ is equal to a diagonal matrix with diagonal entries $t^{a_1}, \ldots, t^{a_d}$, where the sequence of integers $a_1 \geqslant \cdots \geqslant a_d$ depends only on A.

Proof. Let us consider the covering of $\mathbb{P}^1$ by the two open sets $|t| < R$ and $|t| > r$. The matrix A defines a 1-cocycle of this covering with values in $\mathrm{GL}_d(\mathcal{O}_{\mathbb{P}^1})$, hence a vector bundle. The corollary is then a simple translation of Theorem 4.4. □

4.6 Remarks.

(1) Corollary 4.5 was originally proved by G.D. Birkhoff (see [Bir13a]) as a tool in the study of singular points of differential equations. At the same time (cf. [Bir13b]), this author proved results on the normal form of systems with an irregular singularity, normal form which takes his name and that

we will consider in §IV.5. He then realized that the same method had been used by Hilbert and Plemelj to solve the Riemann-Hilbert problem. Theorem IV.2.2, which we will see later, expresses this unity. The proof of Birkhoff is taken up in [Sib90, §3.4].

In 1957, Theorem 4.4, in a more general situation where the structure group of the bundle is not necessarily GL_d, was proved by A. Grothendieck in [Gro57]. The proof given below essentially follows that of Grothendieck.

(2) A consequence of Theorem 4.4 is that any holomorphic bundle on $\mathbb{P}^1$ is trivializable when restricted to U_0 and to U_∞ and that it is isomorphic to an *algebraic* bundle, that is, it admits a cocycle given by a matrix of rational fractions. This result can be generalized: any holomorphic bundle on a closed complex analytic submanifold of the projective space $\mathbb{P}^n$ (see Example §0.2.2) can be defined by an algebraic cocycle (see [Ser56]).

(3) The reader interested in variants and generalizations of Corollary 4.5 is referred to [PS86, Chap. 8].

Proof (of Theorem 4.4, existence). It is done by induction on the rank d of E, the case $d = 1$ being given by Theorem 4.1. Let k_0 be the integer such that

$$H^0(\mathbb{P}^1, \mathscr{E}(k_0)) \neq 0 \quad \text{and} \quad H^0_-(\mathbb{P}^1, \mathscr{E}(k_0 - 1)) = 0.$$

There exists thus a section of $\mathscr{E}(k_0)$ which does not vanish on $\mathbb{P}^1$. It provides an injective homomorphism $\mathscr{O}_{\mathbb{P}^1} \hookrightarrow \mathscr{E}(k_0)$ and the quotient sheaf $\mathscr{F}$ is *locally free* of rank $d-1$ (such would not be the case if the section would vanish). By induction, there exists a sequence $b_2 \geqslant \cdots \geqslant b_d$ such that $\mathscr{F} \simeq \bigoplus_{j=1}^{d-1} \mathscr{O}_{\mathbb{P}^1}(b_i)$. We will show that $\mathscr{E}(k_0) \simeq \mathscr{O}_{\mathbb{P}^1} \oplus \mathscr{F}$.

In order to do that, it is enough, as in Theorem 2.8, to check that the b_i are $\leqslant 1$. We will show more: the b_i are $\leqslant 0$. Setting then $a_i = b_i - k_0$ for $i = 2, \ldots, d$ and $a_1 = -k_0$, we obtain the desired existence.

Let us consider the exact sequence of locally free sheaves that we have yet obtained, tensored with $\mathscr{O}_{\mathbb{P}^1}(-1)$,

$$0 \longrightarrow \mathscr{O}_{\mathbb{P}^1}(-1) \longrightarrow \mathscr{E}(k_0 - 1) \longrightarrow \mathscr{F}(-1) \longrightarrow 0.$$

It gives rise to a long exact sequence, the part which we are interested in being

$$0 := H^0(\mathbb{P}^1, \mathscr{E}(k_0 - 1)) \longrightarrow H^0(\mathbb{P}^1, \mathscr{F}(-1)) \longrightarrow H^1(\mathbb{P}^1, \mathscr{O}_{\mathbb{P}^1}(-1)) = 0,$$

(after Theorem 2.10). Therefore, $H^0(\mathbb{P}^1, \mathscr{F}(-1)) = 0$. But

$$H^0(\mathbb{P}^1, \mathscr{F}(-1)) = \bigoplus_{i=2}^{d} H^0(\mathbb{P}^1, \mathscr{O}_{\mathbb{P}^1}(b_i - 1))$$

and hence we have $b_i - 1 \leqslant -1$ for any $i = 2, \ldots, d$ (cf. Exercise 2.3).

Proof of Theorem 4.4, uniqueness. Assume that we have two decompositions corresponding to two distinct sequences $a_1 \geqslant \cdots \geqslant a_d$ and $a_1' \geqslant \cdots \geqslant a_d'$. Let $j \in \{1, \ldots, d\}$ be such that $a_i = a_i'$ for $i < j$ and $a_j \neq a_j'$, for instance $a_j > a_j'$. Then $a_i' - a_j = a_i - a_j \geqslant 0$ for $i < j$ and $a_i' - a_j < 0$ for $i \geqslant j$. If we have an isomorphism

$$\bigoplus_i \mathscr{O}_{\mathbb{P}^1}(a_i) \simeq \bigoplus_i \mathscr{O}_{\mathbb{P}^1}(a_i')$$

we deduce, by tensoring with $\mathscr{O}_{\mathbb{P}^1}(-a_j)$, an isomorphism

$$\bigoplus_i \mathscr{O}_{\mathbb{P}^1}(a_i' - a_j) \simeq \bigoplus_i \mathscr{O}_{\mathbb{P}^1}(a_i - a_j).$$

However, the dimension of the space of global sections for the left-hand term is strictly smaller than that for the right-hand term, whence a contradiction. $\square$

4.7 Exercise (Degree of a bundle). The *degree* of a bundle E of type $a_1 \geqslant \cdots \geqslant a_d$ on $\mathbb{P}^1$ is by definition the sum $\sum_{i=1}^d a_i$. Prove that a bundle E is trivial if and only if it has degree 0 and satisfies $H^1(\mathbb{P}^1, \mathscr{E}(-1)) = 0$.

4.8 Exercise (The Harder-Narasimhan filtration, case of $\mathbb{P}^1$). For any $a \in \mathbb{Z}$, the subbundle $F^a \mathscr{E} \subset \mathscr{E}$ is defined as the direct sum of components isomorphic to $\mathscr{O}_{\mathbb{P}^1}(a_i)$ with $a_i \geqslant a$.

(1) Prove $F^{a+1}\mathscr{E} \subset F^a\mathscr{E}$, that the quotient F^a/F^{a+1} vanishes if a is not equal to one of the a_i and that it is quasitrivial of weight a. We thus have

$$\{0\} = F^\infty \mathscr{E} \subsetneq F^{a_1} \mathscr{E} \subset \cdots \subset F^{a_d} \mathscr{E} = \mathscr{E}.$$

(2) Prove that a filtration of $\mathscr{E}$ by vector subbundles satisfying the properties above is unique.

(3) We call the *slope* of a nonzero bundle $\mathscr{E}$ the quotient $\mu(\mathscr{E}) = \deg(\mathscr{E})/\operatorname{rk}(\mathscr{E})$, with $\deg(\mathscr{E}) = \sum_{i=1}^d a_i$. Prove that

$$\mu(F^{a+1}\mathscr{E}) \geqslant \mu(F^a\mathscr{E})$$

and consider the case when there is equality.

(4) We say that $\mathscr{E}$ is *semistable* if any nonzero subbundle $\mathscr{F}$ of $\mathscr{E}$ satisfies $\mu(\mathscr{F}) \leqslant \mu(\mathscr{E})$. Deduce that $\mathscr{E}$ is semistable if and only if it is quasitrivial.

It can be suggestive to visualize the decomposition given by Theorem 4.4 as a polygon (see Figure I.2). This polygon is the graph of the function

$$k \in \{0, \ldots, d\} \longmapsto \sum_{i=1}^k a_i.$$

The vertices of this polygon consist of the pairs (k, ℓ_k), where k is such that $F^{a_{k+1}}/F^{a_k} \neq 0$ and ℓ_k is the degree of F^{a_k}.

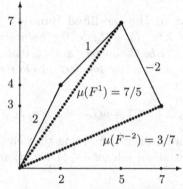

Fig. I.2. The polygon of $\mathcal{O}(2)^2 \oplus \mathcal{O}(1)^3 \oplus \mathcal{O}(-2)^2$.

4.b Application to meromorphic bundles

4.9 Corollary (Meromorphic bundles are trivial). *Any meromorphic bundle $\mathcal{M}$ of rank d on $\mathbb{P}^1$ with poles on some nonempty finite set $\Sigma = \{m_0, \ldots, m_p\}$ is isomorphic to the trivial meromorphic bundle $\mathcal{O}_{\mathbb{P}^1}(*\Sigma)^d$.*

Proof. After Proposition 0.8.4, there exists a lattice $\mathcal{E} \subset \mathcal{M}$. Then $\mathcal{M} = \mathcal{E}(*\Sigma)$. By applying Birkhoff-Grothendieck theorem 4.4 to $\mathcal{E}$, we are reduced to showing that, for any $k \in \mathbb{Z}$, the meromorphic bundle $\mathcal{O}_{\mathbb{P}^1}(k)(*\Sigma)$ is trivial; let us identify $\mathcal{O}_{\mathbb{P}^1}(k)$ with $\mathcal{O}_{\mathbb{P}^1}(km_0)$; we then have in an evident way $\mathcal{O}_{\mathbb{P}^1}(km_0)(*\Sigma) = \mathcal{O}_{\mathbb{P}^1}(*\Sigma)$, whence the result. □

4.c Modification of the type of a bundle

It is clear that, if a bundle $\mathcal{E}$ has type $a_1 \geqslant \cdots \geqslant a_d$, then the bundle $\mathcal{E}(k)$ has type $a_1 + k \geqslant \cdots \geqslant a_d + k$. We will now consider the effect on the type of some modifications on the bundle.

Let us pick $m^o \in \mathbb{P}^1$ and let L^o be a line in the fibre E_{m^o}. Let us define the subsheaf $\mathcal{E}(L^o m^o)$ of $\mathcal{E}(m^o) := \mathcal{E} \otimes_{\mathcal{O}_{\mathbb{P}^1}} \mathcal{O}_{\mathbb{P}^1}(1 \cdot m^o) \simeq \mathcal{E}(1)$: this subsheaf is equal to the sheaf $\mathcal{E}(m^o)$ on the open set $\mathbb{P}^1 \smallsetminus \{m^o\}$; let t be a local coordinate vanishing at m^o; the germ at m^o of $\mathcal{E}(L^o m^o)$ consists of meromorphic sections σ of $\mathcal{E}_{m^o}$ which have at most a simple pole at m^o (i.e., sections of $\mathcal{E}(m^o)_{m^o}$) *and whose residue at m^o belongs to L^o*, that is, such that $(t\sigma(t))_{|t=0} \in L^o$.

4.10 Exercise.

(1) Check that this condition on the residue does not depend on the choice of a local coordinate.
(2) Prove that, if $e_1, \ldots, e_d$ is a local basis of $\mathcal{E}$ and if L^o is the line containing $e_d(m^o)$, then $e_1, \ldots, e_{d-1}, e_d/t$ is a local basis of $\mathcal{E}(L^o \cdot m^o)$, which is thus a locally free $\mathcal{O}_{\mathbb{P}^1}$-module of rank d.

4.11 Proposition (Type of the modified bundle). *Let* $a_1 \geqslant \cdots \geqslant a_d$
*be the type of E and let $F^a\mathscr{E}$ ($a \in \mathbb{Z}$) be the natural associated filtration
(cf. Exercise 4.8). Let $a(L^o)$ be the integer a such that $L^o \subset (F^aE)_{m^o}$ and
$L^o \not\subset (F^{a+1}E)_{m^o}$. Let $j = j(L^o)$ be the smallest integer i such that $a_i = a(L^o)$.
Then the type of $\mathscr{E}(L^o m^o)$ is*

$$a_1 \geqslant \cdots \geqslant a_{j-1} \geqslant a_j + 1 = a(L^o) + 1 > a_{j+1} \geqslant \cdots \geqslant a_d.$$

Proof. Let us first be more explicit concerning the definition[1] of the integer
$j(L^o)$ from $a(L^o)$: it is the unique integer $j \in \{1, \ldots, d\}$ such that there exists
$\ell \geqslant 0$ with

$$a(L^o) = a_j = \cdots = a_{j+\ell}, \quad a(L^o) < a_{j-1}, \quad a(L^o) > a_{j+\ell+1}.$$

We thus have $a(L^o) + 1 \leqslant a_{j-1}$ (by convention, $a_0 = +\infty$).

If $d = 1$, the result is clear because, in this case, the line L^o is equal to the
fibre E_{m^o} and $\mathscr{E}(L^o m^o) \simeq \mathscr{E}(1)$.

For $d > 1$, let us choose, according to Theorem 4.4, a decomposition $\mathscr{E} = \bigoplus_{i=1}^d \mathscr{L}_i$ where $\mathscr{L}_i \simeq \mathscr{O}(a_i)$. If $L^o = L_{j,m^o}$ for some j, we similarly have
$\mathscr{E}(L^o m^o) = \bigoplus_{i \neq j} \mathscr{L}_i \oplus \mathscr{L}_j(1)$ and the result is clear.

In general, with the notation above,

$$L^o \subset F^{a(L^o)}E_{m^o} = L_{1,m^o} \oplus \cdots \oplus L_{j+\ell,m^o}$$

and

$$L^o \not\subset F^{a(L^o)-1}E_{m^o} = L_{1,m^o} \oplus \cdots \oplus L_{j-1,m^o}.$$

We can then modify the decomposition

$$F^{a(L^o)}E_{m^o} = L_{1,m^o} \oplus \cdots \oplus L_{j-1,m^o} \oplus L_{j,m^o} \oplus \cdots \oplus L_{j+\ell,m^o}$$

as

$$F^{a(L^o)}E_{m^o} = L_{1,m^o} \oplus \cdots \oplus L_{j-1,m^o} \oplus L'_{j,m^o} \oplus \cdots \oplus L'_{j+\ell,m^o}$$

in such a way that $L'_{j,m^o} = L^o$. It is therefore enough to show that there exists
a corresponding decomposition

$$(4.12) \qquad \mathscr{E} = \left(\bigoplus_{i=1}^{j-1} \mathscr{L}_i \right) \oplus \left(\bigoplus_{i=j}^{j+\ell} \mathscr{L}'_i \right) \oplus \left(\bigoplus_{i=j+\ell+1}^{d} \mathscr{L}_i \right),$$

as this reduces to the previous situation, and thereby ends the proof.

In order to do so, let us choose an isomorphism $\mathscr{L}_i \simeq \mathscr{O}_{\mathbb{P}^1}(a_i m^o)$, whence
a basis of each L_{i,m^o} ($i \in \{1, \ldots, d\}$); let us choose a basis of each L'_{j+k,m^o}
($k \in \{1, \ldots, \ell\}$); the projection $p_{ik} : L'_{j+k,m^o} \to L_{i,m^o}$ is then the multiplication by some complex number α_{ik}, equal to zero if $i > j + \ell$. We consider the
homomorphism $\mathscr{O}_{\mathbb{P}^1}(a(L^o)m^o) \to \mathscr{O}_{\mathbb{P}^1}(a_i m^o)$ obtained from the natural homomorphism by multiplying it by α_{ik}. We thus get, by direct sum, an injective

---

[1] It is taken from the proof that O. Gabber gives to Theorem IV.2.2.

homomorphism $\mathscr{O}_{\mathbb{P}^1}(a(L^o)m^o) \hookrightarrow \bigoplus_{i=1}^{d} \mathscr{O}_{\mathbb{P}^1}(a_i m^o)$, which defines a subbundle of this direct sum and hence a subbundle $\mathscr{L}'_{j+k}$ of $\mathscr{E}$ ($k \in \{1, \ldots, \ell\}$). One then verifies (exercise) that this family of subbundles satisfies (4.12).   □

4.13 Definition (of the defect). Let $\mathscr{E}$ be a bundle of rank d on $\mathbb{P}^1$ of type $a_1 \geqslant \cdots \geqslant a_d$. The *defect* of E is the integer $\delta(E) = \sum_{i=1}^{d}(a_1 - a_i)$.

The defect is a nonnegative integer and the bundle E has "defect zero" if and only if it is quasitrivial.

4.14 Corollary (Lowering the defect). *If $L^o \subset E_{m^o}$ is not contained in $(F^{a_1})_{m^o}$, we have $\delta(E) > 0$ and $\delta(E(L^o m^o)) = \delta(E) - 1$.*

Proof. This immediately follows from the inequality $a_1 > a_{j(L^o)}$.   □

4.d Algebraic and rational vector bundles

If one uses the Zariski topology instead of the usual topology on $\mathbb{P}^1$, one gets the category of algebraic (instead of "holomorphic") and rational (instead of "meromorphic") vector bundles. The functor which associates to each such bundle its "analytization" is an equivalence between the algebraic categories and the corresponding holomorphic categories (cf. [Ser56]).

In the algebraic framework, the bundles have a somewhat simple description in terms of cocycles. Let us denote as above U_0 the chart of $\mathbb{P}^1$ with coordinate t centered at 0 and U_∞ the chart with coordinate t' centered at ∞.

Recall the correspondence (cf. [Ser55] and §0.10.a) on any affine open set U:

Vector bundles on U $\iff$ Free modules of finite rank over $\mathbb{C}[U]$

where $\mathbb{C}[U]$ denotes the ring of regular functions on U, that is, if for instance U is contained in U_0, the ring of rational fractions in t which have poles in $U_0 \setminus U$.

Any algebraic vector bundle on $\mathbb{P}^1$ can be represented by a cocycle relative to the covering (U_0, U_∞), that is, an element of $\mathrm{GL}_d(\mathbb{C}[t, t^{-1}])$.

The rational bundles which have a pole only at 0 and ∞ are the algebraic bundles on the affine open set $U_0 \cap U_\infty$, that is, the free modules of finite rank over the ring $\mathbb{C}[t, t^{-1}]$ of Laurent polynomials.

4.15 Proposition. *Let $\mathbb{M}$ be a rational bundle of rank d on $\mathbb{P}^1$ with poles at 0 and ∞ (i.e., a free $\mathbb{C}[t, t^{-1}]$-module of rank d).*

(1) *The lattices of $\mathbb{M}$ correspond to the pairs $(\mathbb{E}_0, \mathbb{E}_\infty)$, where $\mathbb{E}_0$ (resp. $\mathbb{E}_\infty$) is a free $\mathbb{C}[t]$ (resp. $\mathbb{C}[t']$)-module of rank d contained in $\mathbb{M}$, such that*

$$\mathbb{C}[t, t^{-1}] \underset{\mathbb{C}[t]}{\otimes} \mathbb{E}_0 = \mathbb{M} \quad and \quad \mathbb{C}[t', t'^{-1}] \underset{\mathbb{C}[t']}{\otimes} \mathbb{E}_\infty = \mathbb{M}.$$

with $t' = t^{-1}$.

(2) *The following properties are equivalent:*
   (a) *the lattice corresponding to* $(\mathbb{E}_0, \mathbb{E}_\infty)$ *is trivial;*
   (b) *we have in* $\mathbb{M}$ *the decomposition* $\mathbb{E}_\infty = (\mathbb{E}_0 \cap \mathbb{E}_\infty) \oplus t'\mathbb{E}_\infty;$
   (c) *we have in* $\mathbb{M}$ *the decomposition* $\mathbb{E}_0 = (\mathbb{E}_0 \cap \mathbb{E}_\infty) \oplus t\mathbb{E}_0;$
   (d) *we have* $\mathbb{M} = \mathbb{E}_0 \oplus t'\mathbb{E}_\infty$ *or also* $\mathbb{M} = \mathbb{E}_\infty \oplus t\mathbb{E}_0.$

Proof. Let us only indicate why one of the decompositions (b) or (c) implies triviality for the lattice $\mathbb{E}$ associated to $(\mathbb{E}_0, \mathbb{E}_\infty)$. The remaining part is left as an exercise.

We first identify $\Gamma(\mathbb{P}^1, \mathbb{E})$ with $\mathbb{E}_0 \cap \mathbb{E}_\infty$. The existence of the decomposition (b) means that the natural restriction mapping $\Gamma(\mathbb{P}^1, \mathbb{E}) \to i_\infty^* \mathbb{E}$ is an isomorphism. Birkhoff-Grothendieck theorem (in its algebraic version), plus Exercise 2.3-(4), enable us to conclude that the bundle $\mathbb{E}$ is necessarily trivial. $\qquad\square$

5 Families of vector bundles on $\mathbb{P}^1$

5.a First properties

Let X be a complex analytic manifold, that we will always assume to be *connected* in the following. A *family* of holomorphic bundles on $\mathbb{P}^1$ parametrized by X is, by definition, a holomorphic bundle E on the manifold $M = \mathbb{P}^1 \times X$. For any $x \in X$, we can then consider the restriction E_x of the bundle E to the submanifold $\mathbb{P}^1 \times \{x\} = \mathbb{P}^1$ (cf. §0.3.3). Each bundle E_x has some type $a_1(x) \geqslant \cdots \geqslant a_d(x)$. As this type consists of integers, one can imagine that it remains constant when x varies in some dense open set of X. Such is indeed the case. More precisely, one can show (see for instance [Bru85]), that there exists a stratification of X with constant type of E_x, namely, there exists a partition of X into complex analytic (not necessarily closed) submanifolds, above each of which the restriction of E has constant type; the closure of each of these submanifolds is a closed analytic (possibly singular) subset of X; the topological boundary of any submanifold of the family is a union of other submanifolds of the family.

Notice that all the "degeneracies" are not possible. Corollary 5.4 (see below) shows for instance that, if at some point of X the type is that of the trivial bundle, the same property holds at any nearby point. Moreover, there is a simple constraint on the type of the bundles E_x for $x \in X$ (cf. for instance [BS76, Chap. 3, Th. 4.12] for the complex analytic case and [Har80] for the algebraic case):

5.1 Proposition (In a deformation, the degree stays constant). *If the manifold X is connected, the degree function* $\deg : x \mapsto \sum_{i=1}^d a_i(x)$ *is constant.* $\qquad\square$

When one replaces the Riemann sphere $\mathbb{P}^1$ with a compact Riemann surface, the *Harder-Narasimhan polygon* of the bundle indicates possible degeneracies (see for instance [Bru85] for a precise study on any compact Riemann surface). For the Riemann sphere, one recovers the polygon of Figure I.2.

5.2 Example (of a nontrivial deformation). The non-split extension (2.7) enables us to construct a family E of bundles of rank 2 on $\mathbb{P}^1$, parametrized by the complex line $X = \mathbb{C}$ with coordinate z, such that $E_0 \simeq \mathscr{O}_{\mathbb{P}^1}(-1) \oplus \mathscr{O}_{\mathbb{P}^1}(1)$ and $E_z \simeq \mathscr{O}_{\mathbb{P}^1} \oplus \mathscr{O}_{\mathbb{P}^1}$ for any $z \in \mathbb{C} \smallsetminus \{0\}$.

Let E' be the trivial bundle of rank 2 on $\mathbb{P}^1 \times \mathbb{C}$. The trivial bundle E'_0 on $\mathbb{P}^1$ contains a subbundle F_0 of rank one isomorphic to $\mathscr{O}_{\mathbb{P}^1}(-1)$. Let then $\mathscr{E}$ be the subsheaf of $\mathscr{E}'$ whose local sections belong to F_0 when restricted to $z = 0$. One can check, as in Exercise 4.10, that $\mathscr{E}$ is locally free of rank 2. As E and E' coincide away from $z = 0$, the bundles E_z are trivial for any $z \in \mathbb{C} \smallsetminus \{0\}$.

One can write $\mathscr{E} = \mathscr{F} + z\mathscr{E}' \subset \mathscr{E}'$, if $\mathscr{F}$ is the pullback bundle of $\mathscr{F}_0$ on $\mathbb{P}^1 \times \mathbb{C}$. Let us set $\mathscr{G}_0 = \mathscr{E}'_0/\mathscr{F}_0 \simeq \mathscr{O}_{\mathbb{P}^1}(1)$ and $\mathscr{G} = \mathscr{E}'/\mathscr{F}$. We will explicitly give a homomorphism $\mathscr{G}_0 \to \mathscr{E}_0$, the image of which only meets $\mathscr{F}_0$ on the zero section, thereby showing that $\mathscr{E}_0 = \mathscr{F}_0 \oplus \mathscr{G}_0$. In order to do that, let us remark that we have a homomorphism

$$\mathscr{G} = \mathscr{E}'/\mathscr{F} \xrightarrow{\ z\ } \mathscr{E}/z\mathscr{E}$$

induced by multiplication by z, as $z\mathscr{E}' \subset \mathscr{E}$ and $z\mathscr{F} \subset z\mathscr{E}$. Moreover, the subsheaf $z\mathscr{G}$ has image 0, as $z^2\mathscr{E}' \subset z\mathscr{E}$. We thus get the desired homomorphism

$$\mathscr{G}_0 = \mathscr{G}/z\mathscr{G} \xrightarrow{\ z\ } \mathscr{E}/z\mathscr{E} = \mathscr{E}_0.$$

5.b Rigidity theorems

We will mainly consider the case where the restriction E_x of E to $\mathbb{P}^1 \times \{x\}$, for x general enough, is isomorphic to the trivial bundle. We will use some fundamental results of the theory of coherent sheaves, for which we refer to [BS76, Chap. 3] or [Fis76]: the semicontinuity of the dimension of fibres of a coherent sheaf and a criterion of local freeness, the coherence of direct images by a proper morphism (statement for a family analogous to the finiteness theorem 3.1) and the proper base change theorem. We will thus assume in this section some familiarity with the theory of coherent sheaves[2].

In particular, we will use the fact that the support of a coherent sheaf $\mathscr{H}$ on X is a closed analytic subset of X and that, at any point x of this support, the fibre $\mathscr{H}/\mathfrak{m}_x\mathscr{H}$ of this sheaf[3], that is, the set of values at x of the germs of sections of $\mathscr{H}$, is not equal to zero (Nakayama).

---

[2] The reader will find a direct proof (i.e., not referring to the theory of coherent sheaves) of Theorem 5.3 and of Corollary 5.6 in [Mal83c].

[3] We denote by $\mathfrak{m}_x$ the sheaf ideals equal to $\mathscr{O}_X$ on $X \smallsetminus \{x\}$ and with germ at x equal to the maximal ideal of $\mathscr{O}_{X,x}$.

If E is a bundle of rank d on $\mathbb{P}^1 \times X$, the degree of its restriction to the fibre $\mathbb{P}^1 \times \{x\}$ does not depend on x, if X is connected, as follows from Proposition 5.1. If we still denote by $\mathcal{O}_{\mathbb{P}^1 \times X}(1)$ the pullback by the first projection of the bundle $\mathcal{O}_{\mathbb{P}^1}(1)$ and $\mathcal{E}(k) = \mathcal{O}_{\mathbb{P}^1 \times X}(k) \otimes \mathcal{E}$, we see that the degree of the restrictions of $\mathcal{E}(-\deg \mathcal{E})$ is zero.

On the other hand, if π denotes the projection $\mathbb{P}^1 \times X \to X$ and if $\mathcal{F}$ is a locally free sheaf on $\mathbb{P}^1 \times X$, the sheaf $\boldsymbol{R}^k \pi_* \mathcal{F}$ ($k \in \mathbb{N}$) is (see [God64, §4.17]) the sheaf associated to the presheaf

$$V \longmapsto H^k(\mathbb{P}^1 \times V, \mathcal{F}).$$

It is a coherent sheaf on X (Grauert's Theorem; see for instance [BS76, Chap. 3, §2]). It is zero for $k \geqslant 2$: indeed, if $\mathcal{F}_x$ is a bundle on $\mathbb{P}^1 \times \{x\}$, we have $H^k(\mathbb{P}^1, \mathcal{F}_x) = 0$ for $k \geqslant 2$, as mentioned in Theorem 3.1; we can then apply [BS76, Cor. 3.11]. It follows from this vanishing that the formation of $\boldsymbol{R}^1 \pi_* \mathcal{F}$ is compatible with base change (see [BS76, Cor. 3.5]). In particular, the fibre at x of $\boldsymbol{R}^1 \pi_* \mathcal{F}$ is equal to $H^1(\mathbb{P}^1, \mathcal{F}_x)$.

5.3 Theorem (The nontriviality divisor). *Let X be a connected complex analytic manifold and let E be a holomorphic bundle of rank d on $\mathbb{P}^1 \times X$ such that, for any $x \in X$, the restriction $\mathbb{P}^1 \times \{x\}$ has degree zero.*

(1) *The support Θ of the sheaf $\boldsymbol{R}^1 \pi_* \mathcal{E}(-1)$ is the set of points $x \in X$ such that the restriction of E to $\mathbb{P}^1 \times \{x\}$ is not trivial.*

(2) *If $\Theta \neq \varnothing$ and $\Theta \neq X$, then Θ is a hypersurface of X.*

Proof. According to the preliminary remarks above, the first point is a straightforward consequence of Exercise 4.7. It remains thus to show that the support Θ, that we will assume nonempty and $\neq X$, is a hypersurface[4]. Let us begin with some remarks on determinant bundles.

Let $\mathcal{F}$ and $\mathcal{G}$ be two locally free sheaves of $\mathcal{O}_X$-modules *having the same rank d* and let $\varphi : \mathcal{F} \to \mathcal{G}$ be a non identically zero homomorphism. In local bases of $\mathcal{F}$ and $\mathcal{G}$, the determinant of its matrix is a holomorphic function, the zero locus of which is the set of points where φ is not an isomorphism. In a more intrinsic way, we call *determinant bundle* of $\mathcal{F}$ the line bundle $\det \mathcal{F} = \wedge^d \mathcal{F}$ and $\det \varphi$ the homomorphism $\det \mathcal{F} \to \det \mathcal{G}$ whose matrix in a local basis of $\det \mathcal{F}$ and $\det \mathcal{G}$ (obtained from a basis of F and G) is the determinant of that of φ. It is also a section of the bundle $\det \mathcal{G} \otimes (\det \mathcal{F})^\vee$, where $(\det \mathcal{F})^\vee$ is the dual bundle of the line bundle $\det \mathcal{F}$ (it is also its inverse in the sense of tensor product, cf. §0.7).

If $\varphi : \mathcal{F} \to \mathcal{G}$ is injective, the cokernel $\mathcal{C}$ of φ is a coherent sheaf on X and its support is the set of zeroes of the section $\det \varphi$: if this set is not equal to X, it is a closed analytic hypersurface of X, empty if and only if φ is everywhere an isomorphism.

---

[4] The proof which follows is inspired by that of [OSS80, p. 214–217]. I learned it from J. Le Potier.

We will therefore show that the sheaf $\boldsymbol{R}^1\pi_*\mathcal{E}(-1)$ admits, in the neighbourhood of any point x^o of Θ, a presentation of the previous kind, that is, that there exists an exact sequence of sheaves on some neighbourhood of x^o in X

$$0 \longrightarrow \mathcal{F} \longrightarrow \mathcal{G} \longrightarrow \boldsymbol{R}^1\pi_*\mathcal{E}(-1) \longrightarrow 0,$$

with $\mathcal{F}$ and $\mathcal{G}$ locally free and having same finite rank.

Let us first show that, up to replacing X with a sufficiently small neighbourhood of x^o, there exist an integer q and a *surjective* homomorphism

$$(*) \qquad \mathcal{E}' := \mathcal{O}_{\mathbb{P}^1 \times X}(q)^d \longrightarrow \mathcal{E}(-1).$$

Indeed, there exists an integer r such that the restriction of $\mathcal{E}(r)$ to $\mathbb{P}^1 \times \{x^o\}$ is generated by its global sections, that is, there exists a surjective homomorphism $\varphi_{x^o} : \mathcal{O}_{\mathbb{P}^1 \times \{x^o\}}^d \to \mathcal{E}_{x^o}(r)$ (if $\mathcal{E}_{x^o}$ has the decomposition $\bigoplus_{i=1}^d \mathcal{O}_{\mathbb{P}^1}(a_i)$, choose r such that $r + a_i \geqslant 0$ for all i).

For such an r, we also have $\boldsymbol{R}^1\pi_*\mathcal{E}(r) = 0$: indeed, we have seen that, in our situation, $\boldsymbol{R}^1\pi_*$ commutes with restriction to x^o; with the previous choice of r, we thus have $\boldsymbol{R}^1\pi_*\mathcal{E}(r)_{x^o} = 0$, by the vanishing theorem 2.10; as $\boldsymbol{R}^1\pi_*\mathcal{E}(r)$ is coherent, Nakayama's Lemma implies that the germ of $\boldsymbol{R}^1\pi_*\mathcal{E}(r)$ at x^o is zero, hence $\boldsymbol{R}^1\pi_*\mathcal{E}(r)$ is identically zero in some neighbourhood of x^o; on such a neighbourhood, π_* commutes then with restriction to x^o; therefore, any section of $\mathcal{E}(r)_{x^o}$ can be locally lifted to a section of $\mathcal{E}(r)$.

In other words, the surjective morphism φ_{x^o} can be lifted, if X is sufficiently small, to a morphism $\varphi : \mathcal{O}_{\mathbb{P}^1 \times X}^d \to \mathcal{E}(r)$. Using once more Nakayama's Lemma at points of $\mathbb{P}^1 \times \{x^o\}$, and up to shrinking X once more, we get the surjectivity of φ. We now tensor φ with $\mathcal{O}_{\mathbb{P}^1 \times X}(q)$, with $q + r = -1$, in order to get the existence and the surjectivity of $(*)$.

Let $\mathcal{K}$ be the kernel of $(*)$. We will apply the long exact sequence of cohomology for the functor π_*, obtained from the exact sequence 0.6.2 by passing to direct limit on neighbourhoods of $x^o \in X$.

Let us first remark that the sheaf $\pi_*\mathcal{E}(-1)$ is zero: its restriction to fibres $\mathbb{P}^1 \times \{x\}$, $x \notin \Theta$, is zero as $H^k(\mathbb{P}^1, \mathcal{O}_{\mathbb{P}^1}(-1)^d) = 0$ for any $k \geqslant 0$ (see [BS76, Cor. 3.11]); on the other hand, it has no local section supported in Θ, as such a section defines a section of the bundle $\mathcal{E}(-1)$ on an open set $\mathbb{P}^1 \times V$ supported in $\pi^{-1}\Theta$, hence is zero.

Moreover, for any coherent sheaf $\mathcal{C}$ on $\mathbb{P}^1$, the cohomology groups $H^k(\mathbb{P}^1, \mathcal{C})$ vanish for $k \geqslant 2$: we have mentioned this above for locally free sheaves (see the proof of Theorem 3.1). The general case of coherent sheaves follows, by using that any coherent sheaf on $\mathbb{P}^1$ admits a finite resolution by locally free sheaves (see for instance [BS76, Chap. 4, Cor. 2.6]).

Therefore, we have $\boldsymbol{R}^k\pi_*\mathcal{H} = 0$ for $k \geqslant 2$, when $\mathcal{H}$ is one of the sheaves $\mathcal{K}$, $\mathcal{E}'$ or $\mathcal{E}(-1)$ ([BS76, Cor. 3.11]).

We finally get

$$\pi_*\mathcal{K} \xrightarrow{\sim} \pi_*\mathcal{E}'$$

and an exact sequence

$$0 \longrightarrow R^1\pi_*\mathscr{K} \longrightarrow R^1\pi_*\mathscr{E}' \longrightarrow R^1\pi_*\mathscr{E}(-1) \longrightarrow 0.$$

Let us note that $\mathscr{G} := R^1\pi_*\mathscr{E}'$ is locally free, as it is coherent (proper direct image of a coherent sheaf) and its fibres have the same rank $\delta = \dim H^1(\mathbb{P}^1, \mathscr{O}_{\mathbb{P}^1}(q)^d)$ (see [BS76, Chap. 3, Lemma 1.6]).

Similarly, $\mathscr{F} := R^1\pi_*\mathscr{K}$ is coherent and locally free of rank equal to δ on $X \smallsetminus \Theta$. Moreover, being a subsheaf of the locally free sheaf $\mathscr{G}$, it has no torsion. If X has dimension one, we conclude that $\mathscr{F}$ is $\mathscr{O}_X$ locally free as, for any $x^o \in X$, the ring $\mathscr{O}_{X,x^o}$ is a principal ideal domain. If X has dimension $n \geqslant 2$ and if $\mathscr{F}$ is not locally free, the dimension of the fibre of $\mathscr{F}$ at x^o is $> \delta$ (see for instance [Fis76, p. 54]). Then there exists, up to shrinking X, a system of local coordinates $(x_1, \ldots, x_n)$ centered at x^o such that the restriction of $\mathscr{F}$ to the x_1-axis has a generic fibre of dimension δ and a fibre at x^o of dimension strictly bigger than δ, that is, this restriction is not locally free. As the formation of $\mathscr{F}$ and $\mathscr{G}$ commutes with base change[5] (properness of π), this is impossible, after the result when $\dim X = 1$. Therefore, $\mathscr{F}$ is locally free, with the same rank as $\mathscr{G}$, and the support of $R^1\pi_*\mathscr{E}(-1)$ is a closed analytic hypersurface of X.   □

The rigidity statement below will be essential. As B. Malgrange remarked, it is basic for the Painlevé's property of some systems of differential equations considered later.

5.4 Corollary (Rigidity of trivial bundles on $\mathbb{P}^1$). *Let E be a bundle of rank d on the product $\mathbb{P}^1 \times X$. We assume that there exists $x^o \in X$ such that $E^o := E_{|\mathbb{P}^1 \times \{x^o\}}$ is trivial. There exists then an open neighbourhood V of x^o in X such that the restriction of E to $\mathbb{P}^1 \times V$ is trivial.*

Proof. After Theorem 5.3-(1), there exists an open neighbourhood W of x^o such that $E_{|\mathbb{P}^1 \times \{x\}}$ is trivial for any $x \in W$. We thus have $H^k(\mathbb{P}^1, \mathscr{E}_{|\mathbb{P}^1 \times \{x\}}) = 0$ for any $k \geqslant 1$; therefore, as indicated before the statement of Theorem 5.3, we have $R^k\pi_*\mathscr{E} = 0$ on W for any $k \geqslant 1$. It follows that the formation of $\pi_*\mathscr{E}$ commutes to base change. Hence, denoting $\mathfrak{m}_{x^o}$ the maximal ideal of $\mathscr{O}_{X,x^o}$, we have $\Gamma(\mathbb{P}^1, \mathscr{E}^o) = \pi_*\mathscr{E}/\mathfrak{m}_{x^o}\pi_*\mathscr{E}$ and this space has dimension d. Let us choose a basis $e_1^o, \ldots, e_d^o$ of it and let $e_1, \ldots, e_d$ be liftings in $(\pi_*\mathscr{E})_{x^o}$. We thus get, for V sufficiently small, a homomorphism

$$e : \mathscr{O}_{\mathbb{P}^1 \times V}^d \longrightarrow \mathscr{E}_{|\mathbb{P}^1 \times V}.$$

In a local basis of $\mathscr{E}_{|\mathbb{P}^1 \times V}$, its matrix $A(t,x)$ satisfies $\det A(t, x^o) \neq 0$ for any t as, E^o being trivial, it admits $e_1^o, \ldots, e_d^o$ as a local basis. It follows that, up to taking a smaller V, we also have $\det A(t,x) \neq 0$ for any t and any $x \in V$. Therefore, e is an isomorphism.   □

---

[5] As $\mathscr{F}$ may not be locally free, we have to use a resolution by locally free $\mathscr{O}_{\mathbb{P}^1 \times V}$-modules to get the assertion for $\mathscr{F}$ from [BS76, Chap. 3, Th. 3.4].

5.5 Remarks.

(1) We also get that $\pi_*\mathscr{E}$ is locally free of rank d on $X \smallsetminus \Theta$.

(2) The following analogy can be helpful in understanding the proof of this corollary. Let H be a finite dimensional $\mathbb{C}$-vector space and let d be an integer $\leqslant \dim H$. The Grassmannian $\mathrm{Gr}(d, H)$ of d-dimensional subspaces of H is equipped with a "tautological" bundle $\mathscr{E}$ whose fibre at the point $[E]$ representing the subspace $E \subset H$ is precisely this subspace (for $d = 1$ and $H = \mathbb{C}^2$, we have $\mathrm{Gr}(1, \mathbb{C}^2) = \mathbb{P}^1$ and $\mathscr{E} = \mathscr{O}_{\mathbb{P}^1}(-1)$). Let us fix a d-dimensional subspace $E^o \subset H$ and a projection $p : H \to E^o$, that is, a decomposition $H = E^o \oplus F^o$. We deduce, for any $E \in \mathrm{Gr}(d, H)$, a morphism $\varphi : E \to E^o$, that we regard as a morphism φ from $\mathscr{E}$ to the trivial bundle $\mathscr{O}_{\mathrm{Gr}(d,H)} \otimes_{\mathbb{C}} E^o$ with fibre E^o. This morphism is an isomorphism away from the locus Θ_p where $\det \varphi = 0$. We regard $\det \varphi$ as a morphism from the line bundle $\det \mathscr{E}$ to the trivial bundle of rank one, that is, a section of the bundle $\det \mathscr{E}^{\vee}$ dual to $\det \mathscr{E}$. Therefore, Θ_p is a hypersurface of $\mathrm{Gr}(d, H)$ (which depends on the choice of p).

Condition (2d) in Proposition 4.15 suggests to extend the reasoning above, to the case of an infinite dimensional space (i.e., take $H = \mathbb{M}$). This is meaningful when H is a separable Hilbert space (cf. [PS86, Chap. 7]), for instance the space of L^2 functions on the unit circle. The proof of [Mal83c] uses these kind of methods.

We can now be more precise concerning Corollary 5.4 in the neighbourhood of a point of Θ:

5.6 Corollary (Meromorphic trivialization). *Let E be a bundle of rank d on $\mathbb{P}^1 \times X$ having degree 0. We assume that the support Θ of $\mathbf{R}^1\pi_*\mathscr{E}(-1)$ is a (possibly empty) hypersurface. Then, for any $x \in X$, there exists an open neighbourhood V of x and a meromorphic trivialization*

$$\mathscr{E}_{|\mathbb{P}^1 \times V}(*\pi^{-1}\Theta) \simeq \mathscr{O}_{\mathbb{P}^1 \times V}^d(*\pi^{-1}\Theta).$$

Proof. It is a consequence of the following lemma:

5.7 Lemma. *Let $\mathscr{F}$ be a $\mathscr{O}_X$-coherent sheaf, locally free of rank d on $X \smallsetminus \Theta$. Then $\mathscr{O}_X(*\Theta) \otimes_{\mathscr{O}_X} \mathscr{F}$ is locally free of rank d as a $\mathscr{O}_X(*\Theta)$-module.*

Indeed, Remark 5.5 enables us to apply this lemma to $\pi_*\mathscr{E}$. Let us pick $x^o \in \Theta$ and let $(e_1, \ldots, e_d)$ be a basis of $\pi_*\mathscr{E}(*\Theta)_{x^o}$. This basis defines thus a homomorphism

$$e : \mathscr{O}_{\mathbb{P}^1 \times V}^d(*\pi^{-1}\Theta) \longrightarrow \mathscr{E}(*\pi^{-1}\Theta)_{|\mathbb{P}^1 \times V}$$

if V is a sufficiently small neighbourhood of x^o. As e induces a basis of each bundle $\mathscr{E}_{|\mathbb{P}^1 \times \{x\}}$ for $x \in V \smallsetminus \Theta$, it follows as above that e is an isomorphism. $\quad\square$

Proof (of Lemma 5.7). Let us pick $x^o \in \Theta$. As $\mathscr{F}$ is coherent, we have

$$\dim \mathscr{F}_{x^o}/\mathfrak{m}_{x^o}\mathscr{F}_{x^o} \geqslant d.$$

Let us choose d independent vectors $\varphi_1^o, \ldots, \varphi_d^o$ in this space. They can be lifted in d independent local sections $\varphi_1, \ldots, \varphi_d$ of $\mathscr{F}$ on a sufficiently small neighbourhood V of x^o. These ones define a homomorphism $\varphi : \mathscr{O}_V^d \to \mathscr{F}_{|V}$. The latter is an isomorphism on $V \smallsetminus \Theta$ as $\mathscr{F}$ is locally free of rank d there and as the determinant of these sections, regarded as a homomorphism from $\mathscr{O}_X$ to $\wedge^d \mathscr{F}$, does not vanish on V. The kernel $\mathscr{K}$ and the cokernel $\mathscr{C}$ of φ are coherent sheaves with support in Θ. If $\theta = 0$ is an equation of Θ in the neighbourhood of x^o, $\mathscr{K}$ and $\mathscr{C}$ are killed by some power of θ. In particular, we have

$$\mathscr{O}_V(*\Theta) \underset{\mathscr{O}_V}{\otimes} \mathscr{K} = 0 \quad \text{and} \quad \mathscr{O}_V(*\Theta) \underset{\mathscr{O}_V}{\otimes} \mathscr{C} = 0.$$

As $\mathscr{O}_V(*\Theta)$ is flat over $\mathscr{O}_V$ (see for instance [Bou98]), we conclude that φ induces an isomorphism $\mathscr{O}_V^d(*\Theta) \xrightarrow{\sim} \mathscr{F}(*\Theta)_{|V}$. $\qquad\square$

Let i_0 and i_∞ be the zero and infinity sections from X to $\mathbb{P}^1 \times X$. Let us consider both natural restriction morphisms

$$\rho_0 : \pi_*\mathscr{E} \longrightarrow i_0^*\mathscr{E} \quad \text{and} \quad \rho_\infty : \pi_*\mathscr{E} \longrightarrow i_\infty^*\mathscr{E}.$$

The mapping ρ_0 is given, for any open set V of X, by

$$\Gamma(\mathbb{P}^1 \times V, \mathscr{E}) \longrightarrow \Gamma(\{0\} \times V, \mathscr{E})$$
$$s(t, x) \longmapsto s(0, x)$$

and ρ_∞ is defined similarly.

Let E be a bundle of rank d on $\mathbb{P}^1 \times X$. We assume that there exists $x^o \in X$ such that $E^o := E_{|\mathbb{P}^1 \times \{x^o\}}$ is trivial. There exists then, after Theorem 5.3, a (possibly empty) hypersurface Θ of X not containing x^o and such that E_x is trivial for any $x \in X \smallsetminus \Theta$.

5.8 Corollary (Canonical identification between the restrictions to 0 and to ∞). *With this assumption, the restriction morphisms ρ_0 and ρ_∞ induce isomorphisms after tensoring with $\mathscr{O}_X(*\Theta)$; in particular, on $X \smallsetminus \Theta$, the vector bundles $\pi_*\mathscr{E}_{|X \smallsetminus \Theta}$, $i_0^*\mathscr{E}_{|X \smallsetminus \Theta}$ and $i_\infty^*\mathscr{E}_{|X \smallsetminus \Theta}$ are isomorphic via ρ_0 and ρ_∞. One also has, after applying the functor π^*, isomorphisms of meromorphic bundles*

$$\mathscr{E}(*\pi^{-1}\Theta), \quad \pi^*i_0^*\mathscr{E}(*\pi^{-1}\Theta) \quad \text{and} \quad \pi^*i_0^*\mathscr{E}(*\pi^{-1}\Theta).$$

Proof. Let us notice first that $i_0^*\mathscr{E}$ and $i_\infty^*\mathscr{E}$ are two vector bundles on X (restrictions of the bundle $\mathscr{E}$). After Corollary 5.6, the bundle $\mathscr{E}(*\pi^{-1}\Theta)$ is a meromorphic bundle trivialized on open sets $\mathbb{P}^1 \times V$. We want to verify that the morphisms ρ_0 and ρ_∞ are isomorphisms. It is enough to check this property locally on X. We may thus assume that $\mathscr{E}(*\pi^{-1}\Theta) = \mathscr{O}_{\mathbb{P}^1 \times V}(*\Theta)^d$ and in this case the result is obvious.

The third part of the corollary is proved in the same way. $\qquad\square$

II

The Riemann-Hilbert correspondence on a Riemann surface

In this chapter, M denotes a Riemann surface, i.e., a complex analytic manifold of dimension one (cf. §0.2) which will always be assumed *connected*. A general presentation of the properties of Riemann surfaces is given in [Rey89]. Their topological properties (that of their fundamental group in particular) are analyzed with details in [Gra84, Mas67].

1 Statement of the problems

Let Σ be a discrete set of points of M and let o be a base point in $M \smallsetminus \Sigma$. If for instance $M = \mathbb{P}^1$, the set $\Sigma = \{m_0, m_1, \ldots, m_p\}$ is finite, as $\mathbb{P}^1$ is compact, and we have $\mathbb{P}^1 \smallsetminus \Sigma = \mathbb{C} \smallsetminus \{m_1, \ldots, m_p\}$ (by taking $\infty = m_0$); therefore, the fundamental group $\pi_1(M \smallsetminus \Sigma, o)$ is the free group having as generators the classes of the loops $\gamma_1, \ldots, \gamma_p$ drawn in Figure II.1 (where the point m_0 is at infinity), or, in a more intrinsic way, the quotient of the free group with $p + 1$ generators $\gamma_0, \ldots, \gamma_p$ by the equivalence relation generated by the relation $\gamma_p \cdots \gamma_0 = 1$.

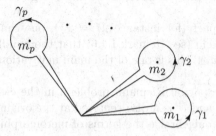

Fig. II.1.

Let us consider some representation $\pi_1(M \smallsetminus \Sigma, o) \rightarrow \mathrm{GL}_d(\mathbb{C})$. For instance, when $M = \mathbb{P}^1$ and $\Sigma = \{m_0, \ldots, m_p\}$, giving such a representation is equivalent to giving p invertible matrices $T_1, \ldots, T_p$ without any other condition or, in an equivalent way, $p + 1$ invertible matrices $T_0, \ldots, T_p$ satisfying $T_p \cdots T_0 = \mathrm{Id}$.

According to Theorem 0.15.8, there exists a locally constant sheaf $\mathscr{F}$ on $M \smallsetminus \Sigma$ corresponding to this representation. Moreover, according to Theorem 0.12.8, there is a holomorphic bundle with connection $(\mathscr{E}, \nabla)$ on $M \smallsetminus \Sigma$ such that $E^\nabla = \mathscr{F}$, namely $\mathscr{E} = \mathcal{O}_M \otimes_{\mathbb{C}_M} \mathscr{F}$. Recall that, as M has dimension one, such a connection is flat; in higher dimension, one should not forget this adjective for the analogous problem.

1.1 The very weak Riemann problem. *Does there exist a meromorphic bundle $(\mathscr{M}, \nabla)$ with (flat) connection on M, whose restriction to $M \smallsetminus \Sigma$ is isomorphic to $(\mathscr{E}, \nabla)$?*

1.2 The weak Riemann problem. *Does there exist a meromorphic bundle $(\mathscr{M}, \nabla)$ with (flat) connection on M which has a regular singularity (cf. §0.14) at each point of Σ and whose restriction to $M \smallsetminus \Sigma$ is isomorphic to $(\mathscr{E}, \nabla)$?*

1.3 The partial Riemann problem. *Given a meromorphic bundle with (flat) connection $({}^U\!\mathscr{M}, {}^U\nabla)$ on some connected open set U of $M \smallsetminus \Sigma$, having poles at a discrete set $\Sigma' \subset U$, assume that the inclusion $U \smallsetminus \Sigma' \hookrightarrow M \smallsetminus (\Sigma \cup \Sigma')$ induces an isomorphism of fundamental groups (for instance $U = M \smallsetminus \Sigma$).*

- *Does there exist a meromorphic bundle with (flat) connection $(\mathscr{M}, \nabla)$ on M, whose restriction to U is isomorphic to $({}^U\!\mathscr{M}, {}^U\nabla)$?*
- *Can one moreover choose $(\mathscr{M}, \nabla)$ with a regular singularity at each point of Σ?*

We will show that the answer to these questions is *positive*. Moreover, the solution is unique when one asks for the regular singularity condition at each point of Σ: we thus get the bijectivity of the *Riemann-Hilbert correspondence* between representations of the fundamental group $\pi_1(M \smallsetminus \Sigma, o)$ and meromorphic bundles with (flat) connection having a regular singularity at each point of Σ.

When M is compact (for instance $M = \mathbb{P}^1$), one can deduce, by using a GAGA[1] type theorem (see Remark I.4.6) that the bundle with connection can be chosen *algebraic*. This is one of the main motivations for these kind of theorems.

A solution to the partial Riemann problem, in the case where $M = \mathbb{P}^1$, $\Sigma = \{\infty\}$, $\Sigma' = \{0\}$ and U is a disc centered at the origin, as in the remark above, enables one to algebraize the germs of meromorphic connections (see §3.c).

When $M = \mathbb{P}^1$, any meromorphic bundle, whose set of poles is nonempty, is isomorphic to the sheaf $\mathcal{O}_{\mathbb{P}^1}(*\Sigma)^d$ (cf. Corollary I.4.9), so that, by taking

---

[1] This means "analytic geometry and algebraic geometry".

2 Local study of regular singularities   85

a basis of this sheaf, one can translate the statements on the connection into statements on meromorphic differential systems.

For the solution to these problems, we will need a precise local study of regular singularities. The question of deforming Birkhoff's problem, which we will consider with details in Chapter VI, will involve a precise local study of irregular singularities, which we will do in §§5 and 6.

In order to do so, we will take a classic point of view from linear algebra: before solving a linear system of differential equations, we give it a "normal" form, in such a way that its resolution becomes easier. We will first introduce the possible normal forms, which we will call "elementary models". Producing the normal form will involve the resolution of a linear system of the same kind to find the matrix of the base change.

2 Local study of regular singularities

In this section, the Riemann surface M is the disc $D = \{t \in \mathbb{C} \mid |t| < r\}$, where r is arbitrarily small, so that 0 is the only possible pole of the meromorphic functions that we consider. More precisely, we consider the germ $(\mathbb{C}, 0)$ of Riemann surface.

2.a Some definitions

Let us take up in a local setting the notions introduced in §0.14. We consider the (germ of) trivial bundle of rank d on the germ $(\mathbb{C}, 0)$, equipped with the connection with matrix $\Omega = A(t)\,dt$, where the entries of A are germs of meromorphic functions with pole at 0.

The ring of germs of meromorphic functions with poles at 0 is the ring $\mathbb{C}\{t\}[1/t]$ of converging meromorphic Laurent series. More precisely, it is a *field*, that we will denote by $\boldsymbol{k}$. The usual action of the derivation makes it a *differential field*, with field of constant (elements killed by the derivation) equal to the field $\mathbb{C}$: a meromorphic Laurent series with zero derivative is constant.

Therefore, a germ of meromorphic bundle with connection is nothing but a $\boldsymbol{k}$-vector space of rank d equipped with a derivation compatible to the derivation of $\boldsymbol{k}$. It amounts to giving a square matrix $A(t)$ of size d with entries in $\boldsymbol{k}$ or also the matrix $\Omega = A(t)\,dt$.

Two $(\boldsymbol{k}, \nabla)$-vector spaces are isomorphic if and only if the corresponding matrices Ω_1 and Ω_2 are related by a meromorphic base change $P \in \mathrm{GL}_d(\boldsymbol{k})$, *via* the relation

$$\Omega_2 = P^{-1}\Omega_1 P + P^{-1}dP$$

and the matrices A_1 and A_2 by the relation

$$(2.1) \qquad A_2(t) = P^{-1}A_1 P + P^{-1}P'.$$

A (k, ∇)-vector space $\mathcal{M}$ of rank d defined by a matrix $A_1(t)$ has a *regular singularity at* 0 if there exists a matrix $P \in \mathrm{GL}_d(k)$ such that the matrix A_2 obtained after the base change (2.1) has at most a *simple pole* at $t = 0$. Otherwise, the singularity is called *irregular*.

A *lattice* of this (k, ∇)-vector space is a free $\mathbb{C}\{t\}$-module $\mathcal{E} \subset \mathcal{M}$ such that

$$k \underset{\mathbb{C}\{t\}}{\otimes} \mathcal{E} = \mathcal{M}.$$

2.b Rank one

The classification of such systems is simple: if $a(t)\, dt$ is the "matrix" of the system, then a system of matrix $b(t)\, dt$ is meromorphically equivalent to the first one if and only if $b(t) = a(t) + p'(t)/p(t)$ for some meromorphic germ $p \in k$.

2.2 Proposition (Regular singularities in rank one).

(1) *A (k, ∇)-vector space of rank one has a regular singularity if and only if, in any basis, the matrix $a(t)\, dt$ of the connection has a simple pole. There exists then a basis in which the matrix takes the form $\alpha\, dt/t$ with $\alpha \in \mathbb{C}$. It admits a horizontal meromorphic section if and only if $\alpha \in \mathbb{Z}$.*

(2) *Two (k, ∇)-vector spaces of rank one with regular singularity, with matrices $a_1(t)\, dt$ and $a_2(t)\, dt$, are isomorphic if and only if $a_1 - a_2$ has a simple pole with integral residue at 0.*

Proof (Sketch).

(1) One shows that a meromorphic function $q(t)$ can be written as p'/p with p meromorphic if and only if q has at most a simple pole at 0 with integral residue.

(2) A homomorphism $(\mathcal{M}_1, \nabla) \to (\mathcal{M}_2, \nabla)$ is an isomorphism if and only if it is nonzero, as $\mathcal{M}_1$ and $\mathcal{M}_2$ have rank one. It is a horizontal section of the (rank one) (k, ∇)-vector space $\mathrm{Hom}_k(\mathcal{M}_1, \mathcal{M}_2)$. The matrix of ∇ on this space is $(a_2 - a_1)\, dt$. $\qquad \square$

Let us pick $\alpha \in \mathbb{C}$ and let e be a basis of some (k, ∇)-vector space of rank one with respect to which the matrix of ∇ is $\alpha\, dt/t$. A horizontal section on an open set U of $D \smallsetminus \{0\}$ takes the form $s(t) = u(t)e$ where u is a holomorphic function on U satisfying Equation (0.12.7), which can be written here as

$$u'(t) + \alpha u(t) = 0.$$

If U is simply connected, Such an equation has a solution, namely $u(t) = e^{-\alpha \log t}$ if log is some determination of the logarithm on U. We denote by $t^{-\alpha}$ this function.

2.3 Exercise (Lattices of (k, ∇)-vector spaces of rank one). Let $\mathcal{M}$ be a (k, ∇)-vector space of rank one and let $\mathcal{E}$ be a lattice. Let e and e' be two bases of $\mathcal{E}$ and let $a(t)\, dt$ and $b(t)\, dt$ be the corresponding "matrices" of the connection.

(1) Prove that there exists a germ p of invertible holomorphic function such that $b(t) = a(t) + p'/p$.

(2) Show that a meromorphic function q can be written as p'/p, with p holomorphic invertible, if and only if q has at most a simple pole (at $t = 0$) with a *nonnegative integral* residue.

(3) Infer from this a classification of lattices of a (k, ∇)-vector space of rank one with regular singularity.

(4) Deduce more generally a classification of lattices of a (k, ∇)-vector space of rank one.

2.c Models in arbitrary rank

We now consider a (k, ∇)-vector space $(\mathcal{M}, \nabla)$ of rank $d \geqslant 1$ equipped with a basis $e = (e_1, \ldots, e_d)$ in which the connection matrix takes the form

$$(2.4) \qquad \qquad \Omega(t) = A\frac{dt}{t}$$

where $A \in M_d(\mathbb{C})$. There exists a constant matrix $P \in \mathrm{GL}_d(\mathbb{C})$ such that $J = P^{-1}AP$ takes the Jordan normal form. The connection matrix in the basis $\varepsilon = e \cdot P$ can be written as

$$\Omega'(t) = J\frac{dt}{t}$$

as the term $P^{-1}P'$ is zero, the matrix P being constant. One can remark that the object we consider is the restriction to a neighbourhood of the origin of a meromorphic bundle with connection on $\mathbb{C}$ with pole at the origin only (because the coefficient of dt in the matrix Ω is a rational fraction with pole at 0 only).

2.5 Definition (Elementary regular models). We define an *elementary regular model* as a (k, ∇)-vector space equipped with a basis in which the connection matrix is written as

$$\Omega(t) = (\alpha \,\mathrm{Id} + N)\frac{dt}{t}$$

where $\alpha \in \mathbb{C}$ and N is a nilpotent matrix. If N is a single Jordan block, we denote it by $\mathcal{N}_{\alpha,d}$.

Note that the connection given in (2.4) is isomorphic to a direct sum of elementary regular models: indeed, this is true after the base change P, as the matrix Ω' is blockdiagonal, each block taking the "elementary" form of Definition 2.5; this is thus also true before the base change.

2.6 Exercise (Horizontal sections of elementary regular models). We assume that the matrix N has a single Jordan block, of size $d \geqslant 1$.

(1) Prove that, on any simply connected open set U of $D \smallsetminus \{0\}$, the horizontal sections s of the elementary regular model with matrix $(\alpha \operatorname{Id} + N) \, dt/t$ are obtained by taking the linear combinations with coefficients in $\mathbb{C}$ of the columns of the matrix

$$t^{-(\alpha \operatorname{Id} + N)} := t^{-\alpha} \left(\operatorname{Id} - N \log t + N^2 \frac{(\log t)^2}{2!} + \cdots + (-N)^d \frac{(\log t)^d}{d!} \right),$$

that is, of the form $s(t) = t^{-(\alpha \operatorname{Id} + N)} \cdot \sigma$, with $\sigma \in \mathbb{C}^d$.

(2) Prove that the monodromy representation (cf. §0.14) defined by the horizontal sections of an elementary regular model associated to the matrix $\alpha \operatorname{Id} + N$ is given by

$$T = \exp\left(-2i\pi(\alpha \operatorname{Id} + N)\right).$$

(3) Prove that the horizontal sections have *moderate growth near the origin*, that is, for any horizontal section s on the neighbourhood of a *closed angular sector* with angle $< 2\pi$, there exist an integer $n \geqslant 0$ and a constant $C > 0$ such that, on this closed sector,

$$\|s(t)\| \leqslant C \, |t|^{-n}.$$

Deduce analogous properties for any model (2.4).

2.7 Exercise (Extensions of elementary regular models). Prove that, if $\alpha - \alpha'$ is not an integer,

(1) any homomorphism of $(\boldsymbol{k}, \nabla)$-vector spaces $\varphi : \mathcal{N}_{\alpha,d} \to \mathcal{N}_{\alpha',d'}$ is zero;
(2) any $(\boldsymbol{k}, \nabla)$-extension of $\mathcal{N}_{\alpha,d}$ by $\mathcal{N}_{\alpha',d'}$ splits.

2.d Classification of $(\boldsymbol{k}, \nabla)$-vector spaces with regular singularity at 0

Let $(\mathcal{M}, \nabla)$ be a $(\boldsymbol{k}, \nabla)$-vector space of rank d with regular singularity. There exists, by definition, a basis $\boldsymbol{e}$ of $\mathcal{M}$ in which the matrix of ∇ is

$$\Omega(t) = A(t) \frac{dt}{t}$$

where $A(t)$ is a matrix of size d with holomorphic entries[2].

2.8 Theorem (Normal form regular singularities). *With these conditions, there exists a matrix $P \in \mathrm{GL}_d(\boldsymbol{k})$ such that, after the base change of matrix P, the matrix Ω' of the connection takes the form*

$$\Omega'(t) = B_0 \frac{dt}{t}$$

where $B_0 \in \mathrm{M}_d(\mathbb{C})$ is constant.

---

[2] If $A(0) = 0$ the singularity is only apparent (in the sense given in Exercise IV.2.3).

We then deduce from the results of §2.c:

2.9 Corollary (Sufficiency of elementary regular models). *Any (k, ∇)-vector space with regular singularity is isomorphic to a direct sum of elementary regular models.* □

2.10 Important remark. The matrix B_0 that one gets is not necessarily equal (or conjugate) to the matrix $A(0)$ (see however Exercise 4.5 for what concerns the characteristic polynomial).

Moreover, the matrix $\exp(-2i\pi B_0)$ only depends, up to conjugation, on $(\mathcal{M}, \nabla)$, as it is the monodromy matrix of horizontal sections (cf. Exercise 2.6-(2)).

The proof of Theorem 2.8 is done in two steps (see for instance [Was65], [Mal74]). One first constructs the matrix $P(t)$ as a formal series, then one proves that this series has a nonzero radius of convergence.

We denote by $\widehat{k}$ the field of meromorphic formal Laurent series, i.e., the series $\sum_{n \geqslant -n_0} a_n t^n$, for which the series $\sum_{n \geqslant 0} a_n t^n$ may have a zero radius of convergence (we also denote by $\mathbb{C}[\![t]\!]$ the ring of such series).

2.11 Proposition. *Let $\widehat{\Omega} = \widehat{A}(t)\, dt/t$, where $\widehat{A}(t)$ has entries in $\mathbb{C}[\![t]\!]$. We moreover assume that any two eigenvalues of the matrix $A(0)$ do not differ by a nonzero integer. There exists then a matrix $\widehat{P} \in \mathrm{GL}_d(\mathbb{C}[\![t]\!])$ such that*

$$\widehat{P}^{-1}\widehat{A}\widehat{P} + t\widehat{P}^{-1}\widehat{P}' = A(0).$$

This proposition is completed by:

2.12 Proposition. *Let $\widehat{\Omega} = \widehat{A}(t)\, dt/t$, where $\widehat{A}(t)$ is a matrix with entries in $\mathbb{C}[\![t]\!]$. There exists a matrix $Q \in \mathrm{GL}_d(\mathbb{C}[t, t^{-1}])$ such that the eigenvalues of $\widehat{B}(0)$, where $\widehat{B}(t)$ is defined by*

$$\widehat{B}(t) := Q^{-1}\widehat{A}Q + tQ^{-1}Q',$$

do not differ by a nonzero integer.

We deduce from both propositions the statement of Theorem 2.8, if we accept matrices $\widehat{P}$ with possibly nonconverging entries. The matrix $A(t)$ we start with has converging entries, as does the (constant) matrix B_0 we get, but the matrix of the base change possibly not. It will thus be necessary, in order to achieve the proof of Theorem 2.8, to show:

2.13 Proposition. *The matrix of the base change $\widehat{P}$ obtained in Proposition 2.11 has converging entries.*

Proof (of Proposition 2.11). It follows that given in [Mal74] by B. Malgrange (see also [Gan59] or [Sab93, §I.5.2]). We search for a matrix $\widehat{P}$ satisfying, if we put $\widehat{A}(t) = A_0 + tA_1 + \cdots$,

$$\widehat{P}^{-1}\widehat{A}\widehat{P} + t\widehat{P}^{-1}\widehat{P}' = A_0.$$

One can regard this relation as a differential equation satisfied by $\widehat{P}$, namely

$$(2.14) \qquad\qquad t\widehat{P}' = \widehat{P}A_0 - \widehat{A}\widehat{P}.$$

We *a priori* set $\widehat{P} = \mathrm{Id} + tP_1 + t^2P_2 + \cdots$, where the P_ℓ are constant matrices to be determined. We will determine them in an inductive way: in degree ℓ, Equation (2.14) is written as

$$(2.15) \qquad \ell P_\ell = P_\ell A_0 - A_0 P_\ell + \Phi_\ell(P_1, \dots, P_{\ell-1}; A_0, \dots, A_\ell),$$

where Φ_ℓ is a matrix depending in a polynomial way on its variables.

2.16 Lemma. *Let $U \in \mathrm{M}_p(\mathbb{C})$ and $V \in \mathrm{M}_q(\mathbb{C})$ be two matrices. Then the following properties are equivalent:*

(1) *for any matrix Y of size $q \times p$ with entries in $\mathbb{C}$, there exists a unique matrix X of the same kind satisfying $XU - VX = Y$;*
(2) *the square matrices U and V have no common eigenvalue.*

Let $\ell \geqslant 1$. Let us assume that we have determined the matrices P_k ($k \leqslant \ell - 1$). The relation (2.15) can be written as in the lemma, with $X = P_\ell$, $Y = \Phi_\ell$, $U = \ell\,\mathrm{Id} - A_0$ and $V = -A_0$. Because of the assumption on A_0, and as $\ell \geqslant 1$, the condition in the lemma is fulfilled, as it means that there does not exist a pair (λ, λ') of eigenvalues of A_0 with $\lambda - \lambda' = \ell$. We can thus determine a matrix P_ℓ which is a solution of (2.15) if we know the matrices P_k for $k \leqslant \ell - 1$. $\qquad\square$

Proof (of the lemma). Let $\varphi : \mathbb{C}^{q \times p} \to \mathbb{C}^{q \times p}$ be the linear map defined by $\varphi(X) = XU - VX$. One shows (for instance assuming first U and V diagonalizable, then using a density argument) that the eigenvalues of φ are exactly the $\lambda_i - \mu_j$, where λ_i belongs to the set of eigenvalues of U and μ_j that of V. Hence φ is bijective if and only if none of the differences $\lambda_i - \mu_j$ is zero. $\qquad\square$

2.17 Remark. With the assumption made in Proposition 2.11, any solution $\widehat{P} \in \mathrm{GL}_d(\boldsymbol{k})$ is in fact in $\mathrm{GL}_d(\mathbb{C}[\![t]\!])$. More generally, if $u \in \boldsymbol{k}^d$ is the solution of a system of the kind $tu'(t) + A(t)u(t) = 0$ with $A \in \mathrm{M}_d(\mathbb{C}[\![t]\!])$ such that the possible integral eigenvalues of $A(0)$ are $\geqslant 0$, then $u(t) \in \mathbb{C}[\![t]\!]^d$. Indeed, if u_ℓ denotes the coefficient of t^ℓ in u, the term $(\ell\,\mathrm{Id} - A(0))u_\ell$ can be expressed in terms of the u_j for $j < \ell$; by induction, we deduce that $u_\ell = 0$ for $\ell < 0$ as, in this case, $\ell\,\mathrm{Id} - A(0)$ is invertible.

Proof (of Proposition 2.13). It is thus a matter of verifying that, if $A(t)$ has entries in the ring $\mathbb{C}\{t\}$ of converging series, so does the matrix $\widehat{P}$. We will not compute the radius of convergence of the series defining $\widehat{P}$: this would lead us to annoying computations. Let us rather remark that Equation (2.14) defines $\widehat{P}$ as a horizontal section of a differential linear system of rank d^2 and that the matrix of this system has at most a simple pole at 0 (the residue of which is the endomorphism $\mathrm{ad}\,A_0$ of $\mathrm{M}_d(\mathbb{C})$, defined by $\mathrm{ad}\,A_0(X) = [A_0, X] = A_0X - XA_0$). The proposition is therefore a particular case of Proposition 2.18 below. $\qquad\square$

2.18 Proposition (Any formal solution is convergent). *Let $A(t)$ be a matrix in $M_d(\mathbb{C}\{t\})$. Any vector $u(t)$ with entries in $\mathbb{C}[\![t]\!]$ which is solution of the system $tu'(t) + A(t)u(t) = 0$ has converging entries (i.e., in $\mathbb{C}\{t\}$).*

Proof. We will use the method of bounding series[3]. Let us write

$$u(t) = \sum_{k=0}^{+\infty} u_k t^k \quad \text{and} \quad A(t) = A_0 + tA_1 + \cdots$$

so that the equation satisfied by u can be written as

$$A_0 u_0 = 0, \quad (\ell \operatorname{Id} - A_0)u_\ell = -\sum_{j=1}^{\ell} A_j u_{\ell-j} \quad (\ell \geqslant 1).$$

Let ℓ_0 be such that $\ell \operatorname{Id} - A_0$ is invertible for any $\ell \geqslant \ell_0$. There exists then[4] a constant $c > 0$ such that $\|(\ell \operatorname{Id} - A_0)^{-1}\| \leqslant c$ for any $\ell \geqslant \ell_0$, if we set, for any matrix $B = (b_{ij})$, $\|B\| = \max_i \sum_j |b_{ij}|$. As a consequence, for $\ell \geqslant \ell_0$,

$$\|u_\ell\| \leqslant c \sum_{j=1}^{\ell} \|A_j\| \, \|u_{\ell-j}\|.$$

Let us set

$$v_\ell = \begin{cases} \|u_\ell\| & \text{for } \ell < \ell_0 \\ c\sum_{j=1}^{\ell} \|A_j\| \, \|u_{\ell-j}\| & \text{for } \ell \geqslant \ell_0. \end{cases}$$

One checks by induction on ℓ that $\|u_\ell\| \leqslant v_\ell$ for any ℓ. Let us then show that the series $\sum_\ell v_\ell t^\ell$ is convergent: this will give the desired result. Let us set $\varphi(t) = \sum_{j=1}^{+\infty} \|A_j\| \, t^j$. This is a converging series.

2.19 Lemma. *We have*

$$\sum_{\ell=0}^{\infty} v_\ell t^\ell = (1 - c\varphi(t))^{-1}\left[\|u_0\| + \sum_{j=1}^{\ell_0} \left(\|u_j\| - c\sum_{i=1}^{j} \|A_i\| \, \|u_{j-i}\| \right)t^j \right].$$

Proof. Exercise. □

As $\varphi(0) = 1$, the function $(1 - c\varphi(t))$ is invertible in the neighbourhood of the origin; moreover, the term between the brackets is a polynomial. The convergence of the series $\sum_{\ell=0}^{\infty} v_\ell t^\ell$ in the neighbourhood of $t = 0$ follows. □

---

[3] At last, some analysis!

[4] Hint: Use that, for a matrix B such that $\|B\| < 1$, the series $\log(\operatorname{Id} - B) := -\sum_{j \geqslant 1} B^j/j$ converges and deduce that, under these conditions, we have $\|(\operatorname{Id} - B)^{-1}\| \leqslant (1 - \|B\|)^{-1}$.

Proof (of Proposition 2.12). Using a base change with constant matrix, we first reduce to the case where the matrix A_0 is blockdiagonal, each block corresponding to some eigenvalue λ_i $(i = 1, \ldots, p)$ of A_0. Let us then write

$$A(t) = \begin{pmatrix} A_0^{(1)} & 0 \\ 0 & A_0^{(2)} \end{pmatrix} + \sum_{j \geqslant 1} A_j t^j$$

where $A_0^{(1)}$ is the block corresponding to the eigenvalue λ_1 and $A_0^{(2)}$ to the eigenvalues $\lambda_j \neq \lambda_1$. Let us set

$$Q(t) = \begin{pmatrix} t\,\mathrm{Id} & 0 \\ 0 & \mathrm{Id} \end{pmatrix}.$$

Then the eigenvalues of the constant part B_0 of the matrix $B = Q^{-1}AQ + tQ^{-1}Q'$ are $\lambda_1 + 1, \lambda_2, \ldots, \lambda_p$. By a finite sequence of such changes, we get a matrix B_0 satisfying the conclusion of Proposition 2.12. □

2.20 Exercise (Levelt normal form [Gan59, Lev61]). It is a matter of giving a normal form to the connection matrix after a *holomorphic* base change, even when the connection matrix does not satisfy the "nonresonance" conditions on eigenvalues, asked in Proposition 2.11.

Let $\Omega = A(t)\,dt/t$ be the connection matrix, with $A(t)$ holomorphic. We can assume that A_0 is blockdiagonal, the blocks corresponding to the distinct eigenvalues. We denote by $D = \mathrm{diag}(\delta_1, \ldots, \delta_d)$ the diagonal matrix, the diagonal entries of which are the integral parts of the real parts of the eigenvalues of A_0. We assume that $\delta_1 \geqslant \delta_2 \geqslant \cdots \geqslant \delta_d$: this is possible up to a permutation of the order of the blocks. We set $\delta = \delta_1 - \delta_d$. We thus have the following properties:

- The endomorphism $\mathrm{ad}\, D = [D, \bullet] : M_d(\mathbb{C}) \to M_d(\mathbb{C})$ is semisimple.
- The matrix A_0 commutes with D and the eigenvalues λ of $A_0 - D$ satisfy $\mathrm{Re}(\lambda) \in [0, 1[$ (hence the only integral eigenvalue of $\mathrm{ad}(A_0 - D)$ is 0).

(1) Prove that there exists a matrix $B_0 = A_0 + B_1 + \cdots + B_\delta \in M_d(\mathbb{C})$ with $\mathrm{ad}\, D(B_i) = -iB_i$ for $i \geqslant 1$, and a matrix $\widehat{P}(t) \in GL_d(\mathbb{C}[\![t]\!])$ such that

(a) $\widehat{P}(t) = P_0 + tP_1 + t^2 P_2 + \cdots$,
(b) $t\widehat{P}'(t) = \widehat{P}(t)B(t) - A(t)\widehat{P}(t)$ with $B(t) = A_0 + tB_1 + t^2 B_2 + \cdots + t^\delta B_\delta = t^{-D}B_0 t^D$.

[Use that $\mathrm{ad}(A_0 - D) + k\,\mathrm{Id}$ commutes with $\mathrm{ad}\, D$, hence preserves the eigenspaces of $\mathrm{ad}\, D$, and induces there an isomorphism if $k \neq 0$; then, decompose the equation of (b) on the eigenspaces of $\mathrm{ad}\, D$.]

(2) Show, by using Proposition 2.18, that the matrix $\widehat{P}$ is convergent. Deduce that there exists a base change $P \in GL_d(\mathbb{C}\{t\})$ after which the connection matrix takes the form

$$B(t)\frac{dt}{t}.$$

(3) By applying the meromorphic base change with matrix $P(t)t^{-D}$ to the matrix $\Omega = A(t)\,dt/t$, show that the new matrix is $(B_0 - D)\,dt/t$ and recover Theorem 2.8. Deduce that the monodromy matrix is $\exp(-2i\pi(B_0 - D))$.

(4) Give an example of a matrix $A(t)$ in $M_2(\mathbb{C}\{t\})$ such that the monodromy of the system with matrix $A(t)\,dt/t$ is not conjugate to $\exp(-2i\pi A(0))$.

2.e Canonical logarithmic lattices

Let $(\mathcal{M}, \nabla)$ be a $(\boldsymbol{k}, \nabla)$-vector space and let $\mathcal{E}$ be a logarithmic lattice. Let e be a basis of $\mathcal{E}$ over $\mathbb{C}\{t\}$ and $\Omega(t) = A(t)\,dt/t$ be the matrix of ∇ in this basis. Then the endomorphism of E_0 having matrix $A(0)$ in the basis e does not depend on the choice of the basis: it is the residue of the connection at the origin (see §0.14.b), that we will denote by $\operatorname{Res}\nabla$, or also by $\operatorname{Res}_{\mathcal{E}}\nabla$ to insist on the lattice which defines it.

We can now be more specific about Exercise 2.6 and construct canonical logarithmic lattices (also called *Deligne lattices*) in any meromorphic bundle with regular singularities. They are obtained by gluing local canonical logarithmic lattices, as constructed below.

The local Deligne lattices enable us to reformulate the classification Theorem 2.8 in terms of an equivalence of categories.

2.21 Corollary (Deligne lattices). *Let $(\mathcal{M}, \nabla)$ be $(\boldsymbol{k}, \nabla)$-vector space with regular singularity.*

(1) *Let $\mathcal{E}$ be a logarithmic lattice of $\mathcal{M}$; if the eigenvalues of the residue $\operatorname{Res}_{\mathcal{E}}\nabla$ of the connection ∇ on $\mathcal{E}$ do not differ by a nonzero integer, there exists a basis of $\mathcal{E}$ over $\mathbb{C}\{t\}$, in which the connection matrix takes the form $A_0\,dt/t$, with A_0 constant.*

(2) *Let σ be a section of the natural projection $\mathbb{C} \to \mathbb{C}/\mathbb{Z}$ (so that two complex numbers in the image $\operatorname{Im}\sigma$ of σ do not differ by a nonzero integer). There exists a unique logarithmic lattice $^{\sigma}\mathcal{V}$ of $(\mathcal{M}, \nabla)$ such that the eigenvalues of $\operatorname{Res}_{\sigma\mathcal{V}}\nabla$ are contained in $\operatorname{Im}\sigma$. Moreover, for any homomorphism $\varphi :$ $(\mathcal{M}, \nabla) \to (\mathcal{M}', \nabla')$,*

$$\varphi(^{\sigma}\mathcal{V}) = {}^{\sigma}\mathcal{V}(\operatorname{Im}\varphi, \nabla) = {}^{\sigma}\mathcal{V}(\mathcal{M}', \nabla') \cap \operatorname{Im}\varphi.$$

(3) *Conversely, let T be an automorphism of $\mathbb{C}^d$ (defining a local system of rank d on some disc punctured D^*). There exists then a unique (up to isomorphism) bundle with meromorphic connection on D with logarithmic pole at 0 for which*
  (a) *the local system it defines on D^* is that associated to T,*
  (b) *the residue at 0 of the connection has eigenvalues in $\operatorname{Im}\sigma$.*

(4) *The functor which associates to any $(\boldsymbol{k}, \nabla)$-vector space $(\mathcal{M}, \nabla)$ with regular singularity the vector space $^{\sigma}H = {}^{\sigma}\mathcal{V}/t\,^{\sigma}\mathcal{V}$ equipped with the automorphism $T = \exp\left(-2i\pi \operatorname{Res}_{\sigma\mathcal{V}}\nabla\right)$ is an equivalence of categories.*

Proof.

(1) This is exactly what gives Propositions 2.11 and 2.13.

(2) The existence of a lattice $^\sigma V$ follows from Proposition 2.12. If $\mathcal{E}$ and $\mathcal{E}'$ are two such lattices, there exists, after (1), a basis e of $\mathcal{E}$ (resp. e' of $\mathcal{E}'$) in which the matrix $A_0 dt/t$ (resp. $A_0' dt/t$) of ∇ has eigenvalues in $\operatorname{Im} \sigma$. By arguing as in Proposition 2.11, we see that the matrix of the base change $P \in \mathrm{GL}_d(\mathbf{k})$ from e to e' is constant and conjugates A_0 with A_0'. Whence the uniqueness.

As the image by φ of a lattice is a lattice of the image of φ, the second assertion follows from uniqueness.

(3) One can decompose in a unique way $\mathbb{C}^d = \bigoplus_\lambda F_\lambda$ (Jordan decomposition of T), where F_λ is a direct sum of spaces $\mathbb{C}[T, T^{-1}]/(T - \lambda)^k$ and we have a corresponding decomposition of the local system associated to T on D^*. The desired bundle will be decomposed similarly and, to any term $\mathbb{C}[T, T^{-1}]/(T - \lambda)^k$, one associates the connection ∇ on the trivial bundle of rank k having matrix $(\alpha \operatorname{Id} + N) \, dt/t$, with $T = \exp -2i\pi(\alpha \operatorname{Id} + N)$, where $-2i\pi\alpha$ is the unique logarithm of λ which belongs to $2i\pi \operatorname{Im} \sigma$, N is a Jordan block of size k and t is a coordinate on D.

If we have two such bundles E and E', the bundle with meromorphic connection $\mathscr{H}om_{\mathcal{O}_D}(\mathcal{E}, \mathcal{E}')$ also has a logarithmic pole at 0 and the eigenvalues of the residue of its connection are obtained as the differences of the eigenvalues of $\operatorname{Res} \nabla'$ and of $\operatorname{Res} \nabla$. Therefore, the only integral difference is 0, by assumption.

As the bundles have the same monodromy on D^*, there exists an isomorphism of the associated locally constant sheaves on D^* and hence we get an invertible horizontal section of $\mathscr{H}om_{\mathcal{O}_D}(\mathcal{E}, \mathcal{E}')$ on D^*. As the singularity of $(\mathscr{H}om_k(\mathcal{M}, \mathcal{M}), \nabla)$ is regular, this section has moderate growth (after Exercise 2.6 and Theorem 2.8), and hence is meromorphic at 0. As the unique integral eigenvalue of the residue of ∇ on $\mathscr{H}om_{\mathcal{O}_D}(\mathcal{E}, \mathcal{E}')$ is $\geqslant 0$, this section is holomorphic at 0 (see Remark 2.17). Its inverse satisfies the same property, whence the existence of an isomorphism between both bundles with meromorphic connection.

(4) Point (3) shows in particular that this functor is essentially surjective. For full faithfulness, let us first remark that, if $\varphi : (\mathcal{M}, \nabla) \to (\mathcal{M}', \nabla')$ satisfies $\varphi(^\sigma V(\mathcal{M})) \subset t \cdot {}^\sigma V(\mathcal{M}')$ then, after (2), we have $^\sigma V(\varphi(\mathcal{M})) = t \cdot {}^\sigma V(\varphi(\mathcal{M}))$ and therefore $\varphi = 0$.

Lastly, let us show that the map induced on the Hom is onto. By choosing suitable bases, we are reduced to showing that, if A_0 and A_0' are two square matrices, of size d and d' respectively, having eigenvalues in $\operatorname{Im} \sigma$, any matrix $P(t)$ of size $d' \times d$ satisfying $PA_0 - A_0'P = tP'$ is constant. This follows from Lemma 2.16, as the only integral eigenvalue of the linear operator $P \mapsto PA_0 - A_0'P$ is 0. $\qquad \square$

2.22 Remark (Algebraization of Deligne lattices). One can express the corollary above in a slightly different way. For any $\alpha \in \mathbb{C}$, let us set

$$\mathbb{M}^\alpha = \{e \in \mathcal{M} \mid \exists\, n, \ (t\nabla_{\partial/\partial t} - \alpha \operatorname{Id})^n e = 0\}.$$

Then, the corollary implies (by first considering an elementary model) that $\mathbb{M}^\alpha$ is a finite dimensional $\mathbb{C}$-vector subspace and multiplication by t induces an isomorphism from $\mathbb{M}^\alpha$ to $\mathbb{M}^{\alpha+1}$. Moreover, $\mathbb{M}^\alpha = 0$ except if $\exp(-2i\pi\alpha)$ is an eigenvalue of T.

Therefore, $\mathbb{M} := \bigoplus_\alpha \mathbb{M}^\alpha$ is a free $\mathbb{C}[t, t^{-1}]$-module contained in $\mathcal{M}$, stable under the action of ∇ and which generates $\mathcal{M}$ over $\boldsymbol{k}$. More precisely, the natural mapping

$$\boldsymbol{k} \underset{\mathbb{C}[t,t^{-1}]}{\otimes} \mathbb{M} \longrightarrow \mathcal{M}$$

is an isomorphism of $(\boldsymbol{k}, \nabla)$-vector spaces.

Let us set $^\sigma V = \bigoplus_{\alpha \in \operatorname{Im}\sigma} \bigoplus_{k \in \mathbb{N}} \mathbb{M}^{\alpha+k}$. Then $^\sigma V$ is the Deligne lattice of $\mathbb{M}$ corresponding to σ and we have

$$\mathbb{C}\{t\} \underset{\mathbb{C}[t]}{\otimes} {}^\sigma V \overset{\sim}{\longrightarrow} {}^\sigma V.$$

Lastly, let us notice that the natural mapping

(2.23)
$$\bigoplus_{\alpha \in \operatorname{Im}\sigma} \mathbb{M}^\alpha \longrightarrow {}^\sigma V / t\, {}^\sigma V$$

is an isomorphism compatible with the action of T, if one defines it on the left-hand term by $\exp(-2i\pi t\nabla_{\partial/\partial t})$. One can thus identify $\mathbb{M}$ to the graded $(\mathbb{C}[t, t^{-1}], \nabla)$-module

$$\operatorname{gr}_{{}^\sigma V} \mathbb{M} := \bigoplus_{k \in \mathbb{Z}} \left(t^k\, {}^\sigma V / t^{k+1}\, {}^\sigma V \right).$$

2.f Adjunction of parameters

The previous results can be extended without any trouble to a situation "with parameters". Although it is in fact not useful for what follows, we will indicate how they are obtained, and we will take this opportunity to introduce useful notions concerning irregular singularities. In the remaining part of this section, the space of parameters X is a complex analytic manifold. Recall the notion of regular singularity (cf. §0.14):

2.24 Definition (Regular singularities with holomorphic parameter). A meromorphic bundle $\mathscr{M}$ on $D \times X$, with poles along $\{0\} \times X$, equipped with a *flat* connection ∇, has regular singularity if, in the neighbourhood of any point $(0, x^o)$ of $\{0\} \times X$, there exists a logarithmic lattice of $(\mathscr{M}, \nabla)$.

Let us denote by $\widehat{\mathscr{O}}_{D\times X}$ the *formal completion* of the sheaf $\mathscr{O}_{D\times X}$ along $\{0\} \times X$. This is a sheaf on $X = \{0\} \times X$. If U is an open set of X, the space of sections $\Gamma(U, \widehat{\mathscr{O}}_{D\times X})$ is the ring $\mathscr{O}_X(U)[\![t]\!]$ of formal series in t with coefficients in the ring $\mathscr{O}_X(U)$. The reader will verify that the presheaf $U \mapsto \mathscr{O}_X(U)[\![t]\!]$ is a sheaf on X. The germ $\widehat{\mathscr{O}}_{D\times X, x^o}$ is the subring of $\mathscr{O}_{X, x^o}[\![t]\!]$ of series $\sum_i a_i(x)t^i$, where the germs of functions $a_i(x) \in \mathscr{O}_{X, x^o}$ are defined on *the same* neighbourhood of x^o, i.e., which, as a series of $x - x^o$, have a radius of convergence bounded from below by the same positive number.

Let $x^o \in X$ and $(x_1, \dots, x_n)$ be a system of local coordinates centered at x^o. Let

$$\Omega = A(t, x)\frac{dt}{t} + \sum_{i=1}^{n} C^{(i)}(t, x)\, dx_i$$

be the connection matrix in some basis e of $\mathscr{M}$ in the neighbourhood of $(0, x^o)$. If the matrices A and $C^{(i)}$ $(i = 1, \dots, n)$ have holomorphic entries, $(\mathscr{M}, \nabla)$ has regular singularity in the neighbourhood of $(0, x^o)$.

2.25 Theorem (Normal form of regular singularities with parameter). *There exists a basis of the germ $\mathscr{M}_{(0, x^o)}$ in which the matrix of ∇ can be written as $B\, dt/t$, where B is constant.*

Proof. Let us first assume that the eigenvalues of the matrix $A(0, x^o)$ do not differ by a nonzero integer. So do the eigenvalues of the matrix $A(0, x)$ for any x in some neighbourhood U of x^o. The endomorphism $\operatorname{ad} A(0, x) + \ell\operatorname{Id}$ is thus invertible on U, with a holomorphic inverse at x, for any $\ell \in \mathbb{Z} \smallsetminus \{0\}$. Formula (2.15) furnishes thus a matrix $\widehat{P} \in \operatorname{GL}_d(\widehat{\mathscr{O}}_{D\times U, x^o})$ such that

$$\widehat{P}^{-1}A(t, x)\widehat{P} + t\widehat{P}^{-1}\partial\widehat{P}/\partial t = A(0, x)$$

for $x \in U$. Let us set $\widehat{\Omega}' = \widehat{P}^{-1}\Omega\widehat{P} + \widehat{P}^{-1}d\widehat{P}$. Then

$$\widehat{\Omega}' = A(0, x)\frac{dt}{t} + \sum_{i=1}^{n} \widehat{C}'^{(i)}(t, x)\, dx_i.$$

As Ω satisfies the integrability condition, so does $\widehat{\Omega}'$ (cf. Exercise 0.12.6). We deduce that $\widehat{C}'^{(i)}(t, x) = \widehat{C}'^{(i)}(0, x) := C'^{(i)}(x)$: indeed, the integrability condition implies that

$$\frac{\partial A(0, x)}{\partial x_i} - t\frac{\partial\widehat{C}'^{(i)}}{\partial t} = [A(0, x), \widehat{C}'^{(i)}]$$

and, setting $\widehat{C}'^{(i)} = \sum_k C_k'^{(i)}(x)t^k$, we get, for $k \neq 0$,

$$(\operatorname{ad} A(0, x) + k\operatorname{Id})(C_k'^{(i)}(x)) = 0;$$

the choice of the open set U shows that $C_k'^{(i)}(x) \equiv 0$ on U for any $k \neq 0$.

As the connection ∇ with matrix $\sum_i C'^{(i)}(x)\, dx_i$ is integrable, and as the residue $A(0,x)$ is horizontal with respect to ∇ (cf. Exercise 0.14.6), there exists a basis in which the matrix of $\widehat{\Omega}'$ is $A(0,x)\, dt/t$.

The argument of Proposition 2.13 can be extended to the situation above, on the open set U. We deduce that $\widehat{P} \in \mathrm{GL}_d(\mathscr{O}_{D\times U,(0,x^o)})$.

Lastly, it is enough to prove the analogue of Proposition 2.12, asking there that the eigenvalues of the matrix $\widehat{B}(0,x^o)$ do not differ by a nonzero integer. The proof is similar. $\qquad\square$

2.26 Deligne lattices. We let the reader extend to the situation "with parameters" the results of §2.e. Let us only indicate that, for any section σ of the projection $\mathbb{C} \to \mathbb{C}/\mathbb{Z}$, there exists a unique logarithmic lattice $^\sigma\!\mathscr{V}$ of the meromorphic bundle $(\mathscr{M}, \nabla)$. The local uniqueness enables us to obtain, by gluing, the global existence from the local existence.

3 Applications

3.a Riemann-Hilbert correspondence

We follow here [Del70]. Let M be a Riemann surface and let Σ be a discrete set of points.

3.1 Theorem (The logarithmic correspondence). *Given a locally constant sheaf $\mathscr{F}$ on $M \smallsetminus \Sigma$, there exists a holomorphic bundle E on M and a connection $\nabla : \mathscr{E} \to \Omega^1_M(\Sigma) \otimes_{\mathscr{O}_M} \mathscr{E}$ with logarithmic poles, such that, on $M \smallsetminus \Sigma$, the locally constant sheaf E^∇ is isomorphic to $\mathscr{F}$.*

Proof. On the open set $M \smallsetminus \Sigma$ we put $\mathscr{E} = \mathscr{O}_{M\smallsetminus\Sigma} \otimes_{\mathbb{C}} \mathscr{F}$. It is thus a matter of finding, for any point $m \in \Sigma$, a sufficiently small open neighbourhood U of m (in particular $U \cap \Sigma = \{m\}$) and a basis e of $\Gamma(U \smallsetminus \{m\}, \mathscr{E})$ in which the connection matrix of ∇ has a logarithmic pole: this basis enables us to define a bundle with connection on U, which coincides with $\mathscr{E}$ on $U \smallsetminus \Sigma$. We then construct the desired bundle by gluing.

Therefore, the problem is local and we can assume that U is a disc D centered at the origin of $\mathbb{C}$. If we restrict to $D \smallsetminus \{0\}$, giving the locally constant sheaf $\mathscr{F} = E^\nabla$ is equivalent to giving the representation of $\pi_1(D \smallsetminus \{0\}, o) = \mathbb{Z}$ in $\mathrm{GL}_d(\mathbb{C})$, that is, an invertible matrix T. We then apply Corollary 2.21-(3). $\qquad\square$

3.2 Corollary (The meromorphic correspondence). *Given a locally constant sheaf $\mathscr{F}$ on $M \smallsetminus \Sigma$, there exists a meromorphic bundle with connection $(\mathscr{M}, \nabla)$ on M with poles at the points of Σ and having regular singularity at these points, such that, on $M \smallsetminus \Sigma$, the locally constant sheaf $\mathscr{M}^\nabla$ is isomorphic to $\mathscr{F}$.*

Proof. It is enough to take $\mathcal{M} = \mathscr{E}(*\Sigma)$, where $(\mathscr{E}, \nabla)$ is given by Theorem 3.1.   □

The correspondence which associates to any meromorphic bundle with connection $(\mathcal{M}, \nabla)$ the locally constant sheaf of horizontal sections $\mathcal{M}^{\nabla}_{|M \smallsetminus \Sigma}$ is functorial: To a homomorphism $\psi : \mathcal{M} \to \mathcal{M}'$ compatible with connections, that is, satisfying $\psi(\nabla s) = \nabla'(\psi(s))$ for any local section s of $\mathcal{M}$, one associates the restricted homomorphism

$$\varphi : \mathcal{M}^{\nabla}_{|M \smallsetminus \Sigma} \longrightarrow \mathcal{M}'^{\nabla'}_{|M \smallsetminus \Sigma}.$$

This correspondence at the level of morphisms is of course compatible with the composition of morphisms.

3.3 Corollary (The Riemann-Hilbert correspondence is an equivalence). *This correspondence induces an equivalence between the category of meromorphic bundles with connection, having poles at the points of Σ and having regular singularity at any point of Σ, and the category of locally constant sheaves of $\mathbb{C}$-vector spaces on $M \smallsetminus \Sigma$.*

Proof. The corollary above shows that the functor is *essentially surjective*. Let us show that it is fully faithful. Let $\varphi : \mathcal{M}^{\nabla}_{|M \smallsetminus \Sigma} \to \mathcal{M}'^{\nabla}_{|M \smallsetminus \Sigma}$ be a homomorphism. It is a matter of showing that there exists a unique homomorphism $\psi : (\mathcal{M}, \nabla) \to (\mathcal{M}', \nabla)$ which induces it. Let us regard φ as a horizontal section of the bundle $(\mathscr{H}om_{\mathcal{O}_M}(\mathcal{M}, \mathcal{M}')_{|M \smallsetminus \Sigma}, \nabla)$. It is a matter of verifying that this horizontal section is the restriction to $M \smallsetminus \Sigma$ of a meromorphic horizontal section of $(\mathscr{H}om_{\mathcal{O}_M}(\mathcal{M}, \mathcal{M}'), \nabla)$. This follows from the lemma below, as $(\mathscr{H}om_{\mathcal{O}_M}(\mathcal{M}, \mathcal{M}'), \nabla)$ has a regular singularity at any point of Σ if $(\mathcal{M}, \nabla)$ and $(\mathcal{M}', \nabla)$ do so.   □

3.4 Lemma. *Let $(\mathcal{N}, \nabla)$ be a meromorphic bundle with connection, with poles at the points of Σ and having regular singularity at any point of Σ. Let s be a horizontal section of $(\mathcal{N}, \nabla)$ on $M \smallsetminus \Sigma$. Then s can be extended in a unique way as a meromorphic section of $\mathcal{N}$ on M.*

Proof. The uniqueness of the extension is clear. Let us show the existence. The problem is local, so that we can assume that M is a disc D centered at 0 in $\mathbb{C}$ and $\Sigma = \{0\}$. After Corollary 2.9, we can assume that $\mathcal{N} = \mathcal{N}_{\alpha,d}$ for some choice of $\alpha \in \mathbb{C}$ and $d \in \mathbb{N}$. There exists a horizontal section on $D \smallsetminus \{0\}$ if and only if $\alpha \in \mathbb{Z}$ (horizontal sections have combinations of functions of the kind $t^{-\alpha}(\log t)^p$ as coordinates) and, if $\alpha \in \mathbb{Z}$, there exists a unique horizontal section (up to a constant), which is meromorphic.   □

3.b Partial Riemann correspondence

We take up the situation of the partial Riemann problem discussed at 1.3. Let us consider the functor which, to any meromorphic bundle with connection $(\mathcal{M}, \nabla)$ on M, having poles at the points of $\Sigma \cup \Sigma'$ and having regular singularity at the points of Σ, associates its restriction $(^U\mathcal{M}, {}^U\nabla)$ to the open set U and similarly for the morphisms of bundles with connection.

3.5 Theorem. *This functor is an equivalence of categories.*

Proof. For the essential surjectivity, we begin by extending $(^U\mathcal{M}, {}^U\nabla)$ to $M \smallsetminus \Sigma$ as a meromorphic bundle with connection: the holomorphic bundle with connection $(^U\mathcal{M}, {}^U\nabla)_{|U\smallsetminus\Sigma'}$ can be extended to $M \smallsetminus (\Sigma \cup \Sigma')$ as a holomorphic bundle with connection, as so does, after the assumption made in Problem 1.3, the locally constant sheaf of its horizontal sections (Corollary 0.15.10); this enables us to construct the desired meromorphic bundle on $M \smallsetminus \Sigma$. Its extension to M can be obtained as in Theorem 3.1. This gives the essential surjectivity of the functor.

Now for full faithfulness. If $U = M \smallsetminus \Sigma$, we argue as in Corollary 3.3. Otherwise, we use the full faithfulness of the restriction functor from $M \smallsetminus \Sigma$ to U, obtained by means of Corollary 0.15.10-(3). $\qquad\qquad\qquad\square$

3.6 Remark. Corollary 3.3 is of course a particular case of Theorem 3.5.

3.c Algebraization of a germ of meromorphic connection

This procedure has been introduced by G. Birkhoff [Bir09]. By applying Theorem 3.5, taking for U some small disc centered at the origin Σ', we get:

3.7 Corollary (Algebraization is an equivalence). *Let $(\mathcal{M}, \nabla)$ be a $\boldsymbol{k}$-vector space with connection. There exists then a meromorphic bundle $(\mathcal{M}, \nabla)$ with connection on $\mathbb{P}^1$, having singularities at most at 0 and ∞, the latter being regular, and with germ at 0 isomorphic to $(\mathcal{M}, \nabla)$. Such a bundle is unique up to isomorphism and this isomorphism is unique if one asks that it induces the identity on $\mathcal{M}$.* $\qquad\qquad\qquad\square$

Let us notice that Theorem 3.5 gives a canonical procedure to obtain this globalization. This corollary has a more concrete expression:

3.8 Corollary (Algebraization of holomorphic differential systems). *Given a matrix $A(t)$ of size d with entries in the field $\boldsymbol{k}$ of meromorphic Laurent series, there exists a matrix $P \in \mathrm{GL}_d(\mathbb{C}\{t\})$ such that $B(t) := P^{-1}AP + P^{-1}P'$ has entries in the ring $\mathbb{C}[t, t^{-1}]$ of Laurent polynomials and such that the connection with matrix $B(t)\,dt$ has a regular singularity at infinity.*

Proof. Let us set $(\mathcal{M}, \nabla) = (\boldsymbol{k}^d, d + A(t)\,dt)$ and let us denote by $\mathcal{E}$ the lattice $\mathbb{C}\{t\}^d$. We can then construct a meromorphic subbundle $\mathcal{E}(*\infty)$ of the meromorphic bundle $\mathcal{M}$ given by Corollary 3.7: the lattice $\mathcal{E}$ defines a lattice $\mathcal{E}_{|D}$ of $\mathcal{M}_{|D}$ on a sufficiently small disc centered at the origin; we glue it with the meromorphic bundle $\mathcal{M}_{|\mathbb{P}^1\smallsetminus\{0\}}$ according to the identification $\mathcal{E}_{|D^*} \simeq \mathcal{M}_{|D^*}$.

on $D^* = D \smallsetminus \{0\}$. This meromorphic bundle has a pole at ∞ only. After Corollary I.4.9, it is isomorphic to $\mathscr{O}_{\mathbb{P}^1}(*\infty)^d$.

Let us take some basis of $\mathscr{E}(*\infty)$, by using this isomorphism. In this basis, the connection matrix of ∇ has entries in the ring $\Gamma(\mathbb{P}^1, \mathscr{O}_{\mathbb{P}^1}(*\{0, \infty\})) = \mathbb{C}[t, t^{-1}]$. The same property holds for the germs of these objects at 0.   □

3.9 Remark (Another algebraization procedure). The base change, obtained by the method above, from the given basis of $\mathcal{M}$ to the basis furnished by the corollary, is not explicit. One can also show that the formal analogue of this corollary, i.e., going from $\widehat{k}$ to $\mathbb{C}[t, t^{-1}]$, is true (cf. [Mal91, Prop. 1.12, p. 49]). This base change is even more explicit, as it is enough to truncate at a sufficiently large order the expansion of $A(t)$. Nevertheless, it is not clear whether this procedure always produces a regular singularity at infinity. On the other hand, the procedure of Corollary 3.8 *a priori* only keeps from A the most polar part, as the matrix P has no pole.

3.10 Exercise. Let $A(t)$ be a matrix with entries in $\mathbb{C}[t, t^{-1}]$ such that the system with matrix $\Omega = A(t) \, dt$ has a regular singularity at infinity. Prove that any germ $u \in \mathbb{C}\{t\}^d$ which is a solution to the system $du + \Omega \cdot u = 0$ is in fact a polynomial, i.e., belongs to $\mathbb{C}[t]^d$.

3.11 Exercise. When $(\mathcal{M}, \nabla)$ has a regular singularity, compare Birkhoff's procedure of Corollary 3.7 with that of Remark 2.22.

4 Complements

4.a How to recognize a regular singularity

Given a germ $(\mathcal{M}, \nabla)$ of meromorphic differential system, with matrix $A(t) \, dt$ in some basis e, where A has entries in k, it is not immediate, in general, to check whether $(\mathcal{M}, \nabla)$ has a regular singularity at 0. Indeed, it could happen, for instance, that A has a double pole and that nevertheless $(\mathcal{M}, \nabla)$ has a regular singularity.

We will give some criteria which happen to be effective. The following should be better regarded as an irregularity criterion; a proof of it is given as an exercise (cf. Exercise 5.9).

4.1 Theorem (An irregularity criterion). *Let us set* $A(t) = B(t)/t^{r+1}$ *with* $B(t)$ *holomorphic,* $B(0) \neq 0$ *and let us assume that* $r \geqslant 1$. *Then, if the matrix* $B(0)$ *is not nilpotent, the system with matrix* $A(t) \, dt$ *has an irregular singularity at* 0.   □

The property of having a regular singularity can be easily checked when the system is defined by a single linear differential equation (cf. [Mal74] for what follows). Let

$$p(t, \partial_t) = a_d(t) \partial_t^d + \cdots + a_0(t)$$

be a differential operator of degree d, with $a_0, \ldots, a_d \in \mathbf{k}$ and $a_d \neq 0$, and let us consider the associated linear differential system of rank d, with matrix

$$(4.2) \qquad \Omega(t) = \begin{pmatrix} 0 & 0 & \cdots & 0 & -a_0/a_d \\ 1 & 0 & \cdots & 0 & -a_1/a_d \\ \vdots & \vdots & \vdots & & \vdots \\ 0 & 0 & \cdots & 1 & -a_{d-1}/a_d \end{pmatrix}$$

Let us denote by $v(a_i)$ the *valuation* of a_i, that is, the integer v_i such that $a_i(t) = t^{v_i} b_i(t)$ with $b_i \in \mathbb{C}\{t\}$ and $b_i(0) \neq 0$.

4.3 Theorem (Fuchs condition, cf. for instance [Mal74]). *The linear system associated to the operator p has a regular singularity if and only if the following condition is satisfied:*

$$\forall i \in \{1, \ldots, d-1\}, \quad i - v(a_i) \leqslant d - v(a_d). \qquad \square$$

One can associate to the operator p a polygon in the plane, which is the convex hull of the quadrants $(i, i - v(a_i)) - \mathbb{N}^2$. The condition means that this polygon is also quadrant, having the point $(d, d - v(a_d))$ as its vertex.

Let $(\mathcal{M}, \nabla)$ be a $(\mathbf{k}, \nabla)$-vector space and let us pick $e \in \mathcal{M}$. We say that e is a *cyclic vector* of $\mathcal{M}$ if

$$e, \nabla_{\partial_t} e, \ldots, \underbrace{\nabla_{\partial_t}(\cdots (\nabla_{\partial_t} e))}_{d-1 \text{ times}}$$

is a basis of $\mathcal{M}$ over $\mathbf{k}$. Any $(\mathbf{k}, \nabla)$-vector space admits a cyclic vector (cf. [Del70]); hence, the construction of a cyclic vector combined with Fuchs criterion gives a method to check the regular singularity property of a system $(\mathcal{M}, \nabla)$.

Another method is given in [Var91], where the author uses a variant of a method due to Turrittin.

4.4 Exercise (Lattices of arbitrarily large order). Set

$$\Omega(t) = \begin{pmatrix} 0 & 0 \\ 1 & 0 \end{pmatrix} \frac{dt}{t} \quad \text{and, for } k \in \mathbb{Z}, \quad P_k(t) = \begin{pmatrix} t^k & 0 \\ 0 & t^{-k} \end{pmatrix}.$$

Compute the order of $\Omega' = P^{-1}\Omega P + P^{-1}dP$. Generalize to higher rank.

4.b Computing the monodromy of horizontal sections

Given a logarithmic lattice of a regular differential system $(\mathcal{M}, \nabla)$, that is, if we choose a basis, a holomorphic matrix $\Omega(t) = A(t)\, dt/t$,

- if $A(t) = A(0)$ is constant, the monodromy of horizontal sections is given by $T = \exp(-2i\pi A(0))$ (cf. Exercise 2.6),

- if $A(t)$ is not constant, the monodromy T is then conjugate to $\exp(-2i\pi B_0)$, where B_0 is constructed in Theorem 2.8. If the eigenvalues of $A(0)$ do not differ by a nonzero integer, one can choose $B_0 = A(0)$ (cf. Propositions 2.11 and 2.13). Otherwise, it can happen that T is not conjugate to $\exp(-2i\pi A(0))$ (see Exercise 2.20-(4)).

4.5 Exercise (The characteristic polynomial of the monodromy). Prove that the characteristic polynomial of T is equal to that of $\exp(-2i\pi A(0))$.

5 Irregular singularities: local study

We will now tackle the analysis of systems with irregular singularities. We will straightly consider the case "with parameters", letting the reader "forget", during a first reading, the parameter space. We will take up both steps that we have followed for regular singularities, namely working first with formal series and then proving convergence. Nevertheless, the second step will not be as simple as in Proposition 2.13.

In the following, the parameter space is an analytic manifold X equipped with coordinates $x_1, \ldots, x_n$ in the neighbourhood of a point x^o. We consider a meromorphic bundle $\mathcal{M}$ on $D \times X$ with poles along $\{0\} \times X$, equipped with a *flat* connection $\nabla : \mathcal{M} \to \Omega^1_{D \times X} \otimes \mathcal{M}$. We will now denote by $\mathcal{M}$ its germ at $(0, x^o) \in D \times X$: this is a module over the ring $\mathcal{O}_{D \times X, (0,x^o)}[t^{-1}] = \mathbb{C}\{t, x_1, \ldots, x_n\}[t^{-1}]$.

5.a Classification in rank one

Let us pick $\varphi \in \mathbb{C}\{t, x_1, \ldots, x_n\}[t^{-1}]$ and let us denote by $\mathcal{E}^\varphi$ the germ

$$(\mathcal{M}, \nabla) = (\mathbb{C}\{t, x_1, \ldots, x_n\}[t^{-1}], d - d\varphi)$$

(this is the system satisfied by the function e^φ). For $\alpha \in \mathbb{C}$, let $\mathcal{N}_{\alpha,0}$ be the germ

$$(\mathcal{M}, \nabla) = (\mathbb{C}\{t, x_1, \ldots, x_n\}[t^{-1}], d + \alpha\, dt/t)$$

(cf. Definition 2.5).

5.1 Proposition (Classification of irregular singularities in rank one). *Any germ $(\mathcal{M}, \nabla)$ of rank one is isomorphic to some germ $\mathcal{E}^\varphi \otimes \mathcal{N}_{\alpha,0}$. Two such germs corresponding to (φ_1, α_1) and (φ_2, α_2) are isomorphic if and only if $\varphi_1 - \varphi_2$ has no pole and $\alpha_1 - \alpha_2 \in \mathbb{Z}$.*

Therefore, the class of φ in $\mathbb{C}\{t, x_1, \ldots, x_n\}[t^{-1}]/\mathbb{C}\{t, x_1, \ldots, x_n\}$ determines the germ $\mathcal{E}^\varphi$. In the following, we will fix this class by considering only germs φ without holomorphic part, i.e., of the form $\sum_{k=1}^r \varphi_k t^{-k}$, where $\varphi_k \in \mathbb{C}\{x_1, \ldots, x_n\}$ for any k.

Proof. A germ $(\mathcal{M}, \nabla)$ of rank one is determined, up to isomorphism, by giving a 1-form ω with coefficients in $\mathbb{C}\{t, x_1, \ldots, x_n\}[t^{-1}]$, satisfying $d\omega = 0$ (integrability condition in rank one). Two forms ω_1 and ω_2 determine the same germ if there exists a function $p(t, x_1, \ldots, x_n) \in \mathbb{C}\{t, x_1, \ldots, x_n\}[t^{-1}]$, with $p(t, 0) \in \boldsymbol{k} \smallsetminus \{0\}$, such that $\omega_2 = \omega_1 + d\log p$.

If ω is such a form, one can write in a unique way $\omega = \omega' + \omega''$, where ω' is the "nonlogarithmic" part of ω and ω'' its logarithmic part: there exists $r \in \mathbb{N}^*$ minimum such that

$$\omega = t^{-r}\Big(a(t, x)\frac{dt}{t} + \sum_{i=1}^{n} b_i(t, x)\, dx_i\Big)$$

with a and b_i holomorphic. The form ω' is obtained by replacing a and the b_i by their Taylor expansion in t up to order $r - 1$.

The condition $d\omega = 0$ can thus be decomposed in $d\omega' = 0$ and $d\omega'' = 0$. One can check, as in Proposition 2.2, that there exist p as above and $\alpha \in \mathbb{C}$ such that $\omega'' = \alpha\, dt/t + d\log p$. On the other hand, the adjunction of a term like $d\log p$ only modifies the logarithmic part ω'' of ω.

In order to end the proof, it is enough to check that there exists a unique φ without holomorphic part, such that $d\varphi = \omega'$. To show this, let us set

$$\omega' = \sum_{k \geqslant 1} t^{-k}\Big(\omega_k + h_k(x)\frac{dt}{t}\Big),$$

where the ω_k only contain the dx_i. Denoting by d_X the differentiation with respect to coordinates x_i only, we obtain

$$d\omega' = \sum_{k \geqslant 1} t^{-k}\Big(d_X\omega_k + (d_X h_k + k\omega_k) \wedge \frac{dt}{t}\Big) = 0.$$

We deduce that, for any $k \geqslant 1$, there exists $\psi_k \in \mathbb{C}\{x_1, \ldots, x_n\}$ with $d_X\psi_k = \omega_k$ (cf. Poincaré Lemma 0.9.7) and there exists $c_k \in \mathbb{C}$ with $h_k = -k(\psi_k + c_k)$. If we put $\varphi_k = \psi_k + c_k$ and $\varphi = \sum_{k=1}^{r} t^{-k}\varphi_k$, we have $\omega' = d\varphi$. Uniqueness is proved similarly. $\qquad\square$

5.b Models in arbitrary rank

As in the regular case, we will first introduce some elementary models, to which we will compare the germ $\mathcal{M}$.

5.2 Definition (Elementary irregular models). We will say that a germ of meromorphic bundle with connection $(\mathcal{M}, \nabla)$ is *elementary* if it is isomorphic to some germ like $(\mathcal{E}^\varphi, \nabla) \otimes (\mathcal{R}, \nabla)$, where $(\mathcal{R}, \nabla)$ has regular singularity along $\{0\} \times X$ (cf. §2.f).

5.3 Exercise (Comparison of elementary models). Let us fix $r \geqslant 1$ and $\varphi = \sum_{k=1}^{r} t^{-k}\varphi_k$ with $\varphi_r \not\equiv 0$.

(1) Compute the Poincaré rank of the meromorphic bundle with connection $(\mathcal{E}^\varphi \otimes \mathcal{R}, \nabla)$ (cf. §0.14).
(2) Prove that there does not exist any nonzero horizontal section of the meromorphic bundle with connection $(\mathcal{E}^\varphi \otimes \mathcal{R}, \nabla)$.
(3) Show that there does not exist any nonzero homomorphism $\mathcal{E}^\varphi \otimes \mathcal{R} \to \mathcal{R}'$, where $\mathcal{R}, \mathcal{R}'$ have regular singularity.

A *model* is a meromorphic bundle with connection $(\mathcal{M}, \nabla)$ isomorphic to a direct sum of elementary models. We will write this direct sum as

$$(5.4) \qquad\qquad \bigoplus_\varphi (\mathcal{E}^\varphi \otimes \mathcal{R}_\varphi)$$

where we assume that the meromorphic bundles with connection $\mathcal{R}_\varphi$ have regular singularity and the $\varphi \in \mathbb{C}\{t, x_1, \ldots, x_n\}[t^{-1}]$ *have no holomorphic part and are pairwise distinct.*

5.5 Exercise (Comparison of elementary models, continuation).

(1) By decomposing each $\mathcal{R}_\varphi$ into elementary regular models, give a simple form for the connection matrix of a model.
(2) Check that any homomorphism between two models is diagonal with respect to the φ-decomposition, and that two models are isomorphic if and only if the corresponding $\mathcal{R}_\varphi$ are isomorphic. Deduce that the decomposition (5.4) is unique.

5.6 Definition (of goodness). We will say that a model (5.4) is *good* if, for all $\varphi \neq \psi$ such that $\mathcal{R}_\varphi, \mathcal{R}_\psi$ are nonzero, the order of the pole along $t = 0$ of $(\varphi - \psi)(t, x)$ does not depend on x being in some neighbourhood of x^o.

If $\varphi - \psi = \sum_{k=1}^r t^{-k}(\varphi - \psi)_k(x)$ with $(\varphi - \psi)_r(x) \not\equiv 0$, the condition means that $(\varphi - \psi)_r(x^o) \neq 0$. It may happen however that the order of φ or ψ is strictly bigger than r. Let us remark that, if a model $(\mathcal{M}, \nabla)$ is good then, for any $\eta \in t^{-1}\mathbb{C}\{x_1, \ldots, x_n\}[t^{-1}]$, the germ $\mathcal{E}^\eta \otimes (\mathcal{M}, \nabla) = (\mathcal{M}, \nabla + d\eta)$ remains a model, which is good.

5.7 Theorem (Formal decomposition). *Let $(\mathcal{M}, \nabla)$ be a germ of meromorphic bundle with connection, equipped with a basis in which the matrix Ω takes the form*

$$\Omega = t^{-r}\left[A(t, x)\frac{dt}{t} + \sum_{i=1}^n C^{(i)}(t, x)\, dx_i\right]$$

with $r \geqslant 1$, A and the $C^{(i)}$ having holomorphic entries, and $A_0 := A(0, x^o)$ being regular semisimple (i.e., with pairwise distinct eigenvalues). There exists then a good model $(\mathcal{M}^{\mathrm{good}}, \nabla)$ and a "formal" isomorphism

$$\widehat{\mathcal{O}}_{D \times X, x^o} \otimes (\mathcal{M}, \nabla) \xrightarrow{\sim} \widehat{\mathcal{O}}_{D \times X, x^o} \otimes (\mathcal{M}^{\mathrm{good}}, \nabla).$$

When the conclusion of the theorem is satisfied, we will say that the germ $(\mathcal{M}, \nabla)$ admits a *good formal decomposition* along $\{0\} \times X$. A more general result due to Turrittin, and for which there exist various approaches (see [Tur55], [Lev75], [Rob80], [Mal79], [BV83], [Var91] for the case "without parameter" and [BV85] for that "with parameters"), says that such a decomposition exists[5], possibly not for $(\mathcal{M}, \nabla)$, but for its pullback by some suitable ramification $t = z^q$ of order $q \geqslant 1$. In the present situation, no ramification is needed. Moreover, we will see that all the components $\mathcal{R}_\varphi$ occurring in the model have rank one, which is not the case in general, even when no ramification is needed.

Proof. Let $\Omega = t^{-r}\left[A(t, x)\dfrac{dt}{t} + \sum_i C^{(i)}(t, x)\,dx_i\right]$ be the matrix of ∇ in some basis of $\mathcal{M}$. Let us set

$$A(t, x) = \sum_{p=0}^{\infty} A_p(x)t^p.$$

As $A(0, x^o)$ is regular semisimple, so is $A_0(x) := A(0, x)$ for any x nearby x^o and there exists a matrix $Q \in \mathrm{GL}_d(\mathscr{O}_{X,x^o})$ such that $(Q^{-1}AQ)(0, x)$ is diagonal with pairwise distinct eigenvalues[6]. We may therefore assume from the beginning that $A_0(x)$ takes the diagonal form $\mathrm{diag}(\lambda_1(x), \ldots, \lambda_d(x))$. The sheaf $\mathrm{M}_d(\mathscr{O}_X)$ of matrices of size d with holomorphic entries admits the decomposition[7]

$$\mathrm{M}_d(\mathscr{O}_X) = \mathrm{Ker\,ad}\,A_0 \oplus \mathrm{Im\,ad}\,A_0,$$

and each term is a locally free sheaf of $\mathscr{O}_X$-modules; moreover,

$$\mathrm{ad}\,A_0 : \mathrm{Im\,ad}\,A_0 \longrightarrow \mathrm{Im\,ad}\,A_0$$

induces an isomorphism: indeed, $\mathrm{Ker\,ad}\,A_0$ consists of diagonal matrices and $\mathrm{Im\,ad}\,A_0$ of matrices having only zeroes on the diagonal.

For $m \in \mathbb{N}$, let us set $P_m = (\mathrm{Id} + t^m T_m)$, where T_m is a matrix of size d with entries in $\mathscr{O}_{X,x^o}$. Let us consider the effect of the base change of matrix P_m on the coefficient of dt/t in the matrix Ω. If we write $\widetilde{\Omega} = P_m^{-1}\Omega P_m + P_m^{-1}dP_m$ as we did for Ω, we have

$$t^{-r}\widetilde{A} = P_m^{-1}(t^{-r}A)P_m + P_m^{-1} \cdot tP_m',$$

which can be written as

$$\widetilde{A} - A = \sum_{p\in\mathbb{N}}\sum_{k\geqslant 0}(-1)^k t^{p+(k+1)m}[T_m, A_p]T_m^k + m\sum_{k\geqslant 0}(-1)^k t^{(k+1)m+r}T_m^k$$

---

[5] Generically on the parameter space; in Theorem 5.7, genericity is implied by the separation of eigenvalues.

[6] Hint: Check that the projection from $X \times \mathbb{C}$ to X induces, on the subset having equation $\det(s\,\mathrm{Id} - A(0, x))$, a *covering* map, in the neighbourhood of the pullback of x^o (i.e., the set of eigenvalues of $A(0, x^o)$) and conclude with Example 0.2.2-(2).

[7] Recall that $\mathrm{ad}\,A_0(B) := [A_0, B]$.

The coefficients $\widetilde{A}_p$ differ from A_p only for $p \geqslant m$ and we have

$$\widetilde{A}_m = A_m + [T_m, A_0].$$

Let us assume that, for any $p < m$, we found T_p such that, after the base change of matrix $P_{<m} = \prod_{0 < q < m} P_q$, the matrix A_p is diagonal for any $p < m$. We then choose T_m in such a way that so is the matrix $\widetilde{A}_m := A_m + [T_m, A_0]$: it is enough to kill the component of A_m in Im ad A_0.

Lastly, after the *formal* base change of matrix $\prod_{m>0} P_m$, we get a decomposition of $\widehat{\mathcal{M}}$ into free $\widehat{\mathcal{O}}[t^{-1}]$-modules, stable by $t^{r+1}\nabla_{\partial_t}$, where the matrix of $t\nabla_{\partial_t}$ is diagonal, with dominant term $t^{-r}A_0(x)$.

It remains to prove that this decomposition is stable by $\nabla_{\partial_{x_\ell}}$ for $\ell = 1, \ldots, n$. In order to do so, we now use the integrability condition. Let us denote by $\widehat{C}^{(\ell)} = \sum_{m \geqslant 0} C_m^{(\ell)}(x)t^m$ the matrix of $t^r\nabla_{\partial_{x_\ell}}$ in the basis constructed above. It is a matter of showing that $C_m^{(\ell)}$ is diagonal and it is enough to check that $[A_0, C_m^{(\ell)}]$ is zero or, what amounts to the same, that $[A_0, C_m^{(\ell)}]$ is also diagonal. We will show this by induction on m. Let us write the integrability relation as

$$t\frac{\partial(t^{-r}\widehat{C}^{(\ell)})}{\partial t} - \frac{\partial(t^{-r}A)}{\partial x_\ell} = [t^{-r}\widehat{C}^{(\ell)}, t^{-r}A],$$

that is, for any m,

$$(m - 2r)C_{m-r}^{(\ell)} - \frac{\partial A_{m-r}}{\partial x_\ell} = \sum_{p+q=m} [C_p^{(\ell)}, A_q].$$

Then $[A_0, C_0^{(\ell)}] = 0$, hence $C_0^{(\ell)}$ is diagonal. For $m \geqslant 1$, $[A_0, C_m^{(\ell)}]$ can be expressed linearly as a function of the $C_p^{(\ell)}$ and of the $\partial A_p/\partial x_\ell$ $(p < m)$, hence is diagonal (by induction). As Ker ad $A_0 \cap$ Im ad $A_0 = \{0\}$, we must have $[A_0, C_m^{(\ell)}] = 0$, hence $C_m^{(\ell)}$ is diagonal.

We have thus proved that $(\widehat{\mathcal{M}}, \widehat{\nabla})$ is isomorphic to some model, all the components of which have rank one, and the exponents φ_i take the form $t^{-r}\lambda_i(x)(1 + \sum_{k=1}^{r-1} u_{i,k}(x)t^k)$, where λ_i is an eigenvalue of $A_0(x)$. By assumption, the differences $\lambda_i(x) - \lambda_j(x)$ do not vanish in the neighbourhood of x^o if $i \neq j$, which implies that the model is good.   □

5.8 Remark. The base change $\widehat{P}$ belongs to $\mathrm{GL}_d(\widehat{\mathcal{O}}_{D \times X, x^o})$, hence has no pole, neither does its inverse, along $\{0\} \times X$.

5.9 Exercise (Decomposition with respect to eigenvalues). We consider a matrix Ω as in Theorem 5.7. We only assume that there exist k holomorphic functions $\lambda_1(x), \ldots, \lambda_k(x)$ in the neighbourhood of x^o and integers $\nu_1, \ldots, \nu_k$ such that, for any x near x^o, the characteristic polynomial of $A(0,x)$ is equal to $\prod_{j=1}^k (s - \lambda_j(x))^{\nu_j}$; we moreover ask that the differences $\lambda_i - \lambda_j$ do not vanish.

(1) Adapt the proof of Theorem 5.7 to show that the system $(\widehat{\mathcal{M}}, \widehat{\nabla})$ is isomorphic to a direct sum of systems of rank ν_i equipped with a connection with pole of order r, for which the corresponding matrix $A^{(i)}(0, x)$ admits only $\lambda_i(x)$ as an eigenvalue.

(2) Prove that, if one of the eigenvalues λ_i is not identically zero, the formal system $(\widehat{\mathcal{M}}, \widehat{\nabla})$ does not have a regular singularity along $\{0\} \times X$ in the neighbourhood of x^o.

(3) Deduce from this a proof of Theorem 4.1 (where the space of parameters is reduced to a point).

5.c The sheaf $\mathscr{A}$

Unlike the case of regular singularities, the formal solutions of a system with an irregular singularity can be divergent[8] and we cannot use the argument of Proposition 2.13 to complete Theorem 5.7 and find an analytic isomorphism with $\mathcal{M}^{\mathrm{good}}$. In fact, we will see later that such an isomorphism may not exist. However, it does exist if one restricts to sectors around the origin. The sectorial analysis will lead us to work with polar coordinates.

If D is an open disc with coordinate t centered at the origin and of radius $r^o > 0$, we will denote by $\widetilde{D}$ the product $[0, r^o[\times S^1$ and by $\pi : \widetilde{D} \to D$ the mapping $(r, e^{i\theta}) \mapsto t = re^{i\theta}$. In the neighbourhood of a point $(0, \theta^o)$ we will use $(r, \theta) \in [0, r^o[\times]\theta^o - \eta, \theta^o + \eta[$ as a coordinate system.

Let $\mathscr{C}^{\infty}_{]-\varepsilon, r^o[\times S^1 \times X}$ be the sheaf of C^{∞} functions on the manifold $]-\varepsilon, r^o[\times S^1 \times X$, for $\varepsilon > 0$. If

$$i : \widetilde{D} \times X \longrightarrow]-\varepsilon, r^o[\times S^1 \times X$$

denotes the inclusion, the pullback sheaf $i^{-1}\mathscr{C}^{\infty}_{]-\varepsilon, r^o[\times S^1 \times X}$, denoted by $\mathscr{C}^{\infty}_{\widetilde{D} \times X}$, is by definition the sheaf of C^{∞} functions on the manifold with boundary $\widetilde{D} \times X$. In other words, a C^{∞} function on $\widetilde{D} \times X$ in the neighbourhood of $(0, \theta^o, x^o)$ is a function which can be extended as a C^{∞} function on a neighbourhood of this point in $]-\varepsilon, r^o[\times S^1 \times X$.

By "working in polar coordinates", we can define the derivations $t\partial/\partial t$, $\bar{t}\partial/\partial\bar{t}$, $\partial/\partial x_i$ and $\partial/\partial\bar{x}_i$ $(i = 1, \ldots, n)$ on $\mathscr{C}^{\infty}_{\widetilde{D} \times X}$: we have

$$t\frac{\partial}{\partial t} = \frac{1}{2}\left(r\frac{\partial}{\partial r} - i\frac{\partial}{\partial\theta}\right), \qquad \bar{t}\frac{\partial}{\partial\bar{t}} = \frac{1}{2}\left(r\frac{\partial}{\partial r} + i\frac{\partial}{\partial\theta}\right).$$

5.10 Definition (The sheaf $\mathscr{A}$). The sheaf of rings $\mathscr{A}_{\widetilde{D} \times X}$ is the subsheaf of $\mathscr{C}^{\infty}_{\widetilde{D} \times X}$ of germs killed by $\bar{t}\partial/\partial\bar{t}$ and $\partial/\partial\bar{x}_i$ $(i = 1, \ldots, n)$.

---

[8] One will find in [Ram94] not only examples but also a survey on the use of divergent series in analysis.

5.11 Remark. The open set $\tilde{D}^* := \tilde{D} \smallsetminus (\{0\} \times S^1)$ coincides with $D^* = D \smallsetminus \{0\}$ and, on $\tilde{D}^* \times X$, the sheaf $\mathscr{A}_{\tilde{D}^* \times X}$ coincides with the sheaf $\mathscr{O}_{D^* \times X}$ of holomorphic functions. A section of $\mathscr{A}_{\tilde{D} \times X}$ on an open set of the kind $\tilde{D} \times V$ is nothing but a holomorphic function on $D \times V$, as it is holomorphic on $D^* \times V$ and locally bounded.

On the sheaf $\mathscr{A}_{\tilde{D} \times X}$ is defined the action of the derivation $\partial/\partial t$ (and not only that of $t\partial/\partial t$): by using the vanishing of $\bar{t}\partial/\partial \bar{t}$, we can set

$$\frac{\partial}{\partial t} = e^{-i\theta} \frac{\partial}{\partial r}.$$

The action of the $\partial/\partial x_i$ on $\mathscr{C}^\infty_{\tilde{D} \times X}$ keeps $\mathscr{A}_{\tilde{D} \times X}$ stable. The sheaf $\mathscr{A}_{\tilde{D} \times X}$ contains the subsheaf $\pi^{-1}\mathscr{O}_{D \times X}$: if f is holomorphic on $D \times X$, $f \circ \pi$ is a section of $\mathscr{A}_{\tilde{D} \times X}$.

The Taylor expansion of a C^∞ germ in (θ^o, x^o) along $r = 0$ can be written as

$$\sum_{k \geqslant 0} f_k(x_1, \ldots, x_n, \theta) r^k,$$

where the f_k are C^∞ functions on some fixed neighbourhood of (θ^o, x^o). The Taylor expansion of a germ of $\mathscr{A}_{\tilde{D} \times X}$ thus takes the form

$$\sum_{k \geqslant 0} f_k(x_1, \ldots, x_n) t^k,$$

where the f_k are holomorphic on some fixed neighbourhood of x^o. In other words, if we denote by $\mathscr{A}_{S^1 \times X}$ the restriction to $\{0\} \times S^1 \times X$ of the sheaf $\mathscr{A}_{\tilde{D} \times X}$, the "Taylor expansion" mapping defines a homomorphism

$$\mathscr{A}_{S^1 \times X, (\theta^o, x^o)} \xrightarrow{\ T\ } \widehat{\mathscr{O}}_{D \times X, x^o}$$

whose kernel[9] is denoted by $\mathscr{A}^{<\{0\} \times X}_{S^1 \times X, (\theta^o, x^o)}$, or also $\mathscr{A}^{<X}_{S^1 \times X, (\theta^o, x^o)}$, if one identifies X to the divisor $\{0\} \times X$ of $D \times X$. By construction, T is compatible with the action of the derivations $\partial/\partial t, \partial/\partial x_1, \ldots, \partial/\partial x_n$. One can remark that it is possible to "divide by t" as many times as wanted the sections of $\mathscr{A}^{<X}_{S^1 \times X}$.

The *Borel-Ritt Lemma* (see for instance [Mal91]) states that T is surjective[10]. In terms of sheaves, we can express this result by saying that the sequence of sheaves on $S^1 \times X$

$$0 \longrightarrow \mathscr{A}^{<X}_{S^1 \times X} \longrightarrow \mathscr{A}_{S^1 \times X} \xrightarrow{\ T\ } \pi^{-1}\widehat{\mathscr{O}}_{D \times X} \longrightarrow 0$$

is *exact* (recall that $\widehat{\mathscr{O}}_{D \times X}$ is a sheaf on $\{0\} \times X$, hence $\pi^{-1}\widehat{\mathscr{O}}_{D \times X}$ is a sheaf on $S^1 \times X$).

---

[9] Here appears the main difference with $\mathscr{O}_{D \times X}$, as the "Taylor expansion" mapping $T : \mathscr{O}_{D \times X} \to \widehat{\mathscr{O}}_{D \times X}$ is injective.

[10] This is the other main difference with $\mathscr{O}_{D \times X}$; between $\mathscr{O}$ and $\mathscr{A}$, the injectivity of T has been replaced by the surjectivity.

5.d Sectorial classification

The sectorial analogue of Proposition 2.13 is known as the "Hukuhara-Turrittin Theorem", although other mathematicians also have contributed, in some particular cases, to its proof (Malmquist [Mal40] for instance). The version with parameters that we indicate below without proof has been obtained by Sibuya (see [Sib62, Sib74], see also [BV89b]).

If $\mathcal{M}$ is a germ at $(0, x^o)$ of a meromorphic bundle with poles along $\{0\} \times X$, we will set

$$\widehat{\mathcal{M}} = \widehat{\mathcal{O}}_{D \times X, x^o} \underset{\mathcal{O}_{D \times X, (0, x^o)}}{\otimes} \mathcal{M}$$

and, for any $e^{i\theta^o} \in S^1$,

$$\widetilde{\mathcal{M}}_{\theta^o} = \mathscr{A}_{S^1 \times X, (\theta^o, x^o)} \underset{\mathcal{O}_{D \times X, (0, x^o)}}{\otimes} \mathcal{M}.$$

Right exactness of the tensor product and Borel-Ritt Lemma show that the natural mapping $\widetilde{\mathcal{M}}_{\theta^o} \to \widehat{\mathcal{M}}$ is onto.

5.12 Theorem (Sectorial decomposition). *Let $\mathcal{M}$ be a germ at $(0, x^o)$ of a meromorphic bundle with poles along $\{0\} \times X$. Let us assume that there exist a good model $\mathcal{M}^{\mathrm{good}}$ and an isomorphism $\widehat{\lambda} : \widehat{\mathcal{M}} \xrightarrow{\sim} \widehat{\mathcal{M}}^{\mathrm{good}}$. There exists then, for any $e^{i\theta^o} \in S^1$, an isomorphism $\widetilde{\lambda}_{\theta^o} : \widetilde{\mathcal{M}}_{\theta^o} \xrightarrow{\sim} \widetilde{\mathcal{M}}_{\theta^o}^{\mathrm{good}}$ lifting $\widehat{\lambda}$, that is, such that the following diagram*

$$
\begin{array}{ccc}
\widetilde{\mathcal{M}}_{\theta^o} & \xrightarrow[\sim]{\widetilde{\lambda}_{\theta^o}} & \widetilde{\mathcal{M}}_{\theta^o}^{\mathrm{good}} \\
\downarrow & & \downarrow \\
\widehat{\mathcal{M}} & \xrightarrow[\widehat{\lambda}]{\sim} & \widehat{\mathcal{M}}^{\mathrm{good}}
\end{array}
$$

commutes.   □

We will apply this theorem to the situation described in Theorem 5.7 of §III.2.

6 The Riemann-Hilbert correspondence in the irregular case

We will now try to give a description of a "topological" nature to the category of germs of meromorphic bundles with connection. For regular singularities, we have seen that this category is equivalent to that of representations of the fundamental group. In the irregular case, a set of nontopological data is furnished by the formal meromorphic bundle with connection that we associate to the bundle we start with. Indeed, the exponents φ which occur in any good model are data of an analytic nature. The data which enable us to recover the meromorphic bundle with connection from the formal meromorphic bundle with connection are called a *Stokes structure*. It is in fact a sheaf on the parameter space. We will see that this sheaf is *locally constant*.

At this stage, many developments are possible, that we will not address: one can define a "wild" fundamental group and identify the category of meromorphic connections to that of the finite dimensional linear representations of this group; one can also define a "differential Galois group" and consider its representations (see for instance [Ber92, LR95, Var96, vdP98, SvdP03] for an introduction to this approach, as well as the references given therein); one can identify the category of meromorphic connections to that of filtered local systems, in the sense of Deligne and give thereby a more geometric flavour to the notion of Stokes structure (see [Mal91, BV89a]); lastly, the theory of multisummability enables one to analyze in a finer way the Stokes phenomenon (see for instance [BBRS91, MR92, LR94, LRP97]).

6.a The Stokes sheaf

Let X be a complex analytic manifold and let $\mathscr{M}^{\mathrm{good}}$ be a meromorphic bundle on $D \times X$ with poles along $\{0\} \times X$, equipped with a flat connection ∇^{good}. We will assume that $(\mathscr{M}^{\mathrm{good}}, \nabla^{\mathrm{good}})$ is a *good model* in the neighbourhood of any point x^o of X, that is, that there exist pairwise distinct germs $\varphi_1, \ldots, \varphi_p \in t^{-1}\mathscr{O}_{X,x^o}[t^{-1}]$ and nonzero germs of systems with regular singularity $\mathscr{R}_{\varphi_1}, \ldots, \mathscr{R}_{\varphi_p}$ along $\{0\} \times X$, such that we have, in the neighbourhood of x^o, an isomorphism $\mathscr{M}^{\mathrm{good}}_{x^o} \simeq \oplus_k (\mathscr{E}^{\varphi_k} \otimes \mathscr{R}_{\varphi_k})$. According to goodness, as the function $\varphi_k - \varphi_\ell$ is not identically zero for $k \neq \ell$, the order of the pole with respect to t of $(\varphi_k - \varphi_\ell)(x,t)$ does not depend on x in some neighbourhood of x^o.

Let us consider on $S^1 \times X$ the sheaf of automorphisms of $\widetilde{\mathscr{M}}^{\mathrm{good}} := \mathscr{A}_{\widetilde{D} \times X} \otimes_{\mathscr{O}_{D \times X}} \mathscr{M}^{\mathrm{good}}$ which are compatible with the connection and are formally equal to the identity, i.e., which induce identity on

$$\widehat{\mathscr{M}}^{\mathrm{good}} := \widehat{\mathscr{O}}_{D \times X} \underset{\mathscr{O}_{D \times X}}{\otimes} \mathscr{M}^{\mathrm{good}}.$$

We denote it by $\mathrm{Aut}^{<\{0\} \times X}(\widetilde{\mathscr{M}}^{\mathrm{good}})$ or, in short, $\mathrm{Aut}^{<X}(\widetilde{\mathscr{M}}^{\mathrm{good}})$ (in this notation, we think of X as the divisor $\{0\} \times X$ in $D \times X$). It is a sheaf of (in general noncommutative) groups on $S^1 \times X$, that we will analyze later. Its local sections are called *Stokes matrices*.

This sheaf enables us to define a sheaf $\mathrm{St}_X(\mathscr{M}^{\mathrm{good}})$ on X, called the *Stokes sheaf*. This is, by definition, the sheaf on X associated to the presheaf

$$U \longmapsto H^1(S^1 \times U, \mathrm{Aut}^{<X}(\widetilde{\mathscr{M}}^{\mathrm{good}})).$$

We will show in §6.c:

6.1 Theorem (The Stokes sheaf is locally constant). *The Stokes sheaf* $\mathcal{S}t_X(\mathcal{M}^{\mathrm{good}})$ *is a locally constant sheaf of pointed sets.*

We deduce that the Stokes sheaf is constant on any simply connected open set.

6.b Classification (statement)

Let $(\mathcal{M}, \nabla)$ be a meromorphic bundle on $D \times X$ with flat connection, having poles along $\{0\} \times X$. We will say that $(\mathcal{M}^{\mathrm{good}}, \nabla^{\mathrm{good}})$ is a *good formal model* for $(\mathcal{M}, \nabla)$ if there exists an isomorphism of sheaves of $\widehat{\mathcal{O}}_{D \times X}$-modules

$$\widehat{f} : (\widehat{\mathcal{M}}, \widehat{\nabla}) \xrightarrow{\sim} (\widehat{\mathcal{M}^{\mathrm{good}}}, \widehat{\nabla^{\mathrm{good}}})$$

compatible with the connections.

We will say that two germs $(\mathcal{M}, \nabla, \widehat{f})$ and $(\mathcal{M}', \nabla', \widehat{f}')$ along $\{0\} \times X$ are *isomorphic* if there exists an isomorphism $g : (\mathcal{M}, \nabla) \xrightarrow{\sim} (\mathcal{M}', \nabla')$ such that, moreover, $\widehat{f} = \widehat{f}' \circ \widehat{g}$. It is important to remark that such an isomorphism is then unique: indeed, in local bases of $\mathcal{M}$ and $\mathcal{M}'$, the Taylor expansion of the matrix of g is equal to that of $\widehat{f}'^{-1}\widehat{f}$ in these bases; the matrix of g is thus uniquely determined.

The good model $(\mathcal{M}^{\mathrm{good}}, \nabla^{\mathrm{good}})$ being fixed, let us consider the presheaf $\mathcal{H}_X$ on X which, to any open set U of X, associates the set $\mathcal{H}_X(U)$ of isomorphism classes of germs $(\mathcal{M}, \nabla, \widehat{f})$ defined on U, equipped with a distinguished element, namely, the class of $(\mathcal{M}^{\mathrm{good}}, \nabla^{\mathrm{good}}, \widehat{\mathrm{Id}})_{|U}$.

6.2 Lemma. *The presheaf* $\mathcal{H}_X$ *is a sheaf.*

Proof. Let $(U_i)_{i \in I}$ be a family of open sets of X and let $(s_i)_{i \in I}$ be a family of sections $s_i \in \mathcal{H}_X(U_i)$, which are compatible on intersections. Let us choose for any i a representative $(\mathcal{M}_i, \nabla_i, \widehat{f}_i)$ of s_i. We wish to construct a germ $(\mathcal{M}, \nabla, \widehat{f})$ on $V = \cup_i U_i$ whose restriction to U_i is isomorphic to $(\mathcal{M}_i, \nabla_i, \widehat{f}_i)$. The uniqueness of isomorphisms enables one to glue them, hence, if such an object exists, it is unique (up to a unique isomorphism). We argue similarly for the existence: for all $i, j \in I$, we have a unique isomorphism

$$g_{ij} : (\mathcal{M}_i, \nabla_i, \widehat{f}_i)_{|U_i \cap U_j} \xrightarrow{\sim} (\mathcal{M}_j, \nabla_j, \widehat{f}_j)_{|U_i \cap U_j}$$

and, by uniqueness, we have $g_{ij}g_{jk} = g_{ik}$ on $U_i \cap U_j \cap U_k$, which enables us to glue the $(\mathcal{M}_i, \nabla_i, \widehat{f}_i)$. $\qquad\square$

According to Theorem 5.12, we can construct a homomorphism of sheaves of pointed sets

$$\mathcal{H}_X \longrightarrow \mathcal{S}t_X(\mathcal{M}^{\mathrm{good}}).$$

Indeed, if $(\mathcal{M}, \nabla, \widehat{f})$ is defined on U, there exist an open covering $\mathfrak{W}$ of $S^1 \times U$ and, for any open set W_i of $\mathfrak{W}$, an isomorphism

$$f_i : (\widetilde{\mathcal{M}}, \widetilde{\nabla})_{|W_i} \xrightarrow{\sim} (\widetilde{\mathcal{M}}^{\text{good}}, \widetilde{\nabla}^{\text{good}})_{|W_i}$$

such that $\widehat{f}_i = \widehat{f}$. Then, $(f_j f_i^{-1})_{i,j}$ is a cocycle of the sheaf $\text{Aut}^{<X}(\widetilde{\mathcal{M}}^{\text{good}})$ relative to covering $\mathfrak{W}$. If f_i' is another lifting of $\widehat{f}$ on W_i, $(f_i' f_i^{-1})_i$ is a 0-cochain of $\text{Aut}^{<X}(\widetilde{\mathcal{M}}^{\text{good}})$ relative to $\mathfrak{W}$ and the cocycles associated to (f_i) and to (f_i') are equivalent *via* the corresponding coboundary. One checks similarly that, if $(\mathcal{M}, \nabla, \widehat{f})$ and $(\mathcal{M}', \nabla', \widehat{f}')$ are isomorphic, the corresponding cocycles define the same cohomology class.

We have thus defined a mapping of pointed sets

$$\Gamma(U, \mathcal{H}_X) \longrightarrow H^1(S^1 \times U, \text{Aut}^{<X}(\widetilde{\mathcal{M}}^{\text{good}}))$$

from which we deduce a homomorphism of sheaves $\mathcal{H}_X \to \text{St}_X(\mathcal{M}^{\text{good}})$, which sends the class of $(\mathcal{M}^{\text{good}}, \nabla^{\text{good}}, \widehat{\text{Id}})_{|U}$ to that of Id.

6.3 Theorem (The Stokes sheaf classifies meromorphic connections with fixed formal type, [Mal83b]). *The homomorphism so defined $\mathcal{H}_X \to \text{St}_X(\mathcal{M}^{\text{good}})$ is an isomorphism of sheaves of pointed sets.*

We deduce from Theorems 6.1 and 6.3 that the sheaf $\mathcal{H}_X$ is locally constant. We also get:

6.4 Corollary (Analytic extension with fixed formal structure). *If X is 1-connected and if $(\mathcal{M}^o, \nabla^o, \widehat{f}^o)$ is a germ of meromorphic bundle on D with pole at 0, equipped with a formal isomorphism $\widehat{f}^o : (\mathcal{M}^o, \nabla^o) \xrightarrow{\sim} i^+(\mathcal{M}^{\text{good}}, \nabla^{\text{good}})$, where i denotes the inclusion $D \times \{x^o\} \hookrightarrow D \times X$, then there exists a meromorphic bundle with connection $(\mathcal{M}, \nabla, \widehat{f})$ equipped with a formal isomorphism $\widehat{f}$ to $(\mathcal{M}^{\text{good}}, \nabla^{\text{good}})$, such that $i^+(\mathcal{M}, \nabla, \widehat{f})$ is isomorphic to $(\mathcal{M}^o, \nabla^o, \widehat{f}^o)$. Such an object is unique up to a unique isomorphism.*

Proof. After Proposition 6.9 below, the germ of the sheaf $\text{St}_X(\mathcal{M}^{\text{good}})$ at x^o is the Stokes set $\text{St}(i^+ \mathcal{M}^{\text{good}})$. Therefore, the initial data $(\mathcal{M}^o, \nabla^o, \widehat{f}^o)$ define an element of $\text{St}_X(\mathcal{M}^{\text{good}})_{x^o}$. As X is 1-connected and as the sheaf $\text{St}_X(\mathcal{M}^{\text{good}})$ is locally constant, there exists a unique section of $\text{St}_X(\mathcal{M}^{\text{good}})$ whose value at x^o is precisely this element (cf. Lemma 0.15.9). The corresponding section of $\mathcal{H}_X$ defines an object $(\mathcal{M}, \nabla, \widehat{f})$, the restriction of which to x^o is isomorphic to $(\mathcal{M}^o, \nabla^o, \widehat{f}^o)$.

Let $(\mathcal{M}, \nabla, \widehat{f})$ and $(\mathcal{M}', \nabla', \widehat{f}')$ be two such objects, with restrictions isomorphic to the object $(\mathcal{M}^o, \nabla^o, \widehat{f}^o)$. According to uniqueness of the section constructed above, their isomorphism classes coincide; in other words, these two objects are isomorphic and, as we have yet seen, this isomorphism is unique. $\square$

6.c Local constancy of the Stokes sheaf

Let us now show Theorem 6.1 by analyzing first with more details the sheaf of Stokes matrices $\mathrm{Aut}^{<X}(\widetilde{\mathscr{M}}^{\mathrm{good}})$. The germs φ_k are polynomials in t^{-1} without constant term, with coefficients in $\mathscr{O}_{X,x^o}$. An automorphism a of $\widetilde{\mathscr{M}}^{\mathrm{good}}$ on a product $I \times U$ of an interval with a sufficiently small neighbourhood of x^o can be decomposed into blocks

$$a_{k\ell} : \widetilde{\mathscr{E}}^{\varphi_k} \otimes \widetilde{\mathscr{R}}_k \longrightarrow \widetilde{\mathscr{E}}^{\varphi_\ell} \otimes \widetilde{\mathscr{R}}_\ell.$$

The term $a_{k\ell}$ thus takes the form $e^{\varphi_k - \varphi_\ell} b_{k\ell}$, where $b_{k\ell}$ is a homomorphism of bundles with connection $\widetilde{\mathscr{R}}_k \to \widetilde{\mathscr{R}}_\ell$. Let us assume that X is a ball. The $\mathbb{C}$-vector space V_k consisting of multivalued horizontal sections of $\mathscr{R}_k$ on $D^* \times X$ is finite dimensional and the space of horizontal sections of $\widetilde{\mathscr{R}}_k$ on an open set like $]0, r[\times I \times X$, with $I \neq S^1$, can be identified to V_k, this identification depending on the choice of a determination of the logarithm on I. With such an identification, $b_{k\ell}$ induces a $\mathbb{C}$-linear map $V_k \to V_\ell$, that we also denote by $b_{k\ell}$. The matrix of $b_{k\ell}$ with respect to bases of V_k, V_ℓ is thus constant, while with respect to $\mathscr{O}_{D \times X}[t^{-1}]$-bases of $\mathscr{R}_k, \mathscr{R}_\ell$, it has moderate growth.

For $k \neq \ell$, let us write

$$(\varphi_k - \varphi_\ell)(t, x) = \frac{\psi_{k\ell}(x)}{t^{n_{k\ell}}} \cdot u_{k\ell}(t, x)$$

with $n_{k\ell} > 0$, $\psi_{k\ell}(x) \in \mathscr{O}^*_{X,x^o}$ (i.e., does not vanish), $u(t, x)$ holomorphic in the neighbourhood of $(0, x^o)$ and $u(0, 0) = 1$. We also choose some C^∞ determination $\eta_{k\ell}(x)$ of the argument of $\psi_{k\ell}$:

$$\psi_{k\ell}(x) = |\psi_{k\ell}(x)| \cdot e^{i\eta_{k\ell}(x)}.$$

Let us remark that $\psi_{\ell k} = -\psi_{k\ell}$.

6.5 Lemma. *Let us pick $e^{i\theta^o} \in S^1$.*

(1) *For $k \neq \ell$, the matrix $e^{\varphi_k - \varphi_\ell} b_{k\ell}$ in $\mathscr{O}_{D \times X}[t^{-1}]$-bases of $\mathscr{E}^{\varphi_k} \otimes \mathscr{R}_k$, $\mathscr{E}^{\varphi_\ell} \otimes \mathscr{R}_\ell$ has entries in $\mathscr{A}^{<X}_{S^1 \times X}$ in the neighbourhood of $(e^{i\theta^o}, x^o)$ if and only if $b_{k\ell} = 0$ or $\cos(n_{k\ell}\theta^o - \eta_{k\ell}) < 0$.*

(2) *The matrix $b_{kk} - \mathrm{Id}$ has entries in $\mathscr{A}^{<X}_{S^1 \times X}$ in the neighbourhood of $(e^{i\theta^o}, x^o)$ if and only if $b_{kk} - \mathrm{Id} = 0$.*

The proof of the lemma is easy and left to the reader: for the first point, it is enough to check that $e^{\varphi_k - \varphi_\ell}$ belongs to $\mathscr{A}^{<X}_{S^1 \times X}$ in the neighbourhood of $(e^{i\theta^o}, x^o)$ if and only if $\mathrm{Re}(\varphi_k - \varphi_\ell) < 0$ on a sufficiently small neighbourhood of $(e^{i\theta^o}, x^o)$, then to express this condition on the leading term of $\varphi_k - \varphi_\ell$. Then, one notices that, as $b_{k\ell}$ has moderate growth, it does not affect the rapid decay or the exponential growth property.   □

Let $V = \bigoplus_k V_k$ be the $\mathbb{C}$-vector space of multivalued horizontal sections of $\bigoplus_k \mathscr{R}_k$ on $D^* \times X$. Let us consider the subsheaf $\mathscr{L}$ of the constant sheaf $\mathrm{Aut}(V)$, the germ of which at $(e^{i\theta^o}, x^o)$ is the space of automorphisms $\mathrm{Id} + (\oplus c_{k\ell})$, with $c_{kk} = 0$ and $c_{k\ell} : V_k \to V_\ell$ nonzero if and only if $\cos(n_{k\ell}\theta^o - \eta_{k\ell}) < 0$. Let us note that any Stokes matrix (local section of $\mathscr{L}$) is *unipotent*: if we suitably order the set of indices (an order which depends on the chosen point θ^o), we can assume that $c_{k\ell} = 0$ for $k < \ell$ and the matrix of $\oplus c_{k\ell}$ is strictly uppertriangular.

We deduce:

6.6 Corollary (The Stokes matrices are constant). *For any open interval $I \neq S^1$, the restrictions to $I \times X$ of the sheaves $\mathrm{Aut}^{<X}(\widetilde{\mathscr{M}}^{\mathrm{good}})$ and $\mathscr{L}$ are isomorphic.*

Let $I \neq S^1$ be an open interval of S^1 such that, for all $k \neq \ell$, the function $t \mapsto \cos(n_{k\ell}t - \eta_{k\ell}(x^o))$ does not vanish at the end points of I, and let V be a neighbourhood of x^o such that the same property holds for $t \mapsto \cos(n_{k\ell}t - \eta_{k\ell}(x))$ for any $x \in V$. If a is any section of $\mathrm{Aut}^{<X}(\widetilde{\mathscr{M}}^{\mathrm{good}})$ on $I \times V$, then $a_{kk} = \mathrm{Id}$ for any k and, for all k, ℓ such that $k \neq \ell$, we have $a_{k\ell} = 0$ or $a_{\ell k} = 0$. If $a_{k\ell} \neq 0$, then $\cos(n_{k\ell}t - \eta_{k\ell}(x^o)) < 0$ for $t \in I$ and in such a case the matrix of $b_{k\ell}$ is constant in bases of V_k and V_ℓ. $\quad\square$

6.7 Corollary (Base change for horizontal sections). *If $i : S^1 \times \{x^o\} \hookrightarrow S^1 \times X$ denotes the inclusion, the natural morphism*

$$i^{-1}\,\mathrm{Aut}^{<X}(\widetilde{\mathscr{M}}^{\mathrm{good}}) \longrightarrow \mathrm{Aut}^{<0}(\widetilde{i^+\mathscr{M}}^{\mathrm{good}})$$

is an isomorphism.

Proof. The question is local, and one can replace the sheaf $i^{-1}\,\mathrm{Aut}^{<X}(\widetilde{\mathscr{M}}^{\mathrm{good}})$ by $\mathscr{L}$. The result is then clear. $\quad\square$

The *manifold of Stokes directions* in $S^1 \times U$, where U is a neighbourhood of x^o on which the φ_k are defined and satisfy the goodness properties[11], is the union of the sets having equation

$$\cos(n_{k\ell}\theta - \eta_{k\ell}(x)) = 0.$$

The restriction of this set to $S^1 \times \{x^o\}$ is a finite set of points, called the *Stokes directions at x^o*. A pair (k, ℓ) contributes to the directions $(\eta_{k\ell} + \pi/2 + j\pi)/n_{k\ell}$, for $j = 0, \ldots, 2n_{k\ell} - 1$.

6.8 Exercise (Straightening the Stokes manifold). In this exercise, we assume that the manifold of Stokes directions is a submanifold of $S^1 \times U$ and that, through a Stokes direction at x^o, passes only one connected component of the manifold of Stokes directions.

---

[11] For a study of the singularities which can occur, in a more general situation, in the Stokes manifold, the reader can refer to [Kos90].

(1) Show that there exists a real analytic mapping Ψ from $S^1 \times U$ into itself, of the form $\Psi(\theta, x) = (\psi(\theta, x), x)$, admitting an inverse mapping of the same kind and sending the component of the manifold of Stokes directions which goes through θ^o at x^o to $\{\theta^o\} \times U$.

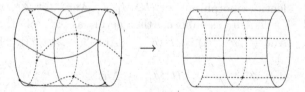

Fig. II.2. The mapping Ψ, with Stokes directions corresponding to a pair (k, ℓ).

(2) Prove that the sheaf $\mathrm{Aut}^{<X}(\widetilde{\mathscr{M}}^{\mathrm{good}})$ is isomorphic to the pullback by the projection $p_1 : S^1 \times U \to S^1$ of its restriction $\mathscr{G}$ to $S^1 \times \{x^o\}$.

When the assumption of this exercise is satisfied, Theorem 6.1 is an immediate consequence of Example 0.5.2 (the existence of a finite good covering will be indicated in §6.e). Otherwise, the argument requires a more precise analysis of the sheaf $\mathcal{St}_X(\mathscr{M}^{\mathrm{good}})$. Let us begin by showing

6.9 Proposition (Base change for the Stokes sheaf). *The germ at x^o of the Stokes sheaf $\mathcal{St}_X(\mathscr{M}^{\mathrm{good}})$ is equal to the Stokes space[12] $H^1(S^1, \mathrm{Aut}^{<0}(\widetilde{i^+\mathscr{M}}^{\mathrm{good}}))$ of the connection $i^+\mathscr{M}^{\mathrm{good}}$.*

Proof. Let U be an open neighbourhood of x^o, and let $\mathscr{U}$ be an open covering of $S^1 \times U$. There exists then an open neighbourhood V of x^o contained in U and a refinement $\mathscr{V}$ of $\mathscr{U}$ such that the covering $\mathscr{V}_{|V}$ of $S^1 \times V$ induced by $\mathscr{V}$ consists of open sets like $I_j \times V$, where I_j is an open interval of S^1, and $I_j \cap I_k \cap I_\ell = \varnothing$ if j, k, ℓ are pairwise distinct (one first refines the restriction of $\mathscr{U}$ to $S^1 \times \{x^o\}$ and then one argues by compactness). It will be moreover convenient to assume that Corollary 6.6 applies to any $I_j \times V$ and $(I_j \cap I_k) \times V$. Consequently, $H^1(S^1 \times U, \mathrm{Aut}^{<X}(\widetilde{\mathscr{M}}^{\mathrm{good}}))$ is the union of the $H^1(\mathscr{V}, \mathrm{Aut}^{<X}(\widetilde{\mathscr{M}}^{\mathrm{good}}))$ over all coverings $\mathscr{V}$ of this kind, and the germ at x^o of the presheaf $W \mapsto H^1(S^1 \times W, \mathrm{Aut}^{<X}(\widetilde{\mathscr{M}}^{\mathrm{good}}))$ (for $W \subset U$) is the union of the germs at x^o of the presheaves $W \mapsto H^1(\mathscr{V}_{|W}, \mathrm{Aut}^{<X}(\widetilde{\mathscr{M}}^{\mathrm{good}}))$.

Let us note that, for $\mathscr{V}$ and V as above,

$$H^1(\mathscr{V}_{|V}, \mathrm{Aut}^{<X}(\widetilde{\mathscr{M}}^{\mathrm{good}})) = H^1(\mathscr{V}_{|\{x^o\}}, \mathrm{Aut}^{<0}(\widetilde{i^+\mathscr{M}}^{\mathrm{good}}))$$

after Corollary 6.6, which says that

$$\Gamma(I_j \times V, \mathrm{Aut}^{<X}(\widetilde{\mathscr{M}}^{\mathrm{good}})) = \Gamma(I_j, \mathrm{Aut}^{<0}(\widetilde{i^+\mathscr{M}}^{\mathrm{good}})),$$

---

[12] The word "space" is used here as this set is equipped with a natural structure of affine space, cf. Example 6.e.

and similarly for intersections $I_j \cap I_k$, which implies the equality of the corresponding Čech complexes. Moreover, for $W \subset V$, the restriction morphism

$$H^1(\mathscr{V}_{|V}, \mathrm{Aut}^{<X}(\widetilde{\mathscr{M}}^{\mathrm{good}})) \longrightarrow H^1(\mathscr{V}_{|W}, \mathrm{Aut}^{<X}(\widetilde{\mathscr{M}}^{\mathrm{good}}))$$

induces the identity morphism on $H^1(\mathscr{V}_{|\{x^o\}}, \mathrm{Aut}^{<0}(\widetilde{i^+ \mathscr{M}}^{\mathrm{good}}))$. Hence, for $\mathscr{V}$ and V chosen as above, the germ at x^o of the presheaf $W \mapsto H^1(\mathscr{V}_{|W}, \mathrm{Aut}^{<X}(\widetilde{\mathscr{M}}^{\mathrm{good}}))$ is $H^1(\mathscr{V}_{|\{x^o\}}, \mathrm{Aut}^{<0}(\widetilde{i^+ \mathscr{M}}^{\mathrm{good}}))$. Therefore,

$$\mathcal{S}t_X(\mathscr{M}^{\mathrm{good}})_{x^o} = \bigcup_{\mathscr{V}} H^1(\mathscr{V}_{|\{x^o\}}, \mathrm{Aut}^{<0}(\widetilde{i^+ \mathscr{M}}^{\mathrm{good}}))$$

$$= H^1(S^1, \mathrm{Aut}^{<0}(\widetilde{i^+ \mathscr{M}}^{\mathrm{good}})). \qquad \square$$

Proof (End of the proof of Theorem 6.1). The previous proof says more precisely that, for any small enough neighbourhood V of x^o and any covering $\mathscr{V}$ of $S^1 \times V$ of the form $(I_j \times V)_j$, where the open intervals I_j satisfy the previous properties, the sheaf associated to the presheaf $W \mapsto H^1(\mathscr{V}_{|W}, \mathrm{Aut}^{<X}(\widetilde{\mathscr{M}}^{\mathrm{good}}))$ is the constant sheaf on V, with germ $H^1(\mathscr{V}_{|\{x\}}, \mathrm{Aut}^{<0}(\widetilde{i^+ \mathscr{M}}^{\mathrm{good}}))$ for any $x \in V$.

In order to conclude, it is enough to show that one can choose a finite covering $(I_j)_j$ of S^1 such that, for any $x \in V$,

$$H^1(\mathscr{V}_{|\{x\}}, \mathrm{Aut}^{<X}(\widetilde{\mathscr{M}}^{\mathrm{good}})) = H^1(S^1 \times \{x\}, \mathrm{Aut}^{<X}(\widetilde{\mathscr{M}}^{\mathrm{good}})).$$

Indeed, the right-hand term is equal to the germ at x of $\mathcal{S}t_X(\mathscr{M}^{\mathrm{good}})$, according to the previous proposition and Corollary 6.7 applied at the point x. The previous equality would imply that the morphism of presheaves on V defined by $H^1(\mathscr{V}_{|W}, \mathrm{Aut}^{<X}(\widetilde{\mathscr{M}}^{\mathrm{good}})) \to H^1(S^1 \times W, \mathrm{Aut}^{<X}(\widetilde{\mathscr{M}}^{\mathrm{good}}))$ induces an isomorphism between the associated sheaves. The proof of this statement will be indicated in §6.e. $\qquad \square$

6.d Classification (proof)

We will now show Theorem 6.3. As the homomorphism $\mathscr{H}_X \to \mathcal{S}t_X(\mathscr{M}^{\mathrm{good}})$ is defined globally, showing that it is an isomorphism is a local problem on X. It is thus a matter, after what we have seen above, of proving that the map which, to the germ $(\mathcal{M}, \nabla, \widehat{f})$ at x^o, associates the element of $H^1(S^1, i^{-1} \mathrm{Aut}^{<X}(\widetilde{\mathscr{M}}^{\mathrm{good}}))$ obtained according to Theorem 5.12, is bijective.

Consider $(\mathcal{M}, \nabla, \widehat{f})$ and $(\mathcal{M}', \nabla', \widehat{f}')$ which define the same element

$$\lambda \in H^1(S^1, i^{-1} \mathrm{Aut}^{<X}(\widetilde{\mathscr{M}}^{\mathrm{good}})).$$

We can assume that there exist a finite covering (I_i) of S^1 and an open neighbourhood V of x^o such that λ is the class of the cocycles $(f_j f_i^{-1})$ and $(f_j' f_i'^{-1})$, where the f_i, f_i' are defined on $I_i \times V$. There exists thus a 0-cochain (g_i) of the sheaf $\operatorname{Aut}^{<X}(\widetilde{\mathscr{M}}^{\text{good}})$ relative to the covering $(I_i \times V)$ of $S^1 \times V$ such that, on $(I_i \cap I_j) \times V$

$$f_j' f_i'^{-1} = g_j f_j f_i^{-1} g_i^{-1}.$$

If we set $\sigma = f_i^{-1} g_i^{-1} f_i'$ on $I_i \times V$, we get a horizontal section, on a sufficiently small open set $[0, r_0[\times S^1 \times V$, of the sheaf $\mathscr{H}om(\widetilde{\mathscr{M}}', \widetilde{\mathscr{M}})$. By making explicit the matrix of such a section in local bases of $\mathscr{M}$ and $\mathscr{M}'$, we see, as in Remark 5.11, that σ is a horizontal section of $\mathscr{H}om(\mathscr{M}', \mathscr{M})$. It is moreover invertible on $D^* \times V$, where D is the disc of radius r_0, hence also on $D \times V$. Lastly, as the g_j are asymptotic to identity, we have $\sigma \circ \widehat{f'} = \widehat{f}$. Therefore $(\mathscr{M}, \nabla, \widehat{f})$ and $(\mathscr{M}', \nabla', \widehat{f'})$ are isomorphic. This gives injectivity for the homomorphism in Theorem 6.3.

Let us now prove surjectivity. We will first give, as in [Mal83b], a necessary and sufficient condition for a class λ in $H^1(S^1, i^{-1}\operatorname{Aut}^{<X}(\mathscr{M}^{\text{good}}))$ to come from an object $(\mathscr{M}, \nabla, \widehat{f})$: such is the case if and only if its image in the set $H^1(S^1, i^{-1}\operatorname{Aut}_{\mathscr{A}}(\mathscr{M}^{\text{good}}))$ is the identity (we take here all the $\mathscr{A}$-linear automorphisms). Indeed, if λ comes from a bundle with connection $(\mathscr{M}, \nabla, \widehat{f})$, there exist a covering (I_i) of S^1 and isomorphisms of connections $f_i : \widetilde{\mathscr{M}} \xrightarrow{\sim} \widetilde{\mathscr{M}}^{\text{good}}$ inducing $\widehat{f}$ such that λ comes from the cocycle (λ_{ij}) with $\lambda_{ij} = f_j f_i^{-1}$ on $I_i \cap I_j$. If we fix a basis of $\mathscr{M}$ and a basis of $\mathscr{M}^{\text{good}}$, the equality of the corresponding matrices shows that (λ_{ij}) is a coboundary of $\operatorname{GL}_d(\mathscr{A}_{\widetilde{D} \times X}[*X]) = \operatorname{Aut}_{\mathscr{A}}(\widetilde{\mathscr{M}}^{\text{good}})$.

Conversely, if for some suitable covering (I_i) the cocycle (λ_{ij}) is a coboundary with values in $\operatorname{Aut}_{\mathscr{A}}(\widetilde{\mathscr{M}}^{\text{good}})$, i.e., $\lambda_{ij} = f_j f_i^{-1}$, we define a new connection ∇ on $\mathscr{M}^{\text{good}}$ by conjugating ∇^{good} by f_i on I_i. Because of the compatibility of λ_{ij} with ∇^{good}, it is globally defined on some sufficiently small neighbourhood $D \times V$ of $(0, x^o)$, hence induces a new structure of $\mathscr{O}_{D \times V}[*\{0\} \times V]$-module with connection on the $\mathscr{O}_{D \times V}[*\{0\} \times V]$-module $\mathscr{M}^{\text{good}}$. Moreover $\widehat{f_i} = \widehat{f_j}$ on $I_i \cap I_j$, so that the formal isomorphisms

$$(\widehat{\widetilde{\mathscr{M}}}^{\text{good}}, \nabla) \xrightarrow{\widehat{f_i}} (\widehat{\widetilde{\mathscr{M}}}^{\text{good}}, \nabla^{\text{good}})$$

can be glued in an isomorphism $\widehat{f} : (\widehat{\widetilde{\mathscr{M}}}^{\text{good}}, \nabla) \xrightarrow{\sim} (\widehat{\widetilde{\mathscr{M}}}^{\text{good}}, \nabla^{\text{good}})$.

Theorem 6.3 is therefore a consequence of the Malgrange-Sibuya Theorem below, for which we refer to [Mal83b, Appendice], [Sib90, Th. 6.4.1], [BV89a, Chap. 4]. $\square$

6.10 Theorem (Malgrange-Sibuya). *The image of the mapping*

$$H^1(S^1, \operatorname{GL}_d^{<X}(i^{-1}\mathscr{A}_{\widetilde{D} \times X})) \longrightarrow H^1(S^1, \operatorname{GL}_d(i^{-1}\mathscr{A}_{\widetilde{D} \times X}))$$

is the identity. $\square$

6.e Some supplementary remarks on the Stokes space

The previous considerations imply that a local knowledge of the Stokes sheaf consists mainly in knowing the fibre of this sheaf. Moreover, according to the base change property seen in Proposition 6.9, we can, for that purpose, forget about the parameter space. Therefore, we now assume that $(\mathcal{M}^{\mathrm{good}}, \nabla^{\mathrm{good}})$ is a good model in the neighbourhood of 0 in $\mathbb{C}$ and we will describe with more details the space $H^1(S^1, \mathrm{Aut}^{<0}(\mathcal{M}^{\mathrm{good}}))$.

(1) The first tool, complementing Exercise 0.5.1, is an analogue of Leray's Theorem 0.6.1 for the nonabelian H^1: let $\mathcal{G}$ be a sheaf of groups on a topological space S; if $\mathfrak{U}$ is a covering of S which is 1-acyclic for $\mathcal{G}$, i.e., such that $H^1(U, \mathcal{G}) = \{\mathrm{Id}\}$ for any open set U of $\mathfrak{U}$, then $H^1(\mathfrak{U}, \mathcal{G}) = H^1(S, \mathcal{G})$; moreover, it is enough, in order that this equality is true, that the restriction mappings $H^1(S, \mathcal{G}) \to H^1(U, \mathcal{G})$ have the identity as image (cf. for instance [BV89a, part II, Cor. 1.2.4]).

(2) Let r be the maximal order of the pole of differences $\varphi_k - \varphi_\ell$, where the φ_j are the exponential factors occurring in the model $\mathcal{M}^{\mathrm{good}}$ (cf. §6.a). Then (cf. [BV89a, part II, Prop. 3.2.3]) any open interval I' of S^1 of length $\leqslant \pi/r$, and whose boundary points are not Stokes directions for $\mathcal{M}^{\mathrm{good}}$, satisfies

$$H^1(I', \mathrm{Aut}^{<0}(\mathcal{M}^{\mathrm{good}})) = \{\mathrm{Id}\}.$$

(3) Let $I' \subset I$ be two open intervals of S^1 such that $I - I'$ does not contain any Stokes direction of $\mathcal{M}^{\mathrm{good}}$. Then the restriction mapping

$$H^1(I, \mathrm{Aut}^{<0}(\mathcal{M}^{\mathrm{good}})) \longrightarrow H^1(I', \mathrm{Aut}^{<0}(\mathcal{M}^{\mathrm{good}}))$$

is bijective (this follows from the fact that the sheaf $\mathrm{Aut}^{<0}(\mathcal{M}^{\mathrm{good}})$ is locally constant on $I - I'$).

We deduce from these results that any element of $H^1(S^1, \mathrm{Aut}^{<0}(\mathcal{M}^{\mathrm{good}}))$ can be represented by a cocycle associated to a covering of S^1 by open intervals of length $\varepsilon + \pi/r$, with $\varepsilon > 0$ sufficiently small (if I' is as in (2), we can enlarge I' by "pushing" the boundary points of I' up to the nearest Stokes directions without changing the vanishing of H^1, after (3)). Here is a way of constructing such a covering (cf. [BJL79]):

- let us call *main directions* the Stokes directions corresponding to pairs (k, ℓ) such that the difference $\varphi_k - \varphi_\ell$ has pole of maximal order r, and *secondary directions* the other ones; let us fix such a pair (k^o, ℓ^o) and a corresponding main direction θ_0^o; the other directions corresponding to (k^o, ℓ^o) are the $\theta_\alpha^o = \theta_0^o + \alpha\pi/r$ $(\alpha = 0, \ldots, 2r - 1)$;
- let us denote by I'_α the open interval of length π/r centered on the main direction θ_α^o (if both boundary points of some I'_α are (main or secondary) Stokes directions, we slightly move the center of each I'_α in the same direction);

- let us "push" the boundary points of each I'_α up to the nearest (main or secondary) directions, getting thus an open interval I_α of length $> \pi/r$;
- lastly, let us consider the covering $\mathfrak{I}$ of S^1 by the intervals I_α.

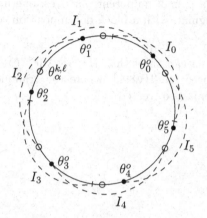

Fig. II.3. An example of a covering.

In the covering by $2r$ intervals constructed in this way, there is no three-by-three intersection, as each interval contains only one Stokes main direction θ^o_α. There are also $2r$ pairwise intersections, each of which has length $\leqslant \pi/r$, as it contains no main direction θ^o_α. Let us also notice that each interval I_α contains, for any k, ℓ, one and only one main direction of type $\theta^{k,\ell}_\beta$.

The set of 1-cocycles $Z^1(\mathfrak{I}, \mathrm{Aut}^{<0}(\mathcal{M}^{\mathrm{good}}))$ is thus the direct product of the groups $G_\alpha = H^0(I_\alpha \cap I_{\alpha+1}, \mathrm{Aut}^{<0}(\mathcal{M}^{\mathrm{good}}))$ (if we agree that $\alpha + 1 = 0$ if $\alpha = 2r-1$). Such a group G_α is made of unipotent matrices. For $i, j = 1, \ldots, p$ and $i \neq j$, the matrices of G_α have some nonzero entry in their i, j-block if and only if $\mathrm{Re}(\varphi_i - \varphi_j) < 0$ on each small sector with direction contained in $I_\alpha \cap I_{\alpha+1}$.

For each $\alpha = 0, \ldots, 2r - 1$, the group $H^0(I_\alpha, \mathrm{Aut}^{<0}(\mathcal{M}^{\mathrm{good}}))$ is made of unipotent matrices. For k, ℓ giving main directions, the k, ℓ-block of these matrices is zero, as I_α contains some Stokes direction of type k, ℓ. For $i \neq j$, the i, j-block is zero if I_α contains the corresponding Stokes direction or if $\mathrm{Re}(\varphi_i - \varphi_j) > 0$ on I_α.

6.11 Example (The Stokes space in the case of a single slope). Let us assume that all Stokes directions are main directions, i.e., that, for all i, j such that $i \neq j$, the difference $\varphi_i - \varphi_j$ has a pole of order r exactly. Then the 0-cochains of the covering $\mathfrak{I}$ reduce to the identity and we have

$$H^1(\mathfrak{I}, \mathrm{Aut}^{<0}(\mathcal{M}^{\mathrm{good}})) = Z^1(\mathfrak{I}, \mathrm{Aut}^{<0}(\mathcal{M}^{\mathrm{good}})).$$

The Stokes space is therefore a product of $2r$ algebraic subgroups of the group of unipotent matrices. It is in particular an affine space, as a group of unipotent matrices can be algebraically identified to its Lie algebra by the logarithm.

In particular, if $r = 1$, one can represent in a unique way an element of the Stokes space by a pair of unipotent matrices, one upper triangular, the other one lower triangular, with a block decomposition corresponding to the decomposition of $\mathcal{M}^{\mathrm{good}}$.

More generally, one can show that $H^1(S^1, \mathrm{Aut}^{<0}(\mathcal{M}^{\mathrm{good}}))$ has a natural structure of affine space (cf. [BV89a], where the "ramified" case, which we did not consider here, is also treated).

III

Lattices

Introduction

We have given in Chapter II a classification of germs of linear systems of one-variable differential equations, up to *meromorphic* equivalence. When the singularities are irregular, we have only considered the simplest case where the dominant part of the connection matrix is semisimple with pairwise distinct eigenvalues.

In this chapter, we will consider the *holomorphic* equivalence, that is, the classification of lattices of a (k, ∇)-vector space $(\mathcal{M}, \nabla)$.

When $\mathcal{M}$ has regular singularity, we can extend the correspondence of §II.2.e

$$(\mathcal{M}, \nabla) \longrightarrow (H, T)$$

to a correspondence "with a lattice"

$$(\mathcal{M}, \nabla, \mathcal{E}) \longrightarrow (H, T, H^\bullet)$$

where $H^\bullet$ is a decreasing filtration of H.

This correspondence shows a particularly nice behaviour when restricted to logarithmic lattices (and to filtrations stable by T on the other side): it reduces then to the Levelt normal form (cf. Exercise II.2.20) and translates in a simple way duality for lattices, or their tensor product[1].

More generally, this correspondence associates to any lattice a *characteristic polynomial*, which is nothing but that of its residue when the lattice is logarithmic. This characteristic polynomial determines that of the monodromy.

One of the key results which will be used later concerns *rigidity* of logarithmic lattices by integrable deformation.

---

[1] One will find in [Sim90] a variant of these results and in Appendix C of [EV86] a generalization of these results for the many variable logarithmic lattices.

The case of irregular singularities is much less clear in general. We however extend the notion of characteristic polynomials to any lattice.

Moreover, some classification can be done when the most polar part of the connection matrix is regular semisimple, and this classification can be extended to integrable deformations of such systems. We will follow [Mal83c] on this question.

In the following, the k- or $\mathbb{C}$-vector spaces are finite dimensional.

1 Lattices of (k, ∇)-vector spaces with regular singularity

In this section, we will assume, implicitly or explicitly, that the (k, ∇)-vector spaces we consider have regular singularity.

Recall (see §II.2.a) that a lattice of a k-vector space $\mathcal{M}$ of finite rank is a free $\mathbb{C}\{t\}$ submodule $\mathcal{E}$ such that $k \otimes_{\mathbb{C}\{t\}} \mathcal{E} = \mathcal{M}$. A lattice is logarithmic (relative to the connection ∇) if it is stable by $t\nabla_{\partial/\partial t}$.

Given a logarithmic lattice, the residue of the connection may not be enough to reconstruct monodromy, when some pair of eigenvalues differ by a nonzero integer (cf. Exercise II.2.20-(4)). In other words, the correspondence which associates to any logarithmic lattice $\mathcal{E}$ of $(\mathcal{M}, \nabla)$ the pair consisting of $\mathcal{E}/t\mathcal{E}$ and $\operatorname{Res} \nabla$ is not an equivalence of categories. The classification of these lattices will introduce auxiliary lattices, to which one compares the given lattice.

1.a Classification of logarithmic lattices

In order to formulate the classification statement, let us fix a canonical lattice (cf. §II.2.e) by choosing σ as being the section of $\mathbb{C} \to \mathbb{C}/\mathbb{Z}$ with values in

$$\{s \in \mathbb{C} \mid -1 < \operatorname{Re}(s) \leqslant 0\}.$$

We will denote by $\mathcal{V}$ the corresponding canonical lattice.

We have seen that the functor which associates to any (k, ∇)-vector space $(\mathcal{M}, \nabla)$ the pair

$$(H, T) = (\mathcal{V}/t\mathcal{V}, \exp(-2i\pi \operatorname{Res}_\mathcal{V} \nabla)),$$

consisting of a vector space and of an automorphism, is an equivalence. How to recover lattices in this description? If $\mathcal{E}$ is a lattice of $(\mathcal{M}, \nabla)$, we will set $\mathcal{E}^k = t^k \mathcal{E}$ for any $k \in \mathbb{Z}$.

1.1 Theorem (The Deligne-Malgrange functor is an equivalence).
The functor F which, to any lattice $\mathcal{E}$ of $(\mathcal{M}, \nabla)$, associates the vector space $H = \mathcal{V}/t\mathcal{V}$, equipped with the endomorphism $T = \exp(-2i\pi \operatorname{Res}_\mathcal{V} \nabla)$ and with the decreasing exhaustive filtration $H^\bullet$ indexed by $\mathbb{Z}$, defined by

$$H^k = \mathcal{E}^k \cap \mathcal{V}/\mathcal{E}^k \cap t\mathcal{V}$$

*is essentially surjective. When restricted to logarithmic lattices and to filtra-
tions stable by T, it induces an equivalence of categories.*

Therefore, giving $(\mathcal{M}, \nabla)$ is equivalent to giving (H, T). The supplementary
datum of a logarithmic lattice $\mathcal{E}$ is equivalent to giving a decreasing filtration
of H stable by T. If $\mathcal{E} = \mathcal{V}$, we have $H^0 = H$ and $H^1 = 0$. Let us notice that
the filtration is *exhaustive* (i.e., $H^k = 0$ for $k \gg 0$ and $H^k = H$ for $k \ll 0$):
indeed, as $\mathcal{E}$ and $\mathcal{V}$ are two lattices of $\mathcal{M}$, they can be compared, that is, there
exist k_0 such that $\mathcal{E}^{k_0} \supset \mathcal{V}$ and k_1 such that $\mathcal{E}^{k_1} \subset t\mathcal{V}$.

Proof. Let us denote by R the endomorphism of H such that $\exp(-2i\pi R) = T$
and whose eigenvalues are in the image of the section σ. Let us set $\mathbb{M} =
\mathbb{C}[t, t^{-1}] \otimes_{\mathbb{C}} H$, equipped with the connection ∇ such that $t\nabla_{\partial_t}(h \otimes p) =
p \otimes R(h) + tp'(t) \otimes h$, for $h \in H$ and $p(t) \in \mathbb{C}[t, t^{-1}]$. Let $H^\bullet$ be a decreasing
exhaustive filtration of H, and let us denote by $\mathbb{E}$ the vector subspace of $\mathbb{M}$
defined by

$$\mathbb{E} = \bigoplus_{k \in \mathbb{Z}} t^{-k} H^k.$$

As $H^\bullet$ is decreasing, $\mathbb{E}$ is a $\mathbb{C}[t]$-submodule of $\mathbb{M}$. Moreover, as $H^\bullet$ is exhaustive, $\mathbb{E}$ is a $\mathbb{C}[t]$-free lattice of $\mathbb{M}$, a basis of which over $\mathbb{C}[t]$ can be obtained
by choosing, for any $\ell \in \mathbb{Z}$, a family e_ℓ in H^ℓ inducing a basis of the quotient
$H^\ell/H^{\ell+1}$ and by considering the family $(t^{-\ell}e_\ell)_{\ell \in \mathbb{Z}}$.

We define in that way a functor $G : (H, T, H^\bullet) \to (\mathcal{M}, \nabla, \mathcal{E})$ by setting

$$(\mathcal{M}, \nabla) = \big(k \otimes_{\mathbb{C}[t,t^{-1}]} \mathbb{M}, \nabla\big) = \big(k \otimes_{\mathbb{C}} H, \nabla\big) \quad \text{and} \quad \mathcal{E} = \mathbb{C}\{t\} \otimes_{\mathbb{C}[t]} \mathbb{E}.$$

By definition, the filtration $^\sigma V_\bullet \mathbb{M}$ is given by $^\sigma V_k \mathbb{M} = t^k \mathbb{C}[t] \otimes_{\mathbb{C}} H$, and
$^\sigma V_k \mathcal{M} = \mathbb{C}\{t\} \otimes_{\mathbb{C}[t]} {}^\sigma V_k \mathbb{M}$. We deduce that

$$F \circ G = \mathrm{Id}.$$

This shows that the functor F is essentially surjective.

If H^k is stable by R (or T) for any k, then $\mathcal{E}$ is logarithmic. Similarly, if $\mathcal{E}$
is logarithmic, then the filtration $H^\bullet$ associated by F is stable by R (or T).

1.2 Lemma. *Let $\mathcal{E}$ be a logarithmic lattice of $(\mathcal{M}, \nabla) = (k \otimes_{\mathbb{C}} H, \nabla)$. There exists a unique decreasing filtration $H^\bullet$ of H, stable by T, such that $(\mathcal{M}, \nabla, \mathcal{E}) =
G(H, T, H^\bullet)$. We have $(H, T, H^\bullet) = F(\mathcal{M}, \nabla, \mathcal{E})$.*

Proof. The second point and the uniqueness of the filtration $H^\bullet$ are clear, as
we have $F \circ G = \mathrm{Id}$. The existence of $H^\bullet$ is nothing but a reformulation of the
existence of the Levelt normal form (Exercise II.2.20). Indeed, from any basis
of $\mathcal{E}$, we can construct (II.2.20-(3)) a basis ε of $\mathcal{M}$ in which the connection
matrix is written as

$$A\frac{dt}{t} = \big[(A_0 - D) + A_1 + \cdots + A_\delta\big]\frac{dt}{t},$$

where D is diagonal with integral eigenvalues, $[D, A_i] = -iA_i$ for any i, and $A_0 - D$ has eigenvalues in the image of σ (the choice of σ is different here than that of Exercise II.2.20, but the argument is the same). If E_j is the eigenspace of D with eigenvalue j, we have $A_i(E_j) \subset E_{j-i}$ for any i, hence A_i is nilpotent for $i \geqslant 1$. It follows that the eigenvalues of A are that of $A_0 - D$. As $(A_0 - D)_{|E_j}$ has eigenvalues in the image of σ, the basis ε is contained in H; it is thus a $\mathbb{C}$-basis of H and A is the matrix of R in this basis. Let us define the decreasing filtration $H^\bullet$ by the formula

$$H^k = \bigoplus_{j \leqslant -k} E_j.$$

This filtration is stable by A, and the basis ε is adapted to the direct sum decomposition $H = \bigoplus_j E_j$. Still according to Exercise II.2.20, we know that the basis $e = \varepsilon \cdot t^D$ of $\mathbb{C}[t, t^{-1}] \otimes_{\mathbb{C}} H$ is a basis of $\mathcal{E}$. This gives $\mathcal{E}$ the desired form. $\qquad \square$

Full faithfulness of the functor F is now clear: set $(H, T, H^\bullet) = F(\mathcal{M}, \nabla, \mathcal{E})$ and $(H', T', H'^\bullet) = F(\mathcal{M}', \nabla', \mathcal{E}')$; if $f : (H, T, H^\bullet) \to (H', T', H'^\bullet)$ is a morphism, there exists, by the Riemann-Hilbert correspondence, a unique morphism $\varphi : (\mathcal{M}, \nabla) \to (\mathcal{M}', \nabla')$ which induces it; it is a matter of proving that φ sends $\mathcal{E}$ into $\mathcal{E}'$; this is immediate, as $\varphi = G(f)$. More precisely, the inverse functor of F is G. $\qquad \square$

1.3 Remarks.

(1) One should not confuse "filtration" and "flag": a filtration consists in giving a flag of vector subspaces and, for any space of the flag, in giving an integer, the set of these defining a decreasing sequence.

(2) If one does not restrict to logarithmic lattices, the functor F is not an equivalence. Let us assume for instance that H has dimension 2, with basis $\varepsilon_1, \varepsilon_2$, and let $\mathcal{E}$ be the lattice of $\mathcal{M} = k \otimes H$ generated by $(e_1, e_2) = (\varepsilon_1, \varepsilon_2) \cdot P(t)$, where $P(t) = \left(\begin{smallmatrix} t & 1 \\ 0 & t \end{smallmatrix} \right)$. If the lattice $\mathcal{E}$ took the form $G(H^\bullet)$, it would admit a basis of the form $(t^{d_1} \varepsilon_1', t^{d_2} \varepsilon_2')$, where $(\varepsilon_1', \varepsilon_2')$ is a basis of H. There would exist thus a holomorphic invertible matrix $Q(t)$, a constant invertible matrix C and a diagonal matrix $D = \mathrm{diag}(d_1, d_2)$, such that $P(t) = C \cdot t^D \cdot Q(t)$. One checks that this is not possible, hence there does not exist a filtration $H^\bullet$ such that $(\mathcal{M}, \nabla, \mathcal{E})$ is isomorphic to $G(H, T, H^\bullet)$.

1.4 Remark (dependence with respect to the section σ). The choice of the section σ with image $\{s \in \mathbb{C} \mid -1 < \mathrm{Re}(s) \leqslant 0\}$ has only been done for convenience. What is the relation between the functors ${}^\sigma F$ for distinct sections σ? Let us notice first that we have a *canonical* isomorphism[2]

(1.5) $$\qquad {}^\sigma H = {}^\sigma \mathcal{V} / t\, {}^\sigma \mathcal{V} \xrightarrow{\ \sim\ } {}^{\sigma'} \mathcal{V} / t\, {}^{\sigma'} \mathcal{V} = {}^{\sigma'} H$$

---

[2] Here, we assume that we have fixed a coordinate t. A change of coordinate would give a nonzero multiple of this isomorphism.

compatible with the action of T. Indeed, Remark II.2.22 shows that the terms are canonically identified to $\bigoplus_{\alpha \in \mathrm{Im}\, \sigma} M^\alpha$ and $\bigoplus_{\alpha' \in \mathrm{Im}\, \sigma'} M^{\alpha'}$. For any $\alpha' \in \mathrm{Im}\, \sigma'$, there exists a unique $\alpha \in \mathrm{Im}\, \sigma$ and a unique $k \in \mathbb{Z}$ such that $\alpha' = \alpha + k$. The isomorphism $M^\alpha \xrightarrow{\sim} M^{\alpha'}$ is that induced by multiplication by t^k.

If $\mathcal{E}$ is a lattice of M, let us denote by $^\sigma H^\bullet$ and $^{\sigma'} H^\bullet$ the filtrations it defines on $^\sigma H$ and $^{\sigma'} H$. A priori, the isomorphism (1.5) does not preserve filtrations. However, when the lattice $\mathcal{E}$ is logarithmic, one can describe precisely the behaviour of filtrations.

Indeed, as $^\sigma H^\bullet$ and $^{\sigma'} H^\bullet$ are stable by T, we have

$$^\sigma H^\bullet = \bigoplus_{\alpha \in \mathrm{Im}\, \sigma} {}^\sigma H^\bullet_\alpha \quad \text{and} \quad {}^{\sigma'} H^\bullet = \bigoplus_{\alpha' \in \mathrm{Im}\, \sigma'} {}^{\sigma'} H^\bullet_{\alpha'}.$$

We then see that, if $\alpha' = \alpha + k$ $(k \in \mathbb{Z})$, the image of $^\sigma H^\bullet_\alpha$ by (1.5) is $^{\sigma'} H^{\bullet + k}_{\alpha'}$.

Let us consider for instance the case of sections σ' and σ with respective images $\{s \mid -1 < \mathrm{Re}(s) \leqslant 0\}$ and $\{s \mid 0 \leqslant \mathrm{Re}(s) < 1\}$, which will occur later on. We can write $^\sigma H = {}^\sigma H_{\neq 1} \oplus {}^\sigma H_1$ and similarly for $^{\sigma'} H$, where the index denotes some eigenvalue of T (and not its logarithm). Then the isomorphism (1.5) is induced by multiplication by t^{-1} on $^\sigma H_{\neq 1}$ and is equal to the identity on $^\sigma H_1 = {}^{\sigma'} H_1$. Therefore, the image by (1.5) of $^\sigma H^\bullet_{\neq 1}$ is $^{\sigma'} H^{\bullet -1}_{\neq 1}$ and that of $^\sigma H^\bullet_1$ is $^{\sigma'} H^\bullet_1$.

Let us give more details concerning the dictionary between lattices and filtrations for a (k, ∇)-vector space (M, ∇) with regular singularity. Let $\mathcal{E}$ be a lattice of M obtained through the functor G: $(M, \nabla, \mathcal{E}) = G(H, T, H^\bullet)$. Let R be the logarithm of $T : H \to H$ whose eigenvalues are in the image of σ, let R_s be its semisimple part and R_n its nilpotent part.

1.6 Proposition.

(1) The lattice $\mathcal{E}$ has order $\leqslant r$ if and only if we have $R(H^k) \subset H^{k-r}$ for all k.

(2) The lattice $\mathcal{E}$ is logarithmic if and only if the filtration is stable by R (or T). If such is the case, the residue $\mathrm{Res}_\mathcal{E} \nabla$ is semisimple if and only if we have $R_n(H^k) \subset H^{k+1}$ for any k.

Proof. The connection on $\mathbb{E}$ can be written as

$$\nabla \left(\sum_k t^k e_{-k} \right) = \left[\sum_k k t^k e_{-k} + \sum_k t^k R e_{-k} \right] \frac{dt}{t}.$$

We deduce the two first assertions. For the last one, we identify $\mathbb{E}/t\mathbb{E}$ with the direct sum $\bigoplus_k (H^k/H^{k+1})$. With this identification, the operator $\mathrm{Res}_\mathcal{E} \nabla$ acts as $R - k\,\mathrm{Id}$ on H^k/H^{k+1}. It is thus semisimple if and only if R_n induces the zero endomorphism on H^k/H^{k+1} for any k. $\qquad \square$

1.b Behaviour with respect to duality

Let us denote by $\mathcal{M}^* = \mathrm{Hom}_{\boldsymbol{k}}(\mathcal{M}, \boldsymbol{k})$ the dual space of the $\boldsymbol{k}$-vector space $\mathcal{M}$, equipped with its natural connection ∇^* (cf. §0.11.b). If $\mathcal{E}$ is a lattice of $\mathcal{M}$, let us denote by $\mathcal{E}^* = \mathrm{Hom}_{\mathbb{C}\{t\}}(\mathcal{E}, \mathbb{C}\{t\})$ the dual module. We can identify in an evident way $\mathcal{E}^*$ to the set of $\varphi \in \mathcal{M}^*$ which send $\mathcal{E}$ into $\mathbb{C}\{t\}$. This shows that $\mathcal{E}^*$ is naturally a lattice of $\mathcal{M}^*$.

Let us apply this to the Deligne lattice $\mathcal{V}$. The residue of ∇^* on $\mathcal{V}^*$ is equal to $-\,^t\mathrm{Res}_{\mathcal{V}} \nabla$ (cf. Exercise 0.14.5), so that its eigenvalues have their real part in $[0, 1[$.

The natural mapping

$$\mathcal{V} \underset{\mathbb{C}\{t\}}{\otimes} \mathcal{V}^* \longrightarrow \mathbb{C}\{t\}$$

$$e \otimes \varphi \longmapsto \varphi(e)$$

induces a nondegenerate $\mathbb{C}$-bilinear mapping

$$(\mathcal{V}/t\mathcal{V}) \underset{\mathbb{C}}{\otimes} (\mathcal{V}^*/t\mathcal{V}^*) \longrightarrow \mathbb{C} = \mathbb{C}\{t\}/t\,\mathbb{C}\{t\},$$

whence an isomorphism

$$(1.7) \qquad (\mathcal{V}^*/t\mathcal{V}^*, T^*) \simeq (\mathrm{Hom}_{\mathbb{C}}(\mathcal{V}/t\mathcal{V}, \mathbb{C}), \,^tT^{-1}).$$

On the other hand, the isomorphism (1.5) identifies the pair $(\mathcal{V}^*/t\mathcal{V}^*, T^*)$ with the pair $(\mathcal{V}(\mathcal{M}^*)/t\mathcal{V}(\mathcal{M}^*), T)$. We thus have a canonical identification (if we still denote by F the functor $(\mathcal{M}, \nabla) \mapsto (H, T)$, and by H^* the dual of H)

$$F(\mathcal{M}^*, \nabla^*) = (H^*, \,^tT^{-1}).$$

We can now extend this identification by introducing a lattice $\mathcal{E}$. However, as we have to use the isomorphism (1.5), we assume that $\mathcal{E}$ is logarithmic. Here, the choice of the section σ simplifies the presentation of the result. As in Remark 1.4, let us denote by H_λ the generalized eigenspace of T for some eigenvalue λ and $H_{\neq 1} = \bigoplus_{\lambda \neq 1} H_\lambda$.

1.8 Proposition (The Deligne-Malgrange functor and duality).

Let $\mathcal{E}$ be a logarithmic lattice of $(\mathcal{M}, \nabla)$. Then the canonical identification $F(\mathcal{M}^*, \nabla^*) = (H^*, \,^tT^{-1})$ can be extended to a canonical identification

$$F(\mathcal{M}^*, \nabla^*, \mathcal{E}^*) = (H^*, \,^tT^{-1}, H'^\bullet)$$

with $H'^k_{\neq 1} = (H^{-k}_{\neq 1})^\perp$ and $H'^k_1 = (H^{-k+1}_1)^\perp$.

Proof. Let us first show that, without assuming that the lattice $\mathcal{E}$ is logarithmic,

$$(1.9) \qquad \forall k \in \mathbb{Z}, \quad \{\varphi \in \mathcal{M}^* \mid \varphi(\mathcal{E}^{-k} \cap \mathcal{V}) \subset t\,\mathbb{C}\{t\}\} = (t^{k+1}\mathcal{E}^*) + t\mathcal{V}^*.$$

The inclusion $\supset$ is immediate. In order to show the other one, let us choose, for any $\ell \in \mathbb{Z}$, a basis of the $\mathbb{C}$-vector space

$$\frac{\mathcal{E}^\ell \cap \mathcal{V}}{(\mathcal{E}^{\ell+1} \cap \mathcal{V}) + (\mathcal{E}^\ell \cap t\mathcal{V})}$$

and let us lift it as a family $e_\ell \subset \mathcal{E}^\ell \cap \mathcal{V}$. The (finite) union of the e_ℓ forms a $\mathbb{C}\{t\}$-basis of $\mathcal{V}$, after Nakayama's lemma.

Let φ be such that $\varphi(\mathcal{E}^{-k} \cap \mathcal{V}) \subset t\mathbb{C}\{t\}$. As, for $\ell \geqslant -k$, we have $\mathcal{E}^\ell \cap \mathcal{V} \subset \mathcal{E}^{-k} \cap \mathcal{V}$, it follows that $\varphi(e_\ell) \subset t\mathbb{C}\{t\}$. For $\ell < -k$, we have

$$\mathcal{E}^\ell \cap \mathcal{V} = (t^{\ell+k}\mathcal{E}^{-k}) \cap \mathcal{V} = t^{\ell+k}(\mathcal{E}^{-k} \cap t^{-(\ell+k)}\mathcal{V}) \subset t^{\ell+k}(\mathcal{E}^{-k} \cap \mathcal{V})$$

and therefore $\varphi(e_\ell) \subset t^{\ell+k+1}\mathbb{C}\{t\}$.

We then write $\varphi = \psi + \eta$, where

$$\psi(e_\ell) = \begin{cases} \varphi(e_\ell) & \text{if } \ell + k \geqslant 0 \\ 0 & \text{if } \ell + k < 0 \end{cases} \quad \text{and} \quad \eta(e_\ell) = \begin{cases} 0 & \text{if } \ell + k \geqslant 0 \\ \varphi(e_\ell) & \text{if } \ell + k < 0. \end{cases}$$

We have $\psi(\mathcal{V}) \subset t\mathbb{C}\{t\}$, hence $\psi \in t\mathcal{V}^*$. On the other hand, we see, still using Nakayama's lemma, that the union of the $t^m e_{-k-m}$ ($m \in \mathbb{Z}$) forms a $\mathbb{C}\{t\}$-basis of $\mathcal{E}^{-k}$. We have $\eta(t^m e_{-k-m}) = 0$ for $m \leqslant 0$ and, for $m > 0$, we have

$$\eta(t^m e_{-k-m}) = \varphi(t^m e_{-k-m}) = t^m \varphi(e_{-k-m}) \subset t^m \cdot t^{-m+1}\mathbb{C}\{t\},$$

which proves (1.9).

The equality (1.9) exactly means that $(\mathcal{E}^*)^{k+1} \cap \mathcal{V}^*/(\mathcal{E}^*)^{k+1} \cap t\mathcal{V}^*$ is the orthogonal space of the space $\mathcal{E}^{-k} \cap \mathcal{V}/\mathcal{E}^{-k} \cap t\mathcal{V}$ through duality (1.7).

When $\mathcal{E}$ is logarithmic, these spaces are stable under the action of monodromy and can be decomposed relatively to its eigenvalues. Remark 1.4 gives the conclusion. $\qquad\square$

1.10 Definition (of a bilinear form). A *bilinear form* on $(\mathcal{M}, \nabla)$ is a k-linear mapping

$$\langle\, ,\, \rangle : \mathcal{M} \underset{k}{\otimes} \mathcal{M} \longrightarrow k$$

compatible with the connection, that is, such that, for all $e, e' \in \mathcal{M}$ we have

$$\nabla\langle e, e' \rangle = \langle \nabla e, e' \rangle + \langle e, \nabla e' \rangle.$$

Giving such a form is equivalent to giving a morphism of (k, ∇)-vector spaces

$$\mathcal{M} \longrightarrow \mathcal{M}^*, \qquad e \longmapsto (e' \mapsto \langle e, e' \rangle).$$

The bilinear form is said to be *nondegenerate* if and only if this morphism is an isomorphism.

1.11 Exercise. The category whose objects are the $(\boldsymbol{k}, \nabla)$-vector spaces $(\mathcal{M}, \nabla)$ with regular singularity equipped with a bilinear form $\langle\,,\,\rangle$ (resp. nondegenerate, resp. symmetric, etc.) and whose morphisms are those which are compatible with bilinear forms, is equivalent to the category whose objects are the $\mathbb{C}$-vector spaces H equipped with a bilinear form $\langle\,,\,\rangle$ (resp. nondegenerate, resp. symmetric, etc.) and with a self-adjoint automorphism T, and whose morphisms...

Let now $\mathcal{E}$ be a lattice of $\mathcal{M}$. We will say that a nondegenerate bilinear form $\langle\,,\,\rangle$ on $(\mathcal{M}, \nabla)$ has *weight* $w \in \mathbb{Z}$ *relatively to the lattice* $\mathcal{E}$ if the corresponding isomorphism $\mathcal{M} \xrightarrow{\sim} \mathcal{M}^*$ sends $\mathcal{E}$ into $(\mathcal{E}^*)^w = t^w \mathcal{E}^*$. We deduce a nondegenerate bilinear form on $\mathcal{E}/t\mathcal{E}$, obtained by composing the isomorphisms

$$\mathcal{E}/t\mathcal{E} \xrightarrow{\sim} t^w \mathcal{E}^*/t^{w+1}\mathcal{E}^* \xrightarrow{\sim} \mathcal{E}^*/t\mathcal{E}^* = \mathrm{Hom}_{\mathbb{C}}(\mathcal{E}/t\mathcal{E}, \mathbb{C}).$$

If $\mathcal{E}$ is logarithmic, the residue of ∇ on $\mathcal{E}$ satisfies then

$$\mathrm{Res}\,\nabla + {}^t\mathrm{Res}\,\nabla = w\,\mathrm{Id}.$$

We deduce from Proposition 1.8 and from Exercise 1.11 above:

1.12 Corollary (With a bilinear form, the Deligne-Malgrange functor is an equivalence). *The functor F induces an equivalence between*

(1) *the category of objects $(\mathcal{M}, \nabla, \mathcal{E}, \langle\,,\,\rangle)$ consisting of a $(\boldsymbol{k}, \nabla)$-vector space with regular singularity, of a logarithmic lattice and of a nondegenerate bilinear form of weight w relatively to the lattice;*

(2) *the category of objects $(H, T, H^\bullet, \langle\,,\,\rangle)$ consisting of a $\mathbb{C}$-vector space equipped with a nondegenerate bilinear form, of a self-adjoint automorphism T and of a decreasing filtration $H^\bullet$ stable by T such that, for any $k \in \mathbb{Z}$, we have (denoting by $H^\perp$ the orthogonal space of H with respect to $\langle\,,\,\rangle$)*

$$\left(H_{\neq 1}^k\right)^\perp = H_{\neq 1}^{-k-w} \quad and \quad \left(H_1^k\right)^\perp = H_1^{-k-w+1}. \qquad \square$$

1.13 Exercise (With a sesquilinear form, the Deligne-Malgrange functor is an equivalence). We fix some coordinate t and we consider the automorphism of the field $\boldsymbol{k}$ defined by $f(t) \mapsto f(-t)$. If $\mathcal{M}$ is a $\boldsymbol{k}$-vector space, we define the $\boldsymbol{k}$-vector space $\overline{\mathcal{M}}$ by modifying the action of $\boldsymbol{k}$ on $\mathcal{M}$: if $f(t) \in \boldsymbol{k}$ and $m \in \mathcal{M}$ we set $f(t) \cdot m = f(-t)m$. Therefore, $\overline{\mathcal{M}}$ and $\mathcal{M}$ coincide as $\mathbb{C}$-vector spaces. We will denote by $\overline{m}$ the element m "regarded in $\overline{\mathcal{M}}$". In other words, we have $f(t)\overline{m} = f(-t)m$. We define similarly $\overline{\mathcal{E}}$ if $\mathcal{E}$ is a lattice of $\mathcal{M}$. If ∇ is a connection on $\mathcal{M}$, we set $\overline{\nabla}(\overline{m}) = -\nabla(m)$.

- Prove that $\overline{\mathcal{V}}$ is the canonical lattice of $\overline{\mathcal{M}}$, that $H = \mathcal{V}/t\mathcal{V} = \overline{\mathcal{V}}/t\overline{\mathcal{V}}$ and that $\mathrm{Res}_{\overline{\mathcal{E}}}\,\overline{\nabla} = \mathrm{Res}_{\mathcal{E}}\,\nabla$.
- Show that the filtrations associated to $\mathcal{E}$ and $\overline{\mathcal{E}}$ coincide.

A *sesquilinear form*[3] on $(\mathcal{M}, \nabla)$ is a k-linear mapping

$$\langle \, , \, \rangle : \mathcal{M} \underset{k}{\otimes} \overline{\mathcal{M}} \longrightarrow k$$

compatible with the connections.

- Prove that Corollary 1.12 still holds if one replaces in (1) "bilinear form" with "sesquilinear form".

1.c Characteristic polynomial of a lattice at the origin

Characteristic polynomial of a logarithmic lattice. Let $\mathcal{E}$ be a logarithmic lattice of $(\mathcal{M}, \nabla)$. The characteristic polynomial of the residue $\mathrm{Res}_{\mathcal{E}} \nabla$ can be computed with the filtration $H^\bullet$ associated to the lattice: this filtration is stable by $A = \mathrm{Res}_{\mathcal{V}} \nabla$; in a similar way, the filtration induced by $\mathcal{V}^\ell = t^\ell \mathcal{V}$ on $\mathcal{E}/t\mathcal{E}$ is stable by $\mathrm{Res}_{\mathcal{E}} \nabla$. Multiplication by t^k identifies the spaces

$$\frac{\mathcal{V}^{-k} \cap \mathcal{E}}{(\mathcal{V}^{-k+1} \cap \mathcal{E}) + (\mathcal{V}^{-k} \cap t\mathcal{E})} \quad \text{and} \quad \frac{\mathcal{E}^k \cap \mathcal{V}}{(\mathcal{E}^{k+1} \cap \mathcal{V}) + (\mathcal{E}^k \cap t\mathcal{V})}$$

and is compatible with $\mathrm{Res}_{\mathcal{E}} \nabla$ (on the left) and $\mathrm{Res}_{\mathcal{V}} \nabla$ (on the right); lastly, the left-hand term above is the subspace of $\mathcal{E}/t\mathcal{E}$ corresponding to eigenvalues of $\mathrm{Res}_{\mathcal{E}} \nabla$ with real part contained in $\,]-k-1, -k]$.

1.14 Definition (of the characteristic polynomial of a logarithmic lattice). The *characteristic polynomial* $\chi_{\mathcal{E}}(s)$ of the logarithmic lattice $\mathcal{E}$ is the characteristic polynomial $\det(s \,\mathrm{Id} - \mathrm{Res}_{\mathcal{E}} \nabla)$ of the residue of ∇ on $\mathcal{E}$.

We can thus write

$$\chi_{\mathcal{E}}(s) = \prod_{\beta} (s - \beta)^{\nu_\beta}$$

with

(1.15) $\nu_\beta = \dim H_\lambda^k / H_\lambda^{k+1}$ where $\lambda = \exp(-2i\pi\beta)$ and $[-\beta] = k$,

if we denote by $[-\beta]$ the integral part of the real part of $-\beta$ and where H_λ^k is as in §1.b.

General case. When the lattice $\mathcal{E}$ is not logarithmic, the residue is not defined in general (see Remark 0.14.8). We can however take Formula (1.15) as an model to define a characteristic polynomial $\chi_{\mathcal{E}}$. It is nevertheless necessary to give a meaning to the spaces H_λ^k when H^k is not stable by T.

In order to do so, let us begin, when the eigenvalues of T do not have modulus equal to 1, by choosing a total order $\leqslant$ on $\mathbb{C}$ subject to both conditions below:

---

[3] The need of such a notion will be clear in the chapter devoted to Fourier transform, cf. in particular Formula V.(2.5).

- it extends the usual order on $\mathbb{R}$ and
- it satisfies the property $\alpha \leqslant \beta \Leftrightarrow \forall k \in \mathbb{Z},\ \alpha + k \leqslant \beta + k$.

We can for instance use the lexicographic order

$$x + iy \leqslant x' + iy' \iff x < x' \text{ or } x = x' \text{ and } y \leqslant y'.$$

Let us fix such an order. We will denote by $]a, b]$ the subset of $\mathbb{C}$ consisting of the complex numbers s such that $a < s \leqslant b$.

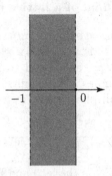

Fig. III.1. The set $]-1, 0]$.

We can then filter H in a decreasing way by setting, for $\alpha \in\ ]-1, 0]$,

$$H^{\alpha} = \bigoplus_{\beta \geqslant \alpha} H_{\exp -2i\pi\beta}.$$

In an analogous way, we can interpolate the filtration $\mathcal{V}^{\bullet}$ by taking, for any $\beta \in \mathbb{C}$, the lattice $\mathcal{V}^{\beta}$ corresponding to the section σ with image $]\beta - 1, \beta]$.

For any $\alpha \in\ ]-1, 0]$, we identify the space $H_{\exp -2i\pi\alpha}$ to the quotient space $\mathcal{V}^{\alpha}/\mathcal{V}^{>\alpha}$, by setting $\mathcal{V}^{>\alpha} = \cup_{\beta>\alpha}\mathcal{V}^{\beta}$.

With the lattice $\mathcal{E}$ we define a decreasing exhaustive filtration of each subspace H_{λ} by setting, for any $k \in \mathbb{Z}$,

$$H_{\lambda}^{k} := (\mathcal{E}^{k} \cap \mathcal{V}^{\alpha})/(\mathcal{E}^{k} \cap \mathcal{V}^{>\alpha}) = (H^{k} \cap H^{\alpha})/(H^{k} \cap H^{>\alpha})$$

(by choosing $\alpha \in\ ]-1, 0]$ such that $\lambda = \exp(-2i\pi\alpha)$ and taking care of the ambiguity of notation in the right-hand term).

The characteristic polynomial $\chi_{\mathcal{E}}(s)$ is then *defined* by Formula (1.15).

1.16 Definition (of the characteristic polynomial of a lattice). The characteristic polynomial $\chi_{\mathcal{E}}(s)$ of the lattice $\mathcal{E}$ is equal to

$$\chi_{\mathcal{E}}(s) = \prod_{\beta \in \mathbb{C}} (s - \beta)^{\nu_{\beta}} \quad \text{with} \quad \nu_{\beta} = \dim \frac{\mathcal{E} \cap \mathcal{V}^{\beta}}{(\mathcal{E} \cap \mathcal{V}^{>\beta}) + (t\mathcal{E} \cap \mathcal{V}^{\beta})}.$$

1.17 Remark. The polynomial $\chi_{\mathcal{E}}(s)$ determines the characteristic polynomial of the monodromy T, which is equal to $\prod_{\beta}(S - \exp(-2i\pi\beta))^{\nu_{\beta}}$.

1.18 Exercise (Behaviour of the characteristic polynomial by duality). Let $(\mathcal{M}^*, \nabla^*)$ be the (k, ∇)-vector space dual to $(\mathcal{M}, \nabla)$.

(1) Prove that, for any $\beta \in \mathbb{C}$, we have

$$V^{\beta}(\mathcal{M}^*) = \mathcal{H}om_{\mathbb{C}\{t\}}(V^{>-(\beta+1)}, \mathbb{C}\{t\}).$$

(2) Deduce that the bilinear form, composition of both mappings

$$V^{\beta}(\mathcal{M}^*) \underset{\mathbb{C}\{t\}}{\otimes} V^{-(\beta+1)}(\mathcal{M}) \longrightarrow t^{-1}\mathbb{C}\{t\} \xrightarrow{\text{Residue}} \mathbb{C}$$

induces a nondegenerate bilinear form

$$\left[V^{\beta}(\mathcal{M}^*)/V^{>\beta}(\mathcal{M}^*)\right] \underset{\mathbb{C}}{\otimes} \left[V^{-(\beta+1)}(\mathcal{M})/V^{>-(\beta+1)}(\mathcal{M})\right] \longrightarrow \mathbb{C}.$$

(3) Show that the orthogonal space of $t^{-1}\mathcal{E} \cap V^{-(\beta+1)}(\mathcal{M})$ is equal to

$$[\mathcal{E}^* \cap V^{>\beta}(\mathcal{M}^*)] + [t\mathcal{E}^* \cap V^{\beta}(\mathcal{M}^*)]$$

and deduce a nondegenerate bilinear form

$$\frac{\mathcal{E}^* \cap V^{\beta}(\mathcal{M}^*)}{[\mathcal{E}^* \cap V^{>\beta}(\mathcal{M}^*)] + [t\mathcal{E}^* \cap V^{\beta}(\mathcal{M}^*)]}$$
$$\underset{\mathbb{C}}{\otimes} \frac{t^{-1}\mathcal{E} \cap V^{-(\beta+1)}(\mathcal{M})}{[\mathcal{E} \cap V^{-(\beta+1)}(\mathcal{M})] + [t^{-1}\mathcal{E} \cap V^{>-(\beta+1)}(\mathcal{M})]}$$
$$\longrightarrow \mathbb{C}.$$

(4) Deduce that $\chi_{\mathcal{E}^*}(s) = (-1)^d\chi_{\mathcal{E}}(-s)$, where $d = \dim_k \mathcal{M}$.
(5) Deduce that, if there exists a nondegenerate bilinear form on $(\mathcal{M}, \nabla)$ of weight w relatively to $\mathcal{E}$, we have $\chi_{\mathcal{E}}(s) = (-1)^d\chi_{\mathcal{E}}(w - s)$.

Prove the analogous statements for $\overline{\mathcal{M}}^*$.

Remark (Various notions of exponents). There exists other ways to associate "exponents" to a lattice $\mathcal{E}$. Let us indicate that of Levelt [Lev61]. E. Corel [Cor99] has shown that these exponents correspond to the characteristic polynomial of the residue of ∇ acting on *the largest logarithmic sublattice of $\mathcal{E}$*. It could also be possible to consider the smallest logarithmic lattice containing $\mathcal{E}$, or also, with Gérard and Levelt [GL76], the lattice obtained from $\mathcal{E}$ by saturating by $t\nabla_{\partial/\partial t}$. These exponents do not necessarily satisfy the duality property of Exercise 1.18, but in loc. cit. E. Corel has shown for them an inequality which enables one, when one applies it to a lattice on $\mathbb{P}^1$, to obtain inequalities analogous to that proved by R. Fuchs for differential equations.

1.d Rigidity of logarithmic lattices

The proposition which follows shows that, locally, logarithmic lattices do not produce any interesting theory of integrable deformations. This statement happens however to be very useful in order to transform local problems into global, hence algebraic, problems.

1.19 Proposition (Rigidity, [Mal83c, Th. 2.1], [Mal86]). *Let (E^o, ∇^o) be a bundle on a disc D equipped with a connection having a logarithmic pole at the origin. Let X be a 1-connected analytic manifold with basepoint x^o. There exists then a unique (up to unique isomorphism) bundle E on $D \times X$ with a flat connection having logarithmic poles along $\{0\} \times X$, such that $(E, \nabla)_{|D \times \{x^o\}} = (E^o, \nabla^o)$.*

1.20 Remarks.

(1) This result can be compared with that of Corollary II.2.21 and §II.2.26, with the difference that the condition on the eigenvalues of the residue is replaced here with the initial condition $(E, \nabla)_{|D \times \{x^o\}} = (E^o, \nabla^o)$, which gives the *strong* uniqueness.

(2) A more precise way to formulate Proposition 1.19 is by saying that the restriction functor, from the category of holomorphic bundles on $D \times X$ with logarithmic connection with pole along $\{0\} \times X$ to the category of bundles on $D \times \{x^o\}$ with logarithmic connection with pole at 0, is an equivalence.

Indeed, if this equivalence is proved, any logarithmic (E, ∇) with restriction equal to (E^o, ∇^o) is isomorphic to $p^+(E^o, \nabla^o)$, if $p : D \times X \to X$ denotes the projection. Full faithfulness shows that there exists a unique isomorphism inducing the identity when restricted to $D \times \{x^o\}$.

(3) If, in Proposition 1.19, we start from a triple $(E^o, \nabla^o, \langle\, , \,\rangle^o)$, where $\langle\, , \,\rangle^o$ is a nondegenerate bilinear or sesquilinear form of weight $w \in \mathbb{Z}$, a reasoning analogous to that done in the proof of this proposition shows the existence and the uniqueness of an extension $(E, \nabla, \langle\, , \,\rangle)$: we indeed regard $\langle\, , \,\rangle$ as a homomorphism $(E, \nabla) \to (E^*, \nabla^*)$ or $(\overline{E}^*, \overline{\nabla}^*)$ and we apply the full faithfulness of the restriction functor to $D \times \{x^o\}$.

Proof (of Proposition 1.19). We will prove that the restriction functor is an equivalence. The essential surjectivity is clear, as we have seen in Remark 1.20-(2): it is enough to take for (E, ∇) the pullback of (E^o, ∇^o) by the projection $p : D \times X \to D$.

Let us show full faithfulness. Any horizontal section of the sheaf $\mathscr{H}om_{\mathcal{O}_D}(\mathscr{E}^o, \mathscr{E}'^o)$ can be lifted in a unique way as a horizontal section of $\mathscr{H}om_{\mathcal{O}_{D \times X}}(\mathscr{E}, \mathscr{E}')_{|D^* \times X}$ as, by assumption of 1-connectedness of X, the inclusion $D^* \times \{x^o\} \hookrightarrow D^* \times X$ induces an isomorphism of fundamental groups. It is a matter of verifying that this section can be extended as a section of $\mathscr{H}om_{\mathcal{O}_{D \times X}}(\mathscr{E}, \mathscr{E}')$. This follows from the existence of the normal form given by Theorem II.2.25: it implies indeed that any horizontal section on $D^* \times X$ is meromorphic along $\{0\} \times X$ and that, if the restriction of such a section to $D \times \{x^o\}$ is holomorphic, the section is also holomorphic along $\{0\} \times X^4$. □

1.21 Remark. One can show in the same way that, if E^o is a bundle on $\mathbb{P}^1$ equipped with a connection ∇^o with logarithmic poles at all points of a subset $\Sigma^o \subset \mathbb{P}^1$ having cardinal $p+1$, there exists a unique (up to a unique isomorphism) bundle E equipped with an integrable connection ∇ with logarithmic poles along $\Sigma^o \times X$ which extends (E^o, ∇^o).

1.22 Remark. Let us now assume that the set Σ^o varies analytically in $\mathbb{P}^1$. In other words, let us be given a smooth hypersurface Σ in $\mathbb{P}^1 \times X$, such that the projection $\Sigma \to X$ is a covering of finite degree $p+1$. For any $x \in X$, the set $\Sigma_x \subset \mathbb{P}^1$ is thus a configuration of $p+1$ distinct points and we have $\Sigma_{x^o} = \Sigma^o$. If X is 1-connected, the statement given in Remark 1.21 still holds in this situation, with the supplementary assumption that the inclusion

$$(\mathbb{P}^1 \smallsetminus \Sigma^o) \times \{x^o\} \longrightarrow (\mathbb{P}^1 \times X) \smallsetminus \Sigma$$

induces an isomorphism of fundamental groups. This assumption is satisfied for instance if the homology group $H_2(X, \mathbb{Z})$ is zero[5]. As X is 1-connected, the covering $\Sigma \to X$ is trivializable and there exists $p+1$ holomorphic mappings $\psi_i : X \to \mathbb{P}^1$ $(i = 0, \ldots, p)$ such that Σ is the union of the graphs of ψ_i. We apply Proposition 1.19 to an open neighbourhood V_i of each component Σ_i isomorphic to $D \times X$ to obtain a bundle (E_i, ∇_i). We get (E, ∇) on $\mathbb{P}^1 \times X$ by gluing the (E_i, ∇_i) with the holomorphic bundle with a flat connection on $\mathbb{P}^1 \times X \smallsetminus \Sigma$ corresponding to the representation of $\pi_1(\mathbb{P}^1 \times X \smallsetminus \Sigma) = \pi_1(\mathbb{P}^1 \smallsetminus \Sigma^o)$ defined by (E^o, ∇^o).

2 Lattices of (k, ∇)-vector spaces with an irregular singularity

2.a Classification of lattices

The classification of (k, ∇)-vector spaces with an irregular singularity is not as simple as for regular singularities (cf. §II.5). Nevertheless, once the latter is known, the classification of lattices reduces to that of formal lattices. Some of these lattices can be classified. We will give some indications below.

---

[4] One will take care that this property can be false for a nonhorizontal meromorphic section.

[5] Indeed, as X is 1-connected, this group is equal to the group $\pi_2(X)$; the homotopy sequence of the fibration $(\mathbb{P}^1 \times X) \smallsetminus \Sigma \to X$ gives the desired result.

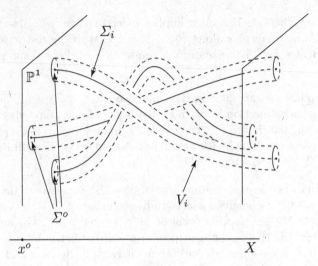

Fig. III.2.

2.1 Lemma (Malgrange). *Let $\mathcal{M}$ be a finite dimensional $\boldsymbol{k}$-vector space. If $\widehat{\mathcal{F}}$ is a lattice of $\widehat{\mathcal{M}} := \widehat{\boldsymbol{k}} \otimes_{\boldsymbol{k}} \mathcal{M}$, then $\mathcal{F} := \widehat{\mathcal{F}} \cap \mathcal{M}$ is a lattice of $\mathcal{M} \subset \widehat{\mathcal{M}}$ and we have $\mathbb{C}[\![t]\!] \otimes_{\mathbb{C}\{t\}} \mathcal{F} = \widehat{\mathcal{F}}$.*

One will find a generalization of this lemma in [Mal96, Prop. 1.2].

Proof. Let us set $\mathcal{F} = \widehat{\mathcal{F}} \cap \mathcal{M}$. The injection $\mathcal{F} \hookrightarrow \widehat{\mathcal{F}}$ induces a homomorphism $\mathbb{C}[\![t]\!] \otimes_{\mathbb{C}\{t\}} \mathcal{F} \to \widehat{\mathcal{F}}$; we will show that it is an isomorphism.

Let us fix a lattice $\mathcal{G}$ of $\mathcal{M}$ and let us consider the lattice $\widehat{\mathcal{G}} = \mathbb{C}[\![t]\!] \otimes_{\mathbb{C}\{t\}} \mathcal{G}$ of $\widehat{\mathcal{M}}$ that it defines. By taking a basis of $\mathcal{G}$ we see that $\mathcal{G} = \widehat{\mathcal{G}} \cap \mathcal{M}$. On the other hand, we can assume, up to changing $\mathcal{G}$ to $t^{-k}\mathcal{G}$ for some suitable k, that $\widehat{\mathcal{G}} \subset \widehat{\mathcal{F}}$.

Let us consider the exact sequences

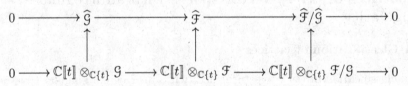

We know that the left up arrow is an isomorphism. It is thus enough to show the same property for the right one. In order to do so, let us remark that $\widehat{\mathcal{M}} = \mathcal{M} + \widehat{\mathcal{G}}$. We deduce that

$$\widehat{\mathcal{F}} = \widehat{\mathcal{F}} \cap (\mathcal{M} + \widehat{\mathcal{G}}) = \mathcal{F} + \widehat{\mathcal{G}},$$

as $\widehat{\mathcal{G}} \subset \widehat{\mathcal{F}}$. We deduce that the natural injection $\mathcal{F}/\mathcal{G} \hookrightarrow \widehat{\mathcal{F}}/\widehat{\mathcal{G}}$ is bijective and, in particular, multiplication by t is nilpotent on $\mathcal{F}/\mathcal{G}$. Therefore, we have $\mathbb{C}[\![t]\!] \otimes_{\mathbb{C}\{t\}} \mathcal{F}/\mathcal{G} = \mathcal{F}/\mathcal{G} = \widehat{\mathcal{F}}/\widehat{\mathcal{G}}$, whence the result. $\square$

Deligne lattices. We will assume in the following[6] that the (k, ∇)-vector space $(\mathcal{M}, \nabla)$ admits a good formal model $(\mathcal{M}^{\mathrm{good}}, \nabla)$. This model can be written as $\bigoplus_\varphi (\mathcal{E}^\varphi \otimes \mathcal{R}_\varphi)$. Let $\mathcal{V}(\mathcal{M}^\ell)$ be the Deligne lattice of $\mathcal{M}^{\mathrm{good}}$, that is, by definition, the lattice $\bigoplus_\varphi (\mathcal{E}^\varphi \otimes \mathcal{V}(\mathcal{R}_\varphi))$, where $\mathcal{V}(\mathcal{R}_\varphi)$ is the Deligne lattice of the (k, ∇)-vector space with regular singularity $\mathcal{R}_\varphi$. We deduce a lattice $\mathcal{V}(\widehat{\mathcal{M}})$. The Deligne lattice of $(\mathcal{M}, \nabla)$ is by definition the lattice (cf. Lemma 2.1)

$$\mathcal{V}(\mathcal{M}) = \mathcal{V}(\widehat{\mathcal{M}}) \cap \mathcal{M}.$$

Formal lattices. Let $\widehat{\mathcal{E}}$ be a lattice of $(\widehat{\mathcal{M}}, \widehat{\nabla})$. This lattice is not necessarily decomposed correspondingly to the components $\widehat{\mathcal{E}}^\varphi \otimes \widehat{\mathcal{R}}_\varphi$ of $(\widehat{\mathcal{M}}, \widehat{\nabla})$. Therefore, in general, the classification of (formal) lattices cannot be reduced to that of lattices in (k, ∇)-vector spaces with regular singularity $\mathcal{R}_\varphi$. However, if the lattice has order equal to the Poincaré rank of $(\mathcal{M}, \nabla)$, that is, if it has *minimal* order among lattices of $(\mathcal{M}, \nabla)$, the decomposition corresponding to eigenvalues of the most polar part (cf. Exercise II.5.9) preserves the lattice (neither the base changes nor their inverses have a pole). In particular, if there is $\mathrm{rk}\,\mathcal{M}$ distinct eigenvalues (as in Theorem II.5.7), the lattice can be decomposed as a direct sum of lattices of rank one: we will come back to these properties in §2.e.

2.b Characteristic polynomial of a lattice at infinity

For want of a general classification of lattices of $(\mathcal{M}, \nabla)$ when the connection has an irregular singularity, we will associate to any lattice a characteristic polynomial. Two kinds of characteristic polynomials can be defined:

(a) By using the notion of Deligne lattice for irregular singularities introduced above, we can proceed as in §1.c. The polynomial that we get in this way determines the characteristic polynomial of the *formal monodromy* of $(\mathcal{M}, \nabla)$.

(b) By algebraizing the germ $(\mathcal{M}, \nabla)$ as in §II.3.c and by using the (regular) Deligne lattice in the neighbourhood of infinity, we can follow an approach analogous to that of §1.c. The characteristic polynomial that we get in this way determines that of the monodromy of $(\mathcal{M}, \nabla)$.

We will follow the latter approach. Let $(\mathbb{M}, \nabla)$ be the free $\mathbb{C}[t, t^{-1}]$-module with connection, with regular singularity at infinity and whose germ at 0 coincides with $(\mathcal{M}, \nabla)$ (cf. §§I.4.d and II.3.c). There exists a unique $\mathbb{C}[t]$-module $\mathbb{E}$ which is a lattice of $\mathbb{M}$ and whose germ at 0 is equal to $\mathcal{E}$. Let $\mathcal{V}'$ be a Deligne lattice of the analytic germ $\mathcal{M}'$ of $\mathbb{M}$ at ∞ (that we have already used above, for instance), such that the eigenvalues of the residue of ∇ have a real part in $]-1, 0]$. Let $\mathbb{V}'$ be the $\mathbb{C}[t']$ submodule of $\mathbb{M}$ whose analytic germ at ∞ is

---

[6] This assumption is made for simplicity; see [Mal83b] for the general case.

equal to $\mathcal{V}'$. We have in an evident way $\mathcal{V}'/t'\mathcal{V}' = \mathbb{V}'/t'\mathbb{V}'$, a space that we will denote by H', equipped with its monodromy T'.

Let us consider the increasing filtration of $\mathbb{M}$ by the $t^{-k}\mathbb{E} = t'^k\mathbb{E}$, for $k \in \mathbb{Z}$. It induces an increasing filtration on H' by the formula

$$G'_k = \mathbb{V}' \cap t'^k\mathbb{E}/(t'\mathbb{V}') \cap t'^k\mathbb{E}.$$

2.2 Lemma. *The filtration $G'_\bullet$ is exhaustive, i.e.,*

$$G'_k = \begin{cases} 0 & \text{if } k \ll 0, \\ H' & \text{if } k \gg 0. \end{cases}$$

Proof. Given $\mathbb{V}'$ and $t'^k\mathbb{E}$, we define a bundle $\mathscr{G}_k$ on $\mathbb{P}^1$ (cf. Proposition I.4.15) and we have

$$\mathscr{G}_k = \mathscr{G} \otimes \mathscr{O}_{\mathbb{P}^1}(k)$$

by setting $\mathscr{G} = \mathscr{G}_0$. We deduce that

$$\mathbb{V}' \cap t'^k\mathbb{E} = H^0(\mathbb{P}^1, \mathscr{G}_k) = 0 \quad \text{for } k \ll 0$$

and therefore $G'_k = 0$ for $k \ll 0$.

On the other hand, we have an isomorphism $t' : \mathbb{V}' \cap t'^{k-1}\mathbb{E} \xrightarrow{\sim} (t'\mathbb{V}') \cap t'^k\mathbb{E}$, so that G'_k is identified to the quotient of $H^0(\mathbb{P}^1, \mathscr{G}_k)$ by $H^0(\mathbb{P}^1, \mathscr{G}_{k-1})$. If m_0 is a point of $\mathbb{P}^1$, the exact sequence

$$0 \longrightarrow H^0(\mathbb{P}^1, \mathscr{G} \otimes \mathscr{O}((k-1)m_0)) \longrightarrow H^0(\mathbb{P}^1, \mathscr{G} \otimes \mathscr{O}(km_0))$$
$$\longrightarrow \mathscr{G}_{m_0} \longrightarrow H^1(\mathbb{P}^1, \mathscr{G} \otimes \mathscr{O}((k-1)m_0))$$

and the fact that H^1 is zero for $k \gg 0$ (see the proof of Corollary I.3.3) show that $\dim G'_k = d$ for $k \gg 0$. $\qquad\square$

In an analogous way, if we fix a total order on $\mathbb{C}$ as in §1.c, we get an increasing exhaustive filtration on each H'_λ by setting

$$G'_k(H'_\lambda) = t^{-k}\mathbb{E} \cap \mathbb{V}'^\alpha/t^{-k}\mathbb{E} \cap \mathbb{V}'^{>\alpha}.$$

2.3 Definition (of the characteristic polynomial at infinity). The characteristic polynomial at infinity $\chi^\infty_{\mathcal{E}}(s)$ is defined by Formula (1.15) where one replaces H'^k_λ with $G'_k(H'_\lambda)$. We thus have

$$\nu_\beta = \dim \frac{\mathbb{E} \cap \mathbb{V}'^\beta}{(\mathbb{E} \cap \mathbb{V}'^{>\beta}) + (t\mathbb{E} \cap \mathbb{V}'^\beta)}.$$

2.4 Remark. This polynomial will reappear later in relation with Birkhoff's problem (cf. Remark IV.5.13-(1)).

2.5 Exercise. Recover the characteristic polynomial of the monodromy T' on the space $H' = \mathbb{V}'/t'\mathbb{V}'$ from $\chi^\infty_{\mathcal{E}}(s)$ (see Remark 1.17).

2.6 Example (Darkness at noon). If $(\mathcal{M}, \nabla)$ has regular singularity and if $\mathcal{E}$ is a logarithmic lattice of it, we have $\chi_{\mathcal{E}}^{\infty}(s) = \chi_{\mathcal{E}}(s)$. Indeed, we can write $\mathrm{M} = \bigoplus_{\beta \in \mathbb{C}} \mathrm{M}^{\beta}$ and $\mathbb{E} = \bigoplus_{\beta \in \mathbb{C}} \mathbb{E}^{\beta}$. The multiplicity ν_{β} of $(s - \beta)$ in $\chi_{\mathcal{E}}(s)$ is equal to $\dim \mathbb{E}^{\beta}$. We remark then that the residue at 0 of ∇ on M^{β} has eigenvalue β, while the residue at ∞ has eigenvalue $-\beta$.

The behaviour by duality is analogous to that of Exercise 1.18:

2.7 Proposition (Behaviour by duality). *If $\mathcal{E}^{*}$ is the lattice dual to $\mathcal{E}$ and $\overline{\mathcal{E}}^{*}$ is its "Hermitian" dual, we have $\chi_{\mathcal{E}^{*}}^{\infty}(s) = \chi_{\overline{\mathcal{E}}^{*}}^{\infty}(s) = (-1)^{d} \chi_{\mathcal{E}}^{\infty}(-s)$.*

Proof. Let us set $\chi_{\mathcal{E}}^{\infty}(s) = \prod_{\beta} (s - \beta)^{\nu_{\beta}}$ and $\chi_{\mathcal{E}^{*}}^{\infty}(s) = \prod_{\beta} (s - \beta)^{\nu_{\beta}^{*}}$. It is a matter of showing that $\nu_{\beta}^{*} = \nu_{-\beta}$. For any $\gamma \in \mathbb{C}$, let $\mathscr{F}^{\gamma}$ be the bundle on $\mathbb{P}^{1}$ constructed by gluing the lattices $\mathcal{E}$ (in the neighbourhood of the origin) and V'^{γ} (in the neighbourhood of infinity). Let us denote similarly by $\mathscr{F}^{*\gamma}$ the bundle constructed with the lattices $\mathcal{E}^{*}$ and $V'^{\gamma}(\mathrm{M}^{*})$. As we have (cf. Exercise 1.18)

$$V'^{\gamma}(\mathrm{M}^{*}) = [V'^{> -(\gamma+1)}]^{*}$$

we also have $\mathscr{F}^{*\gamma} = [\mathscr{F}^{> -(\gamma+1)}]^{*}$. On the other hand, for $k \in \mathbb{Z}$, we have $\mathscr{F}^{\gamma+k} = \mathcal{O}_{\mathbb{P}^{1}}(-k) \otimes \mathscr{F}^{\gamma}$. Serre's duality (cf. [Har80]) and the fact that $\Omega_{\mathbb{P}^{1}}^{1} \simeq \mathcal{O}_{\mathbb{P}^{1}}(-2)$ (cf. Proposition I.2.11) imply then the equalities

$$\dim H^{1}(\mathbb{P}^{1}, \mathscr{F}^{> -(\gamma+1)}) = \dim H^{0}(\mathbb{P}^{1}, \mathscr{F}^{*(\gamma+2)})$$
$$\dim H^{1}(\mathbb{P}^{1}, \mathscr{F}^{-(\gamma+1)}) = \dim H^{0}(\mathbb{P}^{1}, \mathscr{F}^{*>(\gamma+2)}).$$

On the other hand,

$$\dim H^{0}(\mathbb{P}^{1}, \mathscr{F}^{\gamma}) = \dim \mathbb{E} \cap V'^{\gamma} := \delta_{\gamma},$$

so that

$$\nu_{\beta} = (\delta_{\beta} - \delta_{\beta+1}) - (\delta_{>\beta} - \delta_{>\beta+1}).$$

We thus deduce that

$$\nu_{\beta}^{*} = \left(\dim H^{1}(\mathbb{P}^{1}, \mathscr{F}^{> -\beta+1}) - \dim H^{1}(\mathbb{P}^{1}, \mathscr{F}^{> -\beta}) \right)$$
$$- \left(\dim H^{1}(\mathbb{P}^{1}, \mathscr{F}^{-\beta+1}) - \dim H^{1}(\mathbb{P}^{1}, \mathscr{F}^{-\beta}) \right).$$

For any γ we have an exact sequence of sheaves

$$0 \longrightarrow \mathscr{F}^{>\gamma} \longrightarrow \mathscr{F}^{\gamma} \longrightarrow \mathbb{G}^{\gamma} \longrightarrow 0$$

where $\mathbb{G}^{\gamma}$ is the skyscraper sheaf, vanishing on $\mathbb{P}^{1} \setminus \{\infty\}$ and with fibre $V'^{\gamma}/V'^{>\gamma}$ at ∞. This exact sequence induces thus a long exact sequence in cohomology:

$$0 \longrightarrow H^{0}(\mathbb{P}^{1}, \mathscr{F}^{>\gamma}) \longrightarrow H^{0}(\mathbb{P}^{1}, \mathscr{F}^{\gamma}) \longrightarrow \mathbb{G}_{\infty}^{\gamma}$$
$$\longrightarrow H^{1}(\mathbb{P}^{1}, \mathscr{F}^{>\gamma}) \longrightarrow H^{1}(\mathbb{P}^{1}, \mathscr{F}^{\gamma}) \longrightarrow 0.$$

By considering the sequences above for $\gamma = -\beta$ and $\gamma = -\beta+1$, and by using that $\dim \mathbb{G}_{\infty}^{\gamma} = \dim \mathbb{G}_{\infty}^{\gamma+1}$, we deduce the equality $\nu_{\beta}^{*} = \nu_{-\beta}$. $\qquad\square$

2.8 Corollary (Symmetry of the characteristic polynomial at infinity). *If there exists a nondegenerate bilinear or sesquilinear form on $(\mathcal{M}, \nabla)$ of weight w relatively to $\mathcal{E}$, then $\chi_{\mathcal{E}}^{\infty}(s) = (-1)^d \chi_{\mathcal{E}}^{\infty}(w - s)$.* □

2.9 Exercise (Behaviour under tensor product, cf. [Var83]). Let $(\mathcal{M}', \nabla', \mathcal{E}')$ and $(\mathcal{M}'', \nabla'', \mathcal{E}'')$ be two k-vector spaces with connection, equipped with lattices, and let $(\mathcal{M}, \nabla, \mathcal{E})$ be their tensor product:

$$\mathcal{M} = \mathcal{M}' \underset{k}{\otimes} \mathcal{M}'', \quad \mathcal{E} = \mathcal{E}' \underset{\mathbb{C}\{t\}}{\otimes} \mathcal{E}'' \quad \text{and} \quad \nabla = \nabla' \otimes \mathrm{Id}_{\mathcal{M}''} + \mathrm{Id}_{\mathcal{M}'} \otimes \nabla''.$$

(1) Prove that $\mathbb{M} = \mathbb{M}' \otimes_{\mathbb{C}[t,t^{-1}]} \mathbb{M}''$ has regular singularity at infinity.
(2) Show that, if $\mathbb{V}$ denotes the Deligne lattice at infinity,

$$\mathbb{V}(\mathbb{M}) = \mathbb{V}(\mathbb{M}') \otimes_{\mathbb{C}[t']} \mathbb{V}(\mathbb{M}'').$$

(3) Prove that, for any $\beta \in \mathbb{C}$, we have, if we denote by $(\mathbb{V}^\beta)_{\beta \in \mathbb{C}}$ the filtration which interpolates the filtration $(t'^k \mathbb{V})_{k \in \mathbb{Z}}$,

$$\mathbb{V}^\beta(\mathbb{M}) / \mathbb{V}^{>\beta}(\mathbb{M})$$

$$= \bigoplus_{\alpha' \in]-1,0]} \left(\left[\mathbb{V}^{\alpha'}(\mathbb{M}') / \mathbb{V}^{>\alpha'}(\mathbb{M}') \right] \underset{\mathbb{C}}{\otimes} \left[\mathbb{V}^{\beta-\alpha'}(\mathbb{M}'') / \mathbb{V}^{>(\beta-\alpha')}(\mathbb{M}'') \right] \right)$$

(Compute $\mathbb{M}^{\mathrm{reg}} := \bigoplus_{k \in \mathbb{Z}} (t'^k \mathbb{V} / t'^{k+1} \mathbb{V})$ in terms of $\mathbb{M}'^{\mathrm{reg}}$ and $\mathbb{M}''^{\mathrm{reg}}$ by considering the analytic germs of $\mathbb{M}, \mathbb{M}', \mathbb{M}''$ at ∞).
(4) Show that

$$\mathbb{E} \cap \mathbb{V}^\beta(\mathbb{M}) \subset \sum_{\alpha' \in]-1,0]} \left(\left[\mathbb{E}' \cap \mathbb{V}^{\alpha'}(\mathbb{M}') \right] \otimes \left[\mathbb{E}'' \cap \mathbb{V}^{\beta-\alpha'}(\mathbb{M}'') \right] \right).$$

Deduce that, for any $\beta \in \mathbb{C}$, we have

$$\sum_{k \geqslant 0} \nu_{\beta+k} \geqslant \sum_{\alpha' \in]-1,0]} \sum_{i \geqslant 0} \sum_{j \geqslant 0} \nu'_{\alpha'-i} \nu''_{\beta-\alpha'-j}.$$

(5) We set $(\nu' \star \nu'')_\beta = \sum_{\beta'+\beta''} \nu'_{\beta'} \nu''_{\beta''}$. Deduce that, for any β, on a

$$\sum_{k \geqslant 0} \nu_{\beta+k} \geqslant \sum_{k \geqslant 0} (\nu' \star \nu'')_{\beta+k},$$

and

$$N_\beta := \sum_{\gamma \geqslant \beta} \nu_\gamma \geqslant (N' \star N'')_\beta := \sum_{\gamma \geqslant \beta} (\nu' \star \nu'')_\beta.$$

Check that $N_\beta = (N' \star N'')_\beta$ for $\beta \ll 0$ or $\beta \gg 0$.
(6) We moreover assume that $(\mathcal{M}', \nabla')$ and $(\mathcal{M}'', \nabla'')$ are equipped with nondegenerate bilinear (or sesquilinear) forms of weight w' and w'' relatively to $\mathcal{E}'$ and $\mathcal{E}''$. Prove that $(\mathcal{M}, \nabla)$ is also naturally equipped with a nondegenerate bilinear (or sesquilinear) form of weight $w = w' + w''$ relatively to $\mathcal{E}$. Deduce that, for any $\beta \in \mathbb{C}$,

$$\nu_\beta = \nu_{w-\beta} \quad \text{and} \quad (\nu' \star \nu'')_\beta = (\nu' \star \nu'')_{w-\beta}.$$

(7) Keeping the previous assumption, show that $N_\beta = (N' \star N'')_\beta$ for any β and deduce that $\nu_\beta = (\nu' \star \nu'')_\beta$ for any β.

2.c Deformations

The main result of this section is Theorem 2.10 below, which is an analogue of Proposition 1.19 for connections with pole of order one, and the proof of which will occupy the remaining part of this chapter. We restrict ourselves here to the case of poles of order one for simplicity and because only this case will be considered in the following. The reader interested in a more general situation can refer to [Mal83c].

In the following, X denotes a connected analytic manifold of dimension n equipped with a base point x^o and with holomorphic functions $\lambda_1, \ldots, \lambda_d :$ $X \to \mathbb{C}$. We will assume that, for any $x \in X$ and all $i \neq j$, the values $\lambda_i(x)$ and $\lambda_j(x)$ are distinct. We will set $\lambda_i^o = \lambda_i(x^o)$.

2.10 Theorem (cf. [Mal83c]). *Let (E^o, ∇^o) be a bundle on a disc D equipped with a connection with pole of order one at the origin, the "residue" R_0^o of which admits $\lambda_1^o, \ldots, \lambda_d^o$ as eigenvalues. If the parameter space X is 1-connected, there exists a unique (up to isomorphism) holomorphic bundle E on $D \times X$, equipped with a connection having pole of order one along $\{0\} \times X$, such that*

- *for any $x \in X$, the spectrum of $R_0(x)$ is $\{\lambda_1(x), \ldots, \lambda_d(x)\}$;*
- *the restriction $(E, \nabla)_{|D \times \{x^o\}}$ is isomorphic to (E^o, ∇^o).*

2.d Bundles of rank one with connection having a pole of order one

Let (E, ∇) be a holomorphic bundle on $D \times X$ equipped with a *flat* connection ∇. We assume in this paragraph that the bundle E has rank one and, as in Theorem 2.10, that the connection has a pole of order one. Let us begin with the local classification of such objects.

2.11 Proposition. *Let (E, ∇) be a germ at $(0, x^o)$ of bundle of rank one on $D \times X$, with a connection having a pole of order one along $\{0\} \times X$.*

(1) *We can associate to (E, ∇) a unique pair (λ, μ), where $\mu \in \mathbb{C}$ and $\lambda \in \mathscr{O}(X)$, characterized by the property that, in any local basis of E, the polar part of the connection form is written as*

$$\omega_{\text{pol}} = -d\left(\frac{\lambda(x)}{t}\right) + \mu \frac{dt}{t}.$$

(2) *The germ of bundle (E, ∇) admits a non identically zero holomorphic horizontal section if and only if $\lambda \equiv 0$ and $\mu \in -\mathbb{N}$.*

(3) *Two such bundles with associated pairs (λ, μ) and (λ', μ') are locally isomorphic if and only if $\lambda = \lambda'$ and $\mu = \mu'$.*

Proof.

(1) Let ω be the connection form in some local basis of E. The integrability condition reduces to the condition $d\omega = 0$ as E has rank one. Moreover, the polar part of the form ω is not modified by a holomorphic base change, which only adds some exact holomorphic form to ω. The integrability condition applied to the polar part of ω gives the desired expression.

(2) This is analogous to Proposition II.5.1. Let us begin by noticing that there exists a basis of E in which the connection form is reduced to its polar part: indeed, the connection form ω in a given basis can be written as $\omega = \omega_{\mathrm{pol}} + d\varphi$ with φ holomorphic. The base change with matrix $e^{-\varphi}$ gives the existence. In this basis, the coefficient $s(t, x) = \sum_{k \geq 0} s_k(x) t^k$ of a holomorphic horizontal section satisfies the differential equation

$$t\frac{\partial s}{\partial t} + \left(\frac{\lambda(x)}{t} + \mu\right) s = 0,$$

which can be written as

$$\sum_{k \geq 0}(k + \mu)s_k(x)t^k + \lambda(x)\sum_{k \geq 0} s_k(x)t^{k-1} = 0.$$

If $\lambda(x) \not\equiv 0$, we deduce step by step that all coefficients s_k are identically zero. If $\lambda \equiv 0$, we see in the same way that the nonvanishing of at least one coefficient s_k is equivalent to $\mu \in -\mathbb{N}$.

(3) A homomorphism $(E, \nabla) \to (E', \nabla')$ is a holomorphic horizontal section of the bundle $\mathcal{H}om(\mathcal{E}, \mathcal{E}')$. This one still has rank one and its connection still has order one, with form $\omega' - \omega$. The existence of an isomorphism between both bundles with connection implies therefore, after (2), that $\lambda \equiv \lambda'$ and $\mu = \mu'$. Conversely, if this equality is satisfied, both bundles are isomorphic, as there exists for each a basis in which the connection form reduces to its polar part.   □

Let us now consider the global situation: the point (3) of the previous proposition shows the existence of a holomorphic function λ on X and of a complex number μ such that, in any local basis, the polar part ω_{pol} of the connection form is equal to $-d(\lambda(x)/t) + \mu\,dt/t$.

2.12 Proposition. *There exists a locally constant sheaf $\mathcal{E}$ of rank one on X and an isomorphism $\mathcal{E} \simeq \mathcal{O}_{D \times X} \otimes_{\mathbb{C}} \mathcal{E}$ such that the connection ∇ can be written as*

$$\nabla = \widetilde{\nabla} + \omega_{\mathrm{pol}}$$

if $\widetilde{\nabla}$ is the natural holomorphic flat connection on $\mathcal{O}_U \otimes_{\mathbb{C}} \mathcal{E}$.

Proof. It is enough to notice that the local basis of (E, ∇) in which the connection form reduces to its polar part is unique up to a multiplicative constant: indeed, two bases in which the matrix is equal to ω_{pol} are related by multiplication by an invertible holomorphic function ψ such that $d \log \psi = 0$, hence ψ is locally constant.   □

2.13 Remarks.

(1) With the notation of §0.14.c, the endomorphism R_0 is the multiplication by λ, the endomorphism R_∞ is the multiplication by $-\mu$ and $\Phi = -d\lambda$.
(2) If X is 1-connected, it follows from Proposition 2.12 that the bundle E is trivializable.
(3) The sheaf of automorphisms of (E, ∇) is locally constant of rank one on X: as a matter of fact, such an automorphism must preserve the polar part ω_{pol} of the connection matrix of ∇, hence also the connection $\widetilde{\nabla}$ (if one prefers, the natural connection defined by ∇ on $\mathcal{H}om_\mathcal{O}(\mathcal{E}, \mathcal{E})$ has no pole and it is also the natural connection defined by $\widetilde{\nabla}$); as a consequence, it is an automorphism of the flat bundle $(E, \widetilde{\nabla})$.
This sheaf is thus constant if X is 1-connected.

2.14 Exercise. Extend these results to the case of a connection of order $r \geqslant 1$.

2.e Formal structure

When E has rank $\geqslant 1$, the structure of the bundle (E, ∇) is not so simple (this is even so for the associated meromorphic bundle; see §II.5). Nevertheless, if one only considers the associated formal bundle, everything is similar to rank one.

Let $\widehat{\mathcal{O}}$ be the formal completion of the sheaf $\mathcal{O}_{D \times X}$ along $\{0\} \times X$: it is a sheaf on $\{0\} \times X$ whose germ at any point $(0, x^o)$ is the set of formal series $\sum_{i=0}^\infty a_i(x) t^i$, where the a_i are holomorphic functions defined on *the same* neighbourhood of x^o (cf. §II.2.f).

Let E be a holomorphic bundle on $D \times X$ and let $\mathcal{E}$ be the associated sheaf. We will denote by $\widehat{\mathcal{E}}$ the formal bundle associated to $\mathcal{E}$ along $\{0\} \times X$: this is the sheaf $\widehat{\mathcal{O}} \otimes_\mathcal{O} \mathcal{E}$; it is thus a sheaf on $\{0\} \times X$. The notion of formal bundle with meromorphic connection remains meaningful and, if E is equipped with a connection having pole along $\{0\} \times X$, then so is $\widehat{\mathcal{E}}$.

In general, going from (E, ∇) to $(\widehat{\mathcal{E}}, \widehat{\nabla})$ loses much information, if (E, ∇) does not have regular singularity along $\{0\} \times X$.

Nevertheless, we notice that, when E has rank one and ∇ has a pole of order one, giving the formal bundle with connection $(\widehat{\mathcal{E}}, \widehat{\nabla})$ is equivalent to giving a pair (λ, μ) and, in this case, taking the formal bundle is an equivalence of categories.

Let us resume the assumptions of §2.c.

2.15 Theorem (Formal decomposition of a lattice of order one). *If* (E, ∇) *is a holomorphic bundle on* $D \times X$ *with connection of order one along* $\{0\} \times X$ *and which "residue"* $R_0(x)$ *has eigenvalues* $\lambda_i(x)$ *at any* $x \in X$, *then the associated formal bundle with connection can be decomposed in a unique way (up to a permutation) as a direct sum of subbundles of rank one*

$$(\widehat{\mathscr{E}}, \widehat{\nabla}) = \bigoplus_{i=1}^{d} (\widehat{\mathscr{E}}_i, \widehat{\nabla})$$

which are locally pairwise nonisomorphic.

Proof. The restricted bundle $i_0^* E$ can be decomposed as the direct sum of eigenbundles of R_0. It is a matter of lifting this decomposition to $\widehat{\mathscr{E}}$ in a way compatible with $\widehat{\nabla}$.

The local existence of the decomposition follows from Theorem II.5.7 because, as we mentioned in Remark II.5.8, neither the base change that we use not its inverse has a pole. This decomposition also provides complex numbers μ_i.

Moreover, this local decomposition is unique: as a matter of fact, let us consider another decomposition by $(\widehat{\mathscr{E}}_i', \widehat{\nabla})$; the projection $\widehat{\mathscr{E}}_i' \to \widehat{\mathscr{E}}_j$ induced by the local decomposition is zero if $i \neq j$ as $\lambda_i \neq \lambda_j$ (Proposition 2.11-(2)) and, arguing the other way around, we deduce that $\widehat{\mathscr{E}}_i' = \widehat{\mathscr{E}}_i$.

Lastly, the summand $(\widehat{\mathscr{E}}_i, \widehat{\nabla})$ is the unique subbundle of rank one of $(\widehat{\mathscr{E}}, \widehat{\nabla})$ on which the polar part of $\widehat{\nabla}$ is equal to $-d(\lambda_i/t) + \mu_i \, dt/t$: if $(\widehat{\mathscr{E}}_i', \widehat{\nabla})$ is another one, we see as above that the sections of $\widehat{\mathscr{E}}_i'$ have no component on the summands $\widehat{\mathscr{E}}_j$ ($j \neq i$) of the decomposition, hence $\widehat{\mathscr{E}}_i' \subset \widehat{\mathscr{E}}_i$; having fixed the residue μ_i implies $\widehat{\mathscr{E}}_i' = \widehat{\mathscr{E}}_i$.

The uniqueness now gives the *global* existence of the subbundle $(\widehat{\mathscr{E}}_i, \widehat{\nabla})$. Finally, the natural homomorphism $\bigoplus_i (\widehat{\mathscr{E}}_i, \widehat{\nabla}) \to (\widehat{\mathscr{E}}, \widehat{\nabla})$ is locally, hence globally, an isomorphism. $\square$

2.16 Remarks.

(1) The sheaf of automorphisms of $(\widehat{\mathscr{E}}, \widehat{\nabla})$ is locally constant: indeed, an automorphism of $(\widehat{\mathscr{E}}, \widehat{\nabla})$ preserves the decomposition of Theorem 2.15 as the $(\widehat{\mathscr{E}}_i, \widehat{\nabla})$ are pairwise nonisomorphic.

(2) When X is 1-connected, we deduce from what we have seen above that the bundle $\widehat{\mathscr{E}}$ is trivializable and admits a basis in which the matrix $\widehat{\Omega}$ of $\widehat{\nabla}$ can be written as $\mathrm{diag}(\omega_1, \ldots, \omega_d)$, where the ω_i take the form $-d(\lambda_i(x)/t) + \mu_i \, dt/t$. Such a basis is unique up to conjugation by a constant diagonal matrix.

2.f Proof of Theorem 2.10

Step 1: construction of the formal connection. If the formal bundle $(\widehat{E}, \widehat{\nabla})$ exists, it must be trivializable, as X is assumed 1-connected and after Remark 2.16-(2). It admits thus a basis in which the connection matrix is $\operatorname{diag}(\omega_1, \ldots, \omega_d)$, with

$$\omega_i = -d\left(\frac{\lambda_i(x)}{t}\right) + \mu_i \frac{dt}{t},$$

where $\mu_i = \mu_i^o$ is determined by (E^o, ∇^o). We equip the trivial formal bundle with this connection, which is suitable, and we fix a formal isomorphism $\widehat{f}^o$ of $(\widehat{E}^o, \widehat{\nabla}^o)$ with the former.

Step 2: construction of the meromorphic bundle. The meromorphic bundle

$$(\mathcal{M}, \nabla) = (\mathscr{E}(*(\{0\} \times X)), \nabla)$$

can be obtained by means of the Riemann-Hilbert correspondence with parameter of §II.6.b. Indeed, the formal bundle $(\widehat{\mathcal{M}}, \widehat{\nabla}) = (\widehat{\mathscr{E}}(*\{0\} \times X), \widehat{\nabla})$ is determined by Step 1. Moreover, Theorem II.6.1 shows that the Stokes sheaf is locally constant, hence constant as X is 1-connected. There exists therefore a unique section σ of this Stokes sheaf, the value of which at x^o is the element determined by $(E^o, \nabla^o, \widehat{f}^o)$. By means of $(\widehat{\mathcal{M}}, \widehat{\nabla})$ and of the section σ, we can thus construct a meromorphic bundle with connection $(\mathcal{M}, \nabla, \widehat{f})$. Its restriction to $D \times \{x^o\}$ is isomorphic to $(\mathcal{M}^o, \nabla^o, \widehat{f}^o) = (\mathscr{E}(*\{0\}), \nabla^o, \widehat{f}^o)$.

Step 3: construction of the bundle $(\mathscr{E}, \nabla)$. We can proceed in two ways. The simpler method notes that, once the formal decomposition of Theorem 2.15 is obtained, the arguments of §II.6 do not necessitate inverting the variable t and can thus be applied to $(\mathscr{E}, \nabla, \widehat{f})$ as well as to $(\mathcal{M}, \nabla, \widehat{f})$. We can therefore gather Steps 2 and 3.

Another method involves constructing $\mathscr{E}$ as a lattice of the meromorphic bundle $\mathcal{M}$ obtained above and verifying that the connection ∇ has a pole of order one. Moreover, it is enough to construct the germ of $\mathscr{E}$ along $\{0\} \times X$ as, away from this set, $\mathscr{E}$ must be equal to $\mathcal{M}$. It is thus enough to show that $\mathscr{E} := \widehat{\mathscr{E}} \cap \mathcal{M}$ is a lattice of $\mathcal{M}$ which satisfies $\widehat{\mathcal{O}} \otimes_{\mathcal{O}} \mathscr{E} = \widehat{\mathscr{E}}$; as a matter of fact, if this equality is proved, the connection ∇ on $\mathscr{E}$ has a pole of order one along $\{0\} \times X$ as so has its formalized connection.

One will find in [Mal96, Prop. 1.2] a proof of this result, which is an analogue, with parameters, of Lemma 2.1.

Uniqueness. Let us employ the arguments of the first method above, by using of course the simple connectedness of X. Let (E, ∇) and (E', ∇') be two such bundles with connection, which are isomorphic when restricted to $D \times \{x^o\}$ and let g^o such an isomorphism. On the other hand, let $\widehat{f}$ and $\widehat{f}'$ be formal isomorphisms of these bundles with some model $(E^{\mathrm{good}}, \nabla^{\mathrm{good}})$ as in Theorem 2.15. Remark 2.16 shows that the automorphism $\widehat{h}^o = \widehat{f}'^o \circ \widehat{g}^o \circ (\widehat{f}^o)^{-1}$ of the restriction $i^+(\widehat{E}^{\mathrm{good}}, \widehat{\nabla}^{\mathrm{good}})$ can be extended in a unique way as an automorphism $\widehat{h}$ of $(\widehat{E}^{\mathrm{good}}, \widehat{\nabla}^{\mathrm{good}})$. The restrictions to $D \times \{x^o\}$ of $(\mathscr{E}, \nabla, \widehat{h} \circ \widehat{f})$ and $(\mathscr{E}', \nabla', \widehat{f}')$ are then isomorphic *via* g^o. The constancy of the Stokes sheaf implies that $(\mathscr{E}, \nabla, \widehat{h} \circ \widehat{f})$ and $(\mathscr{E}', \nabla', \widehat{f}')$ are isomorphic. $\square$

The Riemann-Hilbert problem and Birkhoff's problem

Introduction

The question of the existence of logarithmic lattices, *trivial* as vector bundles, in a given meromorphic bundle with connection on $\mathbb{P}^1$, is known as the *Riemann-Hilbert problem*[1]. This question takes for granted that the meromorphic bundle has only regular singularities.

There are various generalizations and variants of this question. *Birkhoff's problem* is one for which the meromorphic bundle has only two singularities, one being regular, and where one fixes the lattice in the neighbourhood of the other singularity, which can be regular or irregular.

We can also ask for the existence of a trivial lattice whose order in each singularity is minimum among the orders of the local lattices of the meromorphic bundle at this singularity: it is a straightforward generalization of the Riemann-Hilbert problem to the case of irregular singularities.

The case of bundles on compact Riemann surfaces has also been considered. One then tries to bound from above the number of *apparent singularities* which are needed for making the bundle trivial. One could also, considering Exercise I.4.8, ask for the existence of logarithmic lattices which are semistable as vector bundles (cf. [EV99]).

One will find in [AB94] and [Bol95] a detailed historical review of this question (cf. also [Bea93]). Let us only note here that, although we have long known examples giving a negative answer to Birkhoff's problem (cf. [Gan59, Mas59]), a negative answer to the Riemann-Hilbert problem has only been given quite recently by A. Bolibrukh.

We will give here proof of the existence of a solution to the Riemann-Hilbert problem and to Birkhoff's problem with an assumption of irreducibility, a result which is drawn from Bolibrukh and Kostov. We will not, how-

---

[1] In fact, the original problem 1.1 shows up the monodromy representation, but this amounts to the same, according to the Riemann-Hilbert correspondence II.3.3.

ever, exhibit examples for which the answer is negative, referring for this to [AB94, Bol95] and to the references given therein.

We will also give a criterion, drawn from M. Saito, for a positive answer to Birkhoff's problem. This criterion happens to be more effective, as we will mention in §VII.5, when applied to differential systems associated to some objects coming from algebraic geometry, than Corollary 5.7 below.

1 The Riemann-Hilbert problem

Given a finite set $\{m_1, \ldots, m_p\}$ of distinct points of the complex line, we fix a base point $o \in \mathbb{C} \setminus \{m_1, \ldots, m_p\}$. The fundamental group $\pi_1(\mathbb{C} \setminus \{m_1, \ldots, m_p\}, o)$ is the free group with p generators. Giving a linear representation

$$\rho : \pi_1(\mathbb{C} \setminus \{m_1, \ldots, m_p\}, o) \longrightarrow \mathrm{GL}_d(\mathbb{C})$$

is then equivalent to giving p invertible matrices $T_1, \ldots, T_p \in \mathrm{GL}_d(\mathbb{C})$.

On the other hand, let us consider a differential system

$$\frac{du}{dt} = -A(t) \cdot u,$$

where $A(t)$ is a matrix $d \times d$ with holomorphic entries on $\mathbb{C} \setminus \{m_1, \ldots, m_p\}$ and u is a vector $(u_1, \ldots, u_d)$ of functions on an open subset of $\mathbb{C} \setminus \{m_1, \ldots, m_p\}$.

Theorem 0.12.8 shows that this equation defines a linear representation as above.

1.1 The Riemann-Hilbert problem. *Given p invertible matrices $T_1, \ldots, T_p \in \mathrm{GL}_d(\mathbb{C})$, do there exist matrices $A_1, \ldots, A_p \in \mathrm{M}_d(\mathbb{C})$ such that the representation defined by $T_1, \ldots, T_p$ is isomorphic to that associated to the system*

$$\frac{du}{dt} = \left[\sum_{i=1}^{p} \frac{A_i}{t - m_i} \right] \cdot u \ ?$$

1.2 Remark. One should not believe that the matrices T_j are related to the matrices A_j by relation $T_j = \exp -2i\pi A_j$. Indeed, in the neighbourhood of m_j, the equation can be written as

$$\frac{du}{dt} = A^{(i)}(t) \cdot u$$

with $A^{(i)}(t) - \dfrac{A_i}{t - m_i}$ holomorphic in the neighbourhood of m_i and $\neq 0$ in general. The situation of Exercise II.2.20-(4) may thus happen here.

We will first consider an analogous problem, where representations do not occur. Let $(\mathcal{M}, \nabla)$ be a meromorphic bundle with connection on the Riemann sphere $\mathbb{P}^1$, with poles in $p + 1$ distinct points $m_0, m_1, \ldots, m_p$.

1.3 The algebraic Riemann-Hilbert problem. *We suppose that* $(\mathscr{M}, \nabla)$
has a regular singularity at $m_0, \ldots, m_p$ *(i.e., that there exists in the neighbourhood of each* m_i *a logarithmic lattice).*

(a) Does there exist a global logarithmic *lattice of* $(\mathscr{M}, \nabla)$*?*
(b) Does there exist such a lattice which is moreover quasitrivial*?*
(c) Does there exist such a lattice which is moreover trivial*?*

Let us first translate the questions in terms of differential systems. Recall
that (Corollary I.4.9) when $\Sigma \neq \varnothing$, the meromorphic bundle $\mathscr{M}$ is isomorphic
to $\mathcal{O}_{\mathbb{P}^1}(*\Sigma)^d$. Let Ω be the connection matrix in some basis $e_1, \ldots, e_d$ of global
sections of $\mathscr{M}$, which exists according to this corollary. This matrix has poles
in Σ. The question (c) is then equivalent to

(c') Does there exist an invertible matrix P whose entries are rational functions
with poles contained in Σ, such that the matrix

$$\Omega' := P^{-1}\Omega P + P^{-1}dP$$

has a logarithmic pole at each $m_i \in \Sigma$ and has no other pole in $\mathbb{P}^1$?

Indeed, the lattice $\mathscr{E}$ generated by the basis $(\varepsilon_1, \ldots, \varepsilon_d) = (e_1, \ldots, e_d) \cdot P$
satisfies then Property (c) and, conversely, any basis of a lattice $\mathscr{E}$ satisfying
(c) provides, through some base change, a matrix having the desired form.

Let us choose a coordinate t on $\mathbb{P}^1 \setminus \{m_0\}$ in such a way that $m_0 = \{t = \infty\}$.
The question (c') is then equivalent to the question

(c") Does there exist an invertible matrix P whose entries are rational functions
with poles contained in Σ, such that the matrix $\Omega' := P^{-1}\Omega P + P^{-1}dP$
can be written as

$$(1.4) \qquad \Omega' = \sum_{i=1}^{p} \frac{A_i}{t - m_i}\, dt \ ?$$

We then say that the differential system defined by Ω' is *Fuchsian*. This
equivalence follows from the lemma below applied to the entries of Ω:

1.5 Lemma. *Let f be a holomorphic function on $\mathbb{C} \setminus \{m_1, \ldots, m_p\}$ having
at most a simple pole at each m_i, with residue $f_i \in \mathbb{C}$ at m_i. The differential
form $f(t)\, dt$ admits at most a simple pole at $m_0 = \infty$ if and only if*

$$f(t)\, dt = \sum_{i=1}^{p} \frac{f_i}{t - m_i}\, dt$$

and the residue at infinity is $-\sum_i f_i$.

Proof. Let us consider the 1-form

$$g(t)\,dt = f(t)\,dt - \sum_{i=1}^{p} \frac{f_i}{t - m_i}\,dt.$$

By definition of f_i, this form has no pole on $\mathbb{C}$. Moreover, we have on $U_0 \cap U_\infty$, if we put $t' = 1/t$,

$$\frac{f_i}{t - m_i}\,dt = \frac{-f_i}{1 - m_i t'} \cdot \frac{dt'}{t'}$$

so that this form has a simple pole at infinity, with residue $-f_i$. Therefore, the 1-form $f(t)\,dt$ has at most a simple pole at infinity if and only if the 1-form $g(t)\,dt$ does as well. But saying that $g(t)\,dt$ has at most a simple pole at infinity is equivalent to saying that $g(t)\,dt$ is a holomorphic global section of the sheaf $\Omega^1_{\mathbb{P}^1}(1 \cdot \infty) \simeq \mathcal{O}_{\mathbb{P}^1}(-1)$ (cf. Proposition I.2.11), which is equivalent to $g(t)\,dt = 0$, after Exercise I.2.3. $\square$

We have a simple result giving a constraint on the residue matrices of the solutions to Problem 1.3-(c).

1.6 Proposition. *Let $\mathcal{E}$ be a trivial logarithmic lattice of a meromorphic bundle with connection $(\mathcal{M}, \nabla)$ on $\mathbb{P}^1$. Then $\sum_{i=0}^{p} \operatorname{tr} \operatorname{Res}_{m_i} \nabla = 0$.*

Proof. One chooses a coordinate t on $\mathbb{C}$ such that the point $\{t = \infty\}$ does not belong to Σ. Let e be a basis of $\mathcal{E}$ in which the matrix of ∇ is

$$\Omega(t) = \sum_{i=0}^{p} \frac{A_i}{t - m_i}\,dt.$$

We then have $\sum_i A_i = 0$ and therefore $\sum_i \operatorname{tr}(A_i) = 0$. $\square$

1.7 Remark (The canonical trivialization of a trivializable bundle).
Let M be a connected complex analytic manifold and let $\mathcal{E}$ be a locally free sheaf of rank d of $\mathcal{O}_M$-modules. The restriction mapping provides, for any $m \in M$, a mapping $\Gamma(M, \mathcal{E}) \to \mathcal{E}_m$ which associates to any holomorphic section of $\mathcal{E}$ on M its germ at the point m. We deduce an homomorphism of sheaves

$$\mathcal{O}_M \underset{\mathbb{C}}{\otimes} \Gamma(M, \mathcal{E}) \longrightarrow \mathcal{E}.$$

We say that $\mathcal{E}$ is *generated by its global sections* if this homomorphism is surjective. One can also show that, if M is compact (for instance $M = \mathbb{P}^1$), the bundle E is trivializable (i.e., isomorphic to the trivial bundle of rank d) if and only if this homomorphism is an isomorphism (the reader should prove this: one could use that, if M is compact and connected, one has $\Gamma(M, \mathcal{O}_M) = \mathbb{C}$). This homomorphism furnishes then a specific isomorphism— called canonical—with the trivial bundle of rank d. It identifies thus all the fibres E_m of E to the same vector space $\Gamma(M, \mathcal{E})$.

When $\mathscr{E}$ is a logarithmic lattice of a meromorphic bundle with connection on $\mathbb{P}^1$, having poles at Σ, the endomorphisms $\operatorname{Res}_{m_i} \nabla : E_{m_i} \to E_{m_i}$ which act on distinct fibres of E can, when E is trivializable, be regarded as endomorphisms of the same vector space $\Gamma(\mathbb{P}^1, \mathscr{E})$. The corresponding matrices in some basis of this space are nothing but the matrices A_i considered above (if we set $A_0 = -\sum_{i=1}^{p} A_i$).

1.8 Exercise.

(1) By using an argument analogous to that of Lemma 1.5, prove that, if ω is a logarithmic 1-form on $\mathbb{P}^1$, having poles at $m_0, \ldots, m_p$ and with integral residue at each m_i, there exists a meromorphic function f on $\mathbb{P}^1$, with poles at $m_0, \ldots, m_p$, such that $\omega = df/f$.

(2) Let $T_1, \ldots, T_p$ be matrices with determinant equal to 1. Show that, if there exists a solution $A_1, \ldots, A_p$ to Problem 1.1, there exists another one with matrices $A_1', \ldots, A_p'$ having *trace zero* (i.e., belonging to the Lie algebra $\mathfrak{sl}_d(\mathbb{C})$ of the Lie group $\mathrm{SL}_d(\mathbb{C})$). Check first that the A_i have an integral trace, then take

$$A_i' = A_i - \begin{pmatrix} \operatorname{tr} A_i & & \\ & 0 & \\ & & \ddots & \\ & & & 0 \end{pmatrix}.$$

1.9 Exercise (Residue formula for a meromorphic form).
Let ω be a meromorphic 1-form on $\mathbb{P}^1$ having poles on a set $\Sigma = \{m_0, \ldots, m_p\}$. Considering the proof of Lemma 1.5, prove the residue formula

$$\sum_{i=0}^{p} \operatorname{res}_{m_i} \omega = 0.$$

1.10 Exercise (Residue formula for a meromorphic connection).
Let $\mathscr{E}$ be a bundle of rank d on $\mathbb{P}^1$ of degree $\deg \mathscr{E}$. The degree only depends on the (rank one) bundle $\det \mathscr{E}$. Let $\nabla : \mathscr{E} \to \Omega_{\mathbb{P}^1}^1(*\Sigma) \otimes \mathscr{E}$ be a meromorphic connection on $\mathscr{E}$, having poles at the points of a finite set $\Sigma = \{m_0, \ldots, m_p\}$. It is a matter of computing the degree of $\mathscr{E}$ by a residue formula from the meromorphic connection ∇, namely

$$(*) \qquad \deg \mathscr{E} = -\sum_{i=0}^{p} \operatorname{tr} \operatorname{Res}_{m_i}(\nabla)$$

This computation can be generalized to any compact Riemann surface.

(1) Let e be a local basis of $\mathscr{E}$ in the neighbourhood of a point m_i and let Ω_i be the matrix of ∇ in this basis. Let $\operatorname{res}_{m_i} \Omega_i$ be the residue of Ω_i at m_i, i.e., the coefficient of dz/z in Ω_i, if z is a local coordinate centered at m_i (cf. Remark §0.9.15). Prove that this residue does not depend on the chosen local coordinate and that its trace does not depend on the chosen local basis.

(2) Show that the determinant bundle $\wedge^d \mathscr{E}$ can be naturally equipped with a meromorphic connection, whose matrix in the local basis $e_1 \wedge \cdots \wedge e_d$ at m_i is tr Ω_i. Deduce that both terms in (∗) only depend on $(\det \mathscr{E}, \nabla)$, so that one can assume from the beginning that $\mathscr{E}$ has rank one.

(3) If the bundle $\mathscr{E}$ is trivial, write $\nabla = d + \eta$, where η is a meromorphic 1-form and apply the classic residue theorem (cf. Exercise 1.9).

(4) If $\mathscr{E}$ is not trivial, it is the bundle associated to some divisor, that we can assume to be equal to δm_0, if $\delta = \deg \mathscr{E}$ is the degree of $\mathscr{E}$. Let s be a meromorphic section of $\mathscr{E}$ defining this divisor. Prove that there exists a meromorphic connection ∇'' on $\mathscr{E}$ for which s is a horizontal section, i.e., such that $\nabla'' fs = df \otimes s$ for any holomorphic or meromorphic function f on an open set of $\mathbb{P}^1$. Show that this connection admits $-\delta$ as residue at m_0 (if e is a local holomorphic basis of $\mathscr{E}$ in the neighbourhood of m_0 and if z is a local coordinate there, set $s(z) = z^\delta u(z)e$ with $u(0) \neq 0$; show that, in the basis e,

$$\nabla'' e = \nabla'' \left(s(z)/z^\delta u(z) \right) = d(1/z^\delta u(z)) \otimes s(z) = \frac{-\delta v(z)\,dz}{z} \otimes e$$

with $v(0) \neq 0$ and conclude).

Consider then the connection $\nabla \otimes \mathrm{Id} - \mathrm{Id} \otimes \nabla''$ on the trivial bundle $\mathscr{E} \otimes \mathscr{E}^*$ to prove (∗) in this case.

Let us come back to Problem 1.3 and let us first show a simple result.

1.11 Proposition.

(1) *The answer to Question* (a) *is positive.*
(2) *There exists a trivial lattice of* $(\mathscr{M}, \nabla)$ *which is logarithmic at each* m_i *($i = 1, \ldots, p$), but possibly not at* m_0.
(3) *Questions* (b) *and* (c) *are equivalent.*

Proof.

(1) Adapt the proof of Corollary I.4.9 starting from local lattices which are logarithmic, which is possible as $(\mathscr{M}, \nabla)$ has regular singularity.

(2) Let us start from the logarithmic lattice $\mathscr{E}$ obtained in (a). It is possibly not trivial, but the meromorphic bundle $\mathscr{E}(*m_0)$ is trivializable (Corollary I.4.9).

Let us then choose a basis $e_1, \ldots, e_d$ of the meromorphic bundle $\mathscr{E}(*m_0)$ and let us consider the lattice $\mathscr{E}' := \sum_{k=1}^d \mathscr{O}_{\mathbb{P}^1} e_k$ of $(\mathscr{M}, \nabla)$. By assumption, it coincides with $\mathscr{E}$ on $\mathbb{P}^1 \smallsetminus \{m_0\}$, hence it is logarithmic at $m_1, \ldots, m_p$. By construction also, it is a trivial bundle. Nevertheless, the connection matrix in the basis $e_1, \ldots, e_d$ is possibly not logarithmic at m_0.

(3) Let us assume that we have answered Question 1.3-(b) positively. Then let $\mathscr{E}$ be a quasitrivial logarithmic lattice of $(\mathscr{M}, \nabla)$. On the other hand, let $(\mathscr{O}_{\mathbb{P}^1}(*m_0), d)$ be the trivial meromorphic bundle of rank one having a pole

at m_0 only, equipped with its natural connection d (usual differentiation of meromorphic functions). The lattices $\mathscr{O}_{\mathbb{P}^1}(k \cdot m_0)$ are logarithmic (the reader should check this). We have in an evident way (cf. 0.11.b)

$$(\mathscr{O}_{\mathbb{P}^1}(*m_0), d) \otimes (\mathscr{M}, \nabla) = (\mathscr{M}, \nabla).$$

Therefore, when $\mathscr{E}$ is a logarithmic lattice of $(\mathscr{M}, \nabla)$, the lattices $\mathscr{O}_{\mathbb{P}^1}(k \cdot m_0) \otimes \mathscr{E}$ also are logarithmic (cf. 0.14.b). If $\mathscr{E}$ is quasitrivial, isomorphic to $\mathscr{O}_{\mathbb{P}^1}(\ell)^d$ for some $\ell \in \mathbb{Z}$, the lattice $\mathscr{O}_{\mathbb{P}^1}(-\ell \cdot m_0) \otimes \mathscr{E}$ is trivial. □

1.12 Remark (The Riemann-Hilbert problem with a bilinear form).
Let $w_1, \ldots, w_p$ be integers. Let us set

$$\omega = \sum_{i=1}^{p} \frac{w_i \, dt}{t - m_i}.$$

This is a 1-form on $\mathbb{P}^1$ with logarithmic poles at the points of Σ, with residues w_i at m_i and $-\sum w_i$ at $m_0 = \infty$. The trivial bundle is then a logarithmic lattice of the meromorphic bundle with connection $(\mathscr{O}_{\mathbb{P}^1}(*\Sigma), d + \omega)$. The latter is isomorphic to the bundle $(\mathscr{O}_{\mathbb{P}^1}(*\Sigma), d)$ via multiplication by the rational fraction $\prod(t - m_i)^{-w_i}$.

Let us set $A_1, \ldots, A_p \in M_d(\mathbb{C})$ and $\Omega = \sum_{i=1}^{p} \frac{A_i}{t - m_i} \, dt$. We denote by $(\mathscr{E}, \nabla)$ the trivial bundle $\mathscr{O}_{\mathbb{P}^1}^d$ equipped with the connection ∇ having matrix Ω in the canonical basis. It is a logarithmic lattice of the meromorphic bundle with connection $(\mathscr{M}, \nabla) := (\mathscr{O}_{\mathbb{P}^1}^d(*\Sigma), d + \Omega)$. The matrix of ∇ in another basis of $\mathscr{E}$ takes the form $C^{-1}\Omega C$ with $C \in \mathrm{GL}_d(\mathbb{C})$.

The property

$$(*) \quad \exists C \in \mathrm{GL}_d(\mathbb{C}), \; \forall i = 1, \ldots, p, \qquad C^{-1}A_iC + {}^t(C^{-1}A_iC) = w_i \, \mathrm{Id}$$

is equivalent to the existence of a nondegenerate pairing

$$(\mathscr{E}, \nabla) \otimes (\mathscr{E}, \nabla) \longrightarrow (\mathscr{O}_{\mathbb{P}^1}, d + \omega),$$

that is, of an isomorphism $(\mathscr{E}^*, \nabla^*) \xrightarrow{\sim} (\mathscr{E}, \nabla + \omega \, \mathrm{Id})$. In order that this property is fulfilled, it is *necessary* that we have an isomorphism

$$(\mathscr{M}^*, \nabla^*) \xrightarrow{\sim} (\mathscr{M}, \nabla + \omega \, \mathrm{Id})$$

and thus $(\mathscr{M}^*, \nabla^*) \simeq (\mathscr{M}, \nabla)$. According to the Riemann-Hilbert correspondence, this is equivalent to the existence of an isomorphism of the corresponding local systems.

In other words, given matrices $T_1, \ldots, T_p \in \mathrm{GL}_d(\mathbb{C})$, in order to solve the problem 1.1 with the supplementary condition $A_i + {}^tA_i = w_i \, \mathrm{Id}$, it is *necessary* that the matrices $T_1, \ldots, T_p$ are conjugate, by the *same* matrix, to *orthogonal* matrices (i.e., satisfying ${}^tT = T^{-1}$)

2 Meromorphic bundles with irreducible connection

2.1 Definition (Irreducibility). We will say that a meromorphic bundle with connection $(\mathcal{M}, \nabla)$ is *irreducible* if there exists no proper (i.e., distinct from $\mathcal{M}$ and 0) meromorphic subbundle $\mathcal{N}$, on which ∇ induces a connection, i.e., such that $\nabla(\mathcal{N}) \subset \Omega^1_{\mathbb{P}^1} \otimes_{\mathcal{O}_{\mathbb{P}^1}} \mathcal{N}$.

The first part of the statement which follows is drawn from J. Plemelj [Ple08] (see also [Tre83]) and the second one from A. Bolibrukh and V. Kostov (see for instance [AB94, Chap. 4]). The proof we give below is due to O. Gabber. This theorem will be instrumental in giving positives answers to Riemann-Hilbert and Birkhoff's problems.

2.2 Theorem (Existence of a quasitrivial lattice). *Let $(\mathcal{M}, \nabla)$ be a meromorphic bundle with connection on $\mathbb{P}^1$ having poles at the points $m_0, \dots, m_p$. Let $\mathscr{E}$ be a lattice of $(\mathcal{M}, \nabla)$ of order $r_i \geqslant 0$ at each m_i $(i = 1, \dots, p)$ and admitting a local basis in the neighbourhood of m_0 (coordinate t) in which the matrix of ∇ takes the form $A\,dt/t$ with $A \in \mathrm{M}_d(\mathbb{C})$. Let us assume that one of the two conditions below is fulfilled*

(1) *(Plemelj) the matrix A is semisimple;*
(2) *(Bolibrukh-Kostov) the bundle with connection $(\mathcal{M}, \nabla)$ is irreducible.*

Then there exists a lattice $\mathscr{E}' \subset \mathscr{E}$ which coincides with $\mathscr{E}$ when restricted to $\mathbb{P}^1 \smallsetminus \{m_0\}$, which is logarithmic at m_0 and which is quasitrivial.

Proof (First part: Plemelj). It is a matter of exhibiting a sublattice of $\mathscr{E}$ having defect *zero* (cf. Definition I.4.13). We will make a sequence of modifications like those of Proposition I.4.11 and apply Corollary I.4.14. As the residue $\mathrm{Res}_{m_0} \nabla$ is semisimple, there exists a one-dimensional eigenspace $L^o \subset E_{m_0}$ for this residue which is not contained in $(F^{a_d+1}E)_{m_0}$. The bundle $\mathscr{E}(L^o m_0)$ satisfies the same properties as $\mathscr{E}$ does: if L^o contains for instance the vector $e_d(m_0)$, a local basis of $\mathscr{E}(L^o m_0)$ is $e_1, \dots, e_{d-1}, e_d/t$ and the connection matrix in this basis and with this coordinate is $A'\,dt/t$, where A' is the diagonal matrix $(\alpha_1, \dots, \alpha_{d-1}, \alpha_d + 1)$ if A is the diagonal matrix $(\alpha_1, \dots, \alpha_d)$.

If we have $\delta(E) > 0$, Corollary I.4.14 shows that the defect of $\mathscr{E}(L^o m_0)$ is $< \delta(E)$.

By a simple induction, we get in this way a lattice $\mathscr{E}''$ contained in $\mathscr{E}(\delta(E))$, which is logarithmic at m_0 and which has defect zero, i.e., is quasitrivial. The lattice $\mathscr{E}' = \mathscr{E}''(-\delta(E))$ is then a solution to the problem. $\qquad \square$

2.3 Exercise (Adjunction of an apparent singularity). Let $(\mathcal{M}, \nabla)$ be a meromorphic bundle with connection having pole at $\Sigma = \{m_0, \dots, m_p\}$. Let us pick $m_{p+1} \in \mathbb{P}^1 \smallsetminus \Sigma$. Show that the conclusion of Theorem 2.2 holds for the bundle with connection $(\mathcal{N}, \nabla) := (\mathcal{M}(*m_{p+1}), \nabla)$ and any lattice $\mathscr{E}$ of $\mathcal{N}$ (one can apply Plemelj's Theorem at the point m_{p+1}). We say that m_{p+1} is an *apparent singularity* of the bundle with connection $(\mathcal{N}, \nabla)$.

Proof (Second part: Bolibrukh-Kostov). The main point is that the irreducibility condition provides an a priori upper bound for the *defect* of any lattice $\mathscr{E}''$ which coincides with $\mathscr{E}$ when restricted to $\mathbb{P}^1 \smallsetminus \{m_0\}$ and which is logarithmic at m_0. Let us set $r = -2 + \sum_{i=0}^p (r_i + 1)$ with $r_0 = 0$, and $R = d(d-1)r/2$. Let us begin with a general result, without assuming $r_0 = 0$.

2.4 Proposition (A priori upper bound of the defect). *Let us assume that $(\mathscr{M}, \nabla)$ is irreducible. Then, for any lattice $\mathscr{E}$ having order $r_i \geqslant 0$ at m_i ($i = 0, \ldots, p$), we have the inequality $\delta(E) \leqslant R$.*

Proof. We can assume that $d \geqslant 2$, otherwise both terms of the inequality vanish. If $a_1 \geqslant \cdots \geqslant a_d$ is the type of E, we have an isomorphism $\mathscr{E} \simeq \bigoplus_{k=1}^d \mathscr{O}_{\mathbb{P}^1}(a_k \cdot m_0)$. There is a meromorphic connection d (with pole at m_0 only, this one being logarithmic) on each summand. Let us set $\theta = \nabla - d$. Its components $\theta_{k\ell}$ are $\mathscr{O}_{\mathbb{P}^1}$-linear morphisms

$$\mathscr{O}_{\mathbb{P}^1}(a_k \cdot m_0) \xrightarrow{\ \theta_{k\ell}\ } \Omega^1_{\mathbb{P}^1}\langle \textstyle\sum_{i=0}^p (r_i + 1)m_i \rangle \otimes_{\mathscr{O}_{\mathbb{P}^1}} \mathscr{O}_{\mathbb{P}^1}(a_\ell \cdot m_0).$$

The right-hand term is isomorphic to $\mathscr{O}_{\mathbb{P}^1}(a_\ell + r)$, after Proposition I.2.11. Hence, after Exercise I.2.3, we have

$$\theta_{k\ell} \neq 0 \implies a_k \leqslant a_\ell + r.$$

If there exists k such that $a_k - a_{k+1} > r$, we have $\theta_{j\ell} = 0$ for any $j \leqslant k$ and $\ell \geqslant k+1$ and therefore the subbundle $F^{a_k} E := \bigoplus_{j \leqslant k} \mathscr{O}_{\mathbb{P}^1}(a_j)$ (cf. Exercise I.4.8) is stable by θ and thus by the connection ∇, which is impossible according to the irreducibility assumption. We thus have $a_k - a_{k+1} \leqslant r$ for any $k = 1, \ldots, d-1$ and we immediately deduce the inequality on the defect $\delta(E) = \sum_{k=2}^d (a_1 - a_k)$.

Lastly, let us note that the hypothesis of the proposition implies in particular

$$d = 1 \quad \text{or} \quad r \geqslant 0. \qquad \square$$

2.5 Remark (The semistability condition, cf. [GS95]). Let $(\mathscr{M}, \nabla)$ be a (not necessarily irreducible) meromorphic bundle with connection, having poles at the points of Σ. A lattice $\mathscr{E}$ of $\mathscr{M}$ is said to be *semistable* (as a bundle with meromorphic connection) if any subbundle $\mathscr{F}$ of $\mathscr{E}$ satisfying $\nabla(\mathscr{F}) \subset \Omega^1_{\mathbb{P}^1}(*\Sigma) \otimes \mathscr{F}$ has slope $\mu(\mathscr{F}) \leqslant \mu(\mathscr{E})$ (cf. Exercise I.4.8). The previous proposition shows in fact that the inequality on the defect is satisfied for any semistable lattice of order $r_i \geqslant 0$ at m_i. When $(\mathscr{M}, \nabla)$ is irreducible, all lattices are semistable.

End of the proof of Theorem 2.2. By assumption, there exists a local basis of $\mathscr{E}$ in the neighbourhood of m_0 and a local coordinate t, such that the connection matrix takes the form $A\,dt/t$ with A constant. Up to a base change with constant matrix, we can moreover suppose that $A = \mathrm{Res}_{m_0} \nabla$ takes the

Jordan normal form. Let us denote by $(e_{\alpha,\ell})$ this basis, where α belongs to a finite set indexing the Jordan blocks and, for any α, the integer ℓ belongs to a finite set $I_\alpha = \{1, \dots, \nu_\alpha\}$. If $\lambda(\alpha)$ denotes the eigenvalue associated to the Jordan block α, we have

$$(A - \lambda(\alpha)\,\mathrm{Id})\,(e_{\alpha,\ell}) = \begin{cases} e_{\alpha,\ell-1} & \text{if } \ell \geqslant 2 \\ 0 & \text{if } \ell = 1. \end{cases}$$

Let us set as above $R = d(d-1)\big(-2 + \sum_{i=0}^p (r_i+1)\big)/2$. We choose, for any α, a strictly increasing sequence of integers $N_{\alpha,\ell} \geqslant 0$ ($\ell \in I_\alpha$) such that, for any α and any $\ell \in \{1, \dots, \nu_\alpha\}$, we have $N_{\alpha,\ell} > N_{\alpha,\ell-1} + R$ (if we set $N_{\alpha,0} = 0$).

We then set $e'_{\alpha,\ell} = t^{N_{\alpha,\ell}} e_{\alpha,\ell}$. The $e'_{\alpha,\ell}$ generate a local lattice of $(\mathcal{M}, \nabla)$, which is contained in $\mathscr{E}$ in the neighbourhood of m_0. We can extend it as a global lattice $\mathscr{E}'$ by gluing this local lattice with $\mathscr{E}_{|\mathbb{P}^1 \smallsetminus \{m_0\}}$.

Let us note that, *whatever the choice of the sequences $(N_{\alpha,\ell})$ is, the defect of the lattice $\mathscr{E}'$ is bounded from above by the number R*, after Proposition 2.4.

It is clear that the lattice $\mathscr{E}'$ has order r_i at any $m_i \neq m_0$, as it coincides with $\mathscr{E}$ in the neighbourhood of such a point. We moreover have

$$t\nabla_{\partial/\partial t} e'_{\alpha,\ell} = (\lambda(\alpha) + N_{\alpha,\ell})e'_{\alpha,\ell} + t^{N_{\alpha,\ell} - N_{\alpha,\ell-1}} e'_{\alpha,\ell-1},$$

so that, for the lattice $\mathscr{E}'$, the connection matrix in the neighbourhood of m_0 takes the form

$$\Omega'(t) = \big(D + t^{R+1} C(t)\big)\frac{dt}{t}$$

where D is diagonal and constant and $C(t)$ has holomorphic entries. The situation is now analogous to that of Plemelj's Theorem, up to a perturbation of order R.

The modification of the lattice $\mathscr{E}'$ done in the proof of Plemelj's Theorem replaces here the connection matrix with a matrix $(D_1 + t^R C_1(t))\,dt/t$, with D_1 diagonal, so that we can iterate R times such a modification if needed, the connection matrix keeping at each step the form

$$\big(D_k + t^{R-k+1} C_k(t)\big)\frac{dt}{t}$$

(D_k diagonal). As $\delta(\mathscr{E}') \leqslant R$, we can get, by at most $\delta(\mathscr{E}')$ modifications, a lattice with defect zero. $\qquad\square$

2.6 Corollary (Existence of a trivial lattice). *Let $(\mathcal{M}, \nabla)$ be a meromorphic bundle having Σ as its set of poles and having at m_0 a regular singularity. If one of the following conditions is fulfilled*

(1) *(Plemelj) the monodromy at m_0 is semisimple,*

(2) *(Bolibrukh-Kostov) the representation* $\pi_1(\mathbb{P}^1 \smallsetminus \Sigma, o) \to \mathrm{GL}_d(\mathbb{C})$ *associated to* $(\mathcal{M}, \nabla)$ *is irreducible,*

(3) *the meromorphic bundle* $(\mathcal{M}, \nabla)$ *is irreducible,*

there exists, for any lattice $\mathcal{E}$ *of* $\mathcal{M}$, *a lattice* $\mathcal{E}'$ *of* $\mathcal{M}$ *such that*

(a) $\mathcal{E}'_{|\mathbb{P}^1 \smallsetminus \{m_0\}} = \mathcal{E}_{|\mathbb{P}^1 \smallsetminus \{m_0\}}$,

(b) $\mathcal{E}'$ *is logarithmic at* m_0,

(c) $\mathcal{E}'$ *is trivial.*

Proof. Let $\mathcal{E}$ be a lattice of $\mathcal{M}$. As $(\mathcal{M}, \nabla)$ has a regular singularity at m_0, there exists a logarithmic lattice $\mathcal{E}_0$ on some neighbourhood Δ of m_0. When restricted to $\Delta \smallsetminus \{m_0\}$, it coincides with $\mathcal{E}_{|\Delta \smallsetminus \{m_0\}} = \mathcal{M}_{|\Delta \smallsetminus \{m_0\}}$. We can therefore construct a lattice $\mathcal{E}''$ which fulfills Properties (a) and (b), by gluing $\mathcal{E}_{|\mathbb{P}^1 \smallsetminus \{m_0\}}$ with $\mathcal{E}_0$ on $\Delta \smallsetminus \{m_0\}$, due to the previous identification.

We can moreover, after Exercise II.2.6 or Corollary II.2.21, suppose that, in a suitable local basis of $\mathcal{E}_0$, the matrix of ∇ takes the form $A\,dt/t$ with A constant.

If the monodromy T is semisimple, we can choose for A the semisimple matrix $\log T$ (with respect to any choice of logarithms of eigenvalues of T) and we conclude by means of Plemelj's Theorem and of Proposition 1.11-(3).

If the representation associated to the holomorphic bundle with connection $(\mathcal{M}, \nabla)_{|\mathbb{P}^1 \smallsetminus \Sigma}$ is irreducible, so is that associated to the meromorphic bundle $(\mathcal{M}, \nabla)$ (the reader will verify this, but with some care, as the converse may not be true); therefore, Condition (2) is stronger than Condition (3). We conclude, in cases (2) and (3), by means of the Bolibrukh-Kostov Theorem 2.2-(2) and of Proposition 1.11-(3).   □

3 Application to the Riemann-Hilbert problem

Let $(\mathcal{M}, \nabla)$ be a meromorphic bundle on $\mathbb{P}^1$ having poles at the points of Σ. Let us moreover suppose that $\mathcal{M}$ has a regular singularity at each point of Σ.

3.1 Corollary. *If one of the two following conditions is fulfilled*

- *the monodromy of* $(\mathcal{M}, \nabla)$ *at* m_0 *is semisimple,*
- *the monodromy representation attached to* $(\mathcal{M}, \nabla)_{|\mathbb{P}^1 \smallsetminus \Sigma}$ *is irreducible,*

there exists a solution to Problem 1.3-(c).

Proof. Indeed, we construct a logarithmic lattice $\mathcal{E}$ of $\mathcal{M}$ by gluing local lattices. We apply Corollary 2.6 to this lattice.   □

3.2 Remark. According to the equivalence of categories given by Corollary II.3.3, both following conditions are equivalent when $(\mathcal{M}, \nabla)$ has only regular singularities:

- the monodromy representation attached to $(\mathcal{M}, \nabla)_{|\mathbb{P}^1 \smallsetminus \Sigma}$ is irreducible,

- the meromorphic bundle with connection $(\mathcal{M}, \nabla)$ is irreducible.

The Riemann-Hilbert Problem 1.1 can now be translated as follows. Given a locally constant sheaf $\mathcal{F}$ of rank d on $\mathbb{P}^1 \smallsetminus \Sigma$, does there exist on the trivial bundle E of rank d a connection ∇ with at most logarithmic poles at the points of Σ, such that, on $\mathbb{P}^1 \smallsetminus \Sigma$, we have $E^\nabla \simeq \mathcal{F}$?

We say that $\mathcal{F}$ is irreducible if there exist no locally constant subsheaf $\mathcal{F}' \subset \mathcal{F}$ of rank $d' \in \,]0, d[$. It amounts to asking for the representation attached to $\mathcal{F}$ to be irreducible.

3.3 Theorem. *If one of the following conditions is fulfilled*

- *one of the monodromy matrices $T_0, \ldots, T_p$ is semisimple,*
- *the representation $\pi_1(\mathbb{C} \smallsetminus \{m_1, \ldots, m_p\}, o) \to \mathrm{GL}_d(\mathbb{C})$ is irreducible,*

the Riemann-Hilbert Problem 1.1 has a solution.

Proof. Let us start with an irreducible locally constant sheaf $\mathcal{F}$. Let us associate to it, by the Riemann-Hilbert correspondence II.3.3, a meromorphic bundle with connection $(\mathcal{M}, \nabla)$ having regular singularity at the points of Σ. Under the second hypothesis, it is irreducible in the sense of §2.1, as the Riemann-Hilbert correspondence is an equivalence of categories. We then apply Corollary 3.1 to it in order to conclude. $\square$

3.4 Remark (The position of singularities has influence on the existence). Without any irreducibility or a semisimplicity hypothesis, one can also show (cf. [Dek79]) that the answer to Problem 1.3-(c) is positive when the rank of the bundle is 2 (the answer in the case of rank one is trivially positive). We give below (as an exercise) the solution suggested by A. Bolibrukh, more elegant than the original solution of Dekkers.

Nevertheless, starting from rank $d = 3$, there exist nonirreducible examples for which the problem has no solution. One can refer to [AB94] for a description of some of them.

One should note that the existence or nonexistence of a solution to Problem 1.1 may depend on the position of the points $m_0, \ldots, m_p$ in $\mathbb{P}^1$. Recall that we can, by an automorphism of $\mathbb{P}^1$, put m_0, m_1, m_2 at $\infty, 0, 1$ respectively. The position of the other points is then the analytic invariant of the configuration of points; for instance, the cross-ratio is the analytic invariant of the position of four points in $\mathbb{P}^1$.

However, if there exists a solution to Problem 1.1 for some position of the points $m_0, \ldots, m_p$, there exists also one for any nearby enough position. Indeed, let $\rho^o : \pi_1(\mathbb{P}^1 \smallsetminus \Sigma^o, o) \to \mathrm{GL}_d(\mathbb{C})$ be a representation. Let U^o be a union of discs centered at the points of Σ^o, pairwise disjoint and not containing the point o. If Σ'^o is a configuration of points of $\mathbb{P}^1$ having exactly one point in each component of U^o, there exists a natural isomorphism $\pi_1(\mathbb{P}^1 \smallsetminus \Sigma'^o, o) \simeq \pi_1(\mathbb{P}^1 \smallsetminus \Sigma^o, o)$ obtained by composing the isomorphisms induced by the inclusion morphisms

$$\pi_1(\mathbb{P}^1 \smallsetminus \Sigma'^o, o) \xrightarrow{\sim} \pi_1(\mathbb{P}^1 \smallsetminus U^o, o) \xleftarrow{\sim} \pi_1(\mathbb{P}^1 \smallsetminus \Sigma^o, o).$$

We deduce a representation $\rho : \pi_1(\mathbb{P}^1 \smallsetminus \Sigma'^o, o) \to \mathrm{GL}_d(\mathbb{C})$.

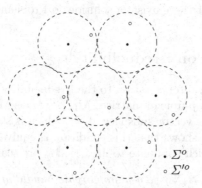

Fig. IV.1. The open set U^o.

Then, if there exists a solution to Riemann's Problem 1.1 for the representation ρ^o, there exists also one for the representation ρ, as soon as U^o is sufficiently small.

Indeed, let X be the product of $p + 1$ copies of $\mathbb{P}^1$ minus the diagonals $m_i = m_j$. In $\mathbb{P}^1 \times X$, let us consider the hypersurfaces (see Fig. III.2)

$$\Sigma_i = \{(m, m_0, \dots, m_p) \mid m = m_i\}.$$

They are smooth and pairwise disjoint. Let Σ be their union. The points of Σ^o define a point $x^o = (m_0^o, \dots, m_p^o)$ of X. There exists by assumption a trivial bundle (E^o, ∇^o) on $\mathbb{P}^1 \times \{x^o\}$ equipped with a logarithmic connection having poles at the points of Σ^o, which associated representation is ρ^o. We can then apply Remarks III.1.22 and III.1.21 to construct a bundle (E, ∇) on $U \times \mathbb{P}^1$ with logarithmic connection along Σ which restricts to (E^o, ∇^o) if $U = \prod U_i^o$ is a contractible open neighbourhood of x^o in X. The rigidity theorem for trivial bundles I.5.4 shows that, up to deleting from U some hypersurface Θ not containing x^o, the bundle E_x is trivial for any $x \in U \smallsetminus \Theta$.

3.5 Exercise (Other examples of solutions to the Riemann-Hilbert Problem 1.1, [Bol95, Dek79]).

(1) Prove that, if the monodromy representation ρ is abelian, i.e., if the matrices $T_i = \rho(\gamma_i)$ pairwise commute, there exists a solution to Problem 1.1 (show first that the logarithms of the matrices T_i pairwise commute).

(2) One now assumes that the representation ρ has rank 2 (i.e., takes values in $\mathrm{GL}_2(\mathbb{C})$).

     (a) Prove that, if the representation ρ is *reducible*, the matrices T_i have a common eigenvector. Deduce that there exists a basis in which each matrix is upper triangular.

(b) Moreover show that if none of the T_i is semisimple, the T_i pairwise commute.
(c) Conclude that there exists a solution to Problem 1.1.

4 Complements on irreducibility

We will give a matrix interpretation to the irreducibility condition.

We will say that a p-uple of matrices $M_1, \ldots, M_p \in M_d(\mathbb{C})$ is *irreducible* if there exists no proper vector subspace of $\mathbb{C}^d$, which is stable by the M_i. The classic *Schur Lemma* shows that this condition is equivalent to the fact that the only matrices which commute to $M_1, \ldots, M_p$ are scalar matrices.

4.1 Theorem. *Let $(\mathscr{M}, \nabla)$ be a meromorphic bundle of rank d with connection on $\mathbb{P}^1$, having poles at Σ and having regular singularity at any $m_i \in \Sigma$. Let us consider the following properties:*

(1) *$(\mathscr{M}, \nabla)$ is irreducible;*
(2) *for any choice of generators of $\pi_1(\mathbb{P}^1 \smallsetminus \Sigma, o)$, the $p+1$-uple consisting of the monodromy matrices $T_0, \ldots, T_p$ is irreducible;*
(3) *if we choose some coordinate t on $\mathbb{P}^1 \smallsetminus \{m_0\}$, there exists a basis $e_1, \ldots, e_d$ of the space $\Gamma(\mathbb{P}^1, \mathscr{M})$ in which the connection matrix Ω takes the form (1.4), where $(A_1, \ldots, A_p)$ is an irreducible p-uple,*
(4) *there exists a logarithmic lattice $\mathscr{E}$ of $(\mathscr{M}, \nabla)$, trivial as a $\mathscr{O}_{\mathbb{P}^1}$-module, such that the endomorphisms $\mathrm{Res}_{m_i} \nabla$ $(i = 1, \ldots, p)$ acting on $\Gamma(\mathbb{P}^1, \mathscr{E})$ (cf. Remark 1.7) form an irreducible p-uple.*

Then we have (1) $\Leftrightarrow$ (2), (3) $\Leftrightarrow$ (4) and, lastly, (1) or (2) $\Rightarrow$ (3) or (4).

Proof. The equivalence between (1) and (2) has been indicated in Remark 3.2. On the other hand, Remark 1.7 shows that (3) and (4) are equivalent.

(1) $\Rightarrow$ (4): if $(\mathscr{M}, \nabla)$ is irreducible, Corollary 3.1 furnishes a trivial logarithmic lattice. If the residues of the connection on this lattice at $m_1, \ldots, m_p$ do not form an irreducible p-uple, any subspace of $\Gamma(\mathbb{P}^1, \mathscr{E})$ invariant by these residues is also invariant by the residue at m_0, as we have $\mathrm{Res}_{m_0} \nabla = -\sum_{i=1}^{p} \mathrm{Res}_{m_i} \nabla$, and we can then construct an eigenbundle of E stable by the connection, which then has logarithmic poles at each m_i on this bundle, as it does so on E. By tensoring this subbundle by $\mathscr{O}(*\Sigma)$, we get a meromorphic subbundle with connection of $(\mathscr{M}, \nabla)$ of rank $< d$. This is in contradiction with the irreducibility assumption. $\qquad\square$

4.2 Remark. The converse implication (4) $\Rightarrow$ (1) is false. The reason is that Condition (4) means that E has no logarithmic *trivial* subbundle stable by the connection, while Condition (1) for E means that there is no logarithmic *not necessarily trivial* subbundle stable (in the meromorphic sense) by the connection (use a Deligne lattice). The exercise below gives a counterexample; I learned it from A. Bolibrukh. One can also refer to [Bol95, Prop. 5.1.1].

4.3 Exercise (According to A. Bolibrukh). We consider three matrices

$$A_1 = \begin{pmatrix} 1/2 & 0 \\ 0 & 0 \end{pmatrix}, \quad A_2 = \begin{pmatrix} -1/2 & 1 \\ 0 & 0 \end{pmatrix}, \quad A_3 = \begin{pmatrix} 0 & -1 \\ 0 & 0 \end{pmatrix}$$

and, on the trivial bundle of rank 2 over $\mathbb{C}$, the 1-form

$$\Omega = \left(\frac{A_1}{t} + \frac{A_2}{t-1} + \frac{A_3}{t+1} \right) dt.$$

(1) Prove that the meromorphic bundle with connection $(\mathcal{M}, \nabla)$ that it defines has a pole at $m_0 = 0$, $m_1 = 1$, $m_2 = -1$ only (and not at infinity) and that it is *reducible* (i.e., nonirreducible).
(2) Set

$$P(t) = \begin{pmatrix} 1 & 0 \\ 3/4t & 1 \end{pmatrix} \in \mathrm{GL}_2(\mathbb{C}[t^{-1}]).$$

Compute the connection matrix after the base change of matrix P.
(3) Show that the new basis of $\mathcal{M}$ generates a trivial logarithmic lattice and compute the residue matrices at the poles m_i.
(4) Prove, by computing the eigenspaces of these matrices, that the latter form an irreducible triple and conclude.

5 Birkhoff's problem

5.a Local analytic and algebraic Birkhoff's problems

Let us begin with the local problem and let us consider a disc centered at the origin of $\mathbb{C}$, equipped with the coordinate[2] τ. Let us be given, on this disc, a meromorphic connection of order $r \geqslant 0$ on the trivial bundle. In some basis of this bundle, the connection matrix takes the form $A(\tau)d\tau$, where $A(\tau)$ is a square matrix of size d such that $\tau^{r+1}A(\tau)$ has holomorphic entries, one of which at least does not vanish $\tau = 0$.

5.1 Local analytic Birkhoff's problem. *Does there exist a matrix $P(\tau)$ in* $\mathrm{GL}_d(\mathcal{O})$, *where $\mathcal{O}$ is the ring of converging series at 0 (i.e., P has holomorphic entries and its determinant does not vanish at 0) such that the matrix $B(\tau) = P^{-1}AP + P^{-1}P'$ can be written as*

$$(5.2) \qquad B(\tau) = \tau^{-(r+1)} B_{-(r+1)} + \cdots + \tau^{-1} B_{-1}$$

where $B_{-1}, \ldots, B_{-(r+1)}$ are constant matrices?

We then say that the matrix B takes Birkhoff's normal form: its Laurent expansion at 0 has no coefficient with positive index.

---

[2] Changing the name of the coordinate will enable us, in the following, to better distinguish the objects and their Fourier transforms.

Remarks.

(1) If one only asks for the matrix $B(\tau)$ to be a rational fraction with pole at 0 at most, that is, if one also accepts a finite number of coefficients B_k with $k \geqslant 0$, the problem has a solution (see Corollary II.3.8). This is the result proved by Birkhoff in [Bir09] (see the proof of Birkhoff in [Sib90, §3.3]).

(2) In the statement of Birkhoff's problem, there is no hypothesis on the germ of the underlying meromorphic bundle: it may or may not have a regular singularity (if it has a regular singularity, the matrix $B_{-(r+1)}$ is nilpotent). It is quite easy (cf. [Gan59, vol. 2], see also [Mas59]) to give examples where the problem has no solution (unlike what Birkhoff thought having proved): when $r = 0$, the problem reduces to asking whether any germ of logarithmic lattice admits a basis in which the connection matrix takes the form B_{-1}/τ; if the monodromy is not semisimple, there exists Levelt normal forms which are not of this kind (choose a nontrivial decreasing filtration of H which is stable by monodromy, cf. §III.1.a).

(3) On the other hand, Turrittin ([Tur63], see also [Sib90, §3.10]) proved that, when the eigenvalues of $A_{-(r+1)}$ are pairwise distinct, there exists a matrix $P \in \mathrm{GL}_d(\mathscr{O}[\tau^{-1}])$ such that B takes the form (5.2) (with the same index r, which is not evident, as P may have a pole at 0).

We will also consider the problem in its algebraic form. In this case, we assume that $A(\tau)$ has entries in the ring $\mathbb{C}[\tau, \tau^{-1}]$ of Laurent polynomials.

5.3 Algebraic Birkhoff's problem. *Let us suppose moreover that the connection with matrix $\Omega = A(\tau)d\tau$ has a regular singularity at infinity. Does there exist a matrix $P \in \mathrm{GL}_d(\mathbb{C}[\tau])$ such that $B(\tau) := P^{-1}AP + P^{-1}P'$ takes the form (5.2)?*

Let us begin by noticing that the algebraic problem is nothing but a particular case of the local problem analytic: as a matter of fact, in the algebraic situation, assume that we have found a matrix $P \in \mathrm{GL}_d(\mathbb{C}\{\tau\})$ giving to the connection matrix the form (5.2). Then P is a germ at 0 of solution of the system $P' = PB - AP$, which is algebraic and has a regular singularity at infinity. This implies that P has polynomial entries (cf. Exercise II.3.10). One can show similarly that P^{-1} has polynomial entries.

Let us also notice that, in the local analytic problem 5.1, we can suppose that the matrix $A(\tau)d\tau$ is as in Corollary 5.3: it is indeed enough to apply the base change given by Corollary II.3.8.

Let us formulate the algebraic problem in an equivalent way:

5.4 Algebraic Birkhoff's problem (bis). *Let $(\mathscr{M}, \nabla)$ be a meromorphic bundle of rank d on $\mathbb{P}^1$, with pole at 0 and ∞ only, equipped with a connection having a regular singularity at ∞ and admitting, in the neighbourhood of 0, a lattice of order $r \geqslant 1$. Does there exist a lattice of $(\mathscr{M}, \nabla)$ which coincides* *with the given lattice in the neighbourhood of* 0, *which is logarithmic at infinity and which is isomorphic to the trivial bundle of rank d?*

Using Zariski topology and Proposition I.4.15, we obtain the following formulation:

5.5 Algebraic Birkhoff's problem (ter). *Let* $(\mathbb{M}, \nabla)$ *be a free* $\mathbb{C}[\tau, \tau^{-1}]$*-module of rank* d *equipped with a connection having a pole at* 0 *and* ∞ *only, with a regular singularity at* ∞. *Let* $\mathbb{E}_0 \subset \mathbb{M}$ *be a free* $\mathbb{C}[\tau]$*-module which is a lattice of order* $r \geqslant 1$. *Does there exist on* U_∞ *a lattice* $\mathbb{E}_\infty \subset \mathbb{M}$, *which is logarithmic at* ∞, *such that we have* $\mathbb{E}_0 = (\mathbb{E}_0 \cap \mathbb{E}_\infty) \oplus \tau \mathbb{E}_0$?

Lastly, we can formulate the equivalence between the local Birkhoff's Problem 5.1 and the algebraic Birkhoff's Problem 5.3 or 5.5 in the proposition below, the (easy) proof of which is left to the reader.

Let $(\mathbb{M}, \nabla)$ be a $(\boldsymbol{k}, \nabla)$-vector space of rank d and let $(\mathscr{M}, \nabla)$ be a meromorphic bundle on $\mathbb{P}^1$, having a pole at 0 and ∞, with a regular singularity at ∞, whose germ $(\mathscr{M}, \nabla)_0$ is isomorphic to $(\mathbb{M}, \nabla)$; we know (cf. Corollary II.3.7) that such a bundle is unique up to isomorphism. Let $\mathcal{E}$ be a lattice of $\mathcal{M}$.

5.6 Proposition (The various Birkhoff's problems are equivalent).

(1) *If* $\boldsymbol{e}$ *is a basis of* $\mathcal{E}$ *in which the matrix of* ∇ *takes the form* (5.2)
   - *the* $\mathbb{C}[\tau, \tau^{-1}]$*-module* $\mathbb{M} := \mathbb{C}[\tau, \tau^{-1}] \cdot \boldsymbol{e} \subset \mathcal{M}$ *is stable by the connection* ∇ *and it has singularity at* 0 *and* ∞ *only, the latter being regular;*
   - *the* $\mathbb{C}[\tau]$*-module* $\mathbb{E}_0 := \mathbb{C}[\tau] \cdot \boldsymbol{e} \subset \mathcal{E}$ *is a lattice on* U_0, *with analytic germ* $\mathcal{E}$ *at the origin;*
   - *the* $\mathbb{C}[\tau^{-1}]$*-module* $\mathbb{E}_\infty := \mathbb{C}[\tau^{-1}] \cdot \boldsymbol{e}$ *is a logarithmic lattice at infinity, which gives a solution to Problem 5.5 for* $\mathbb{E}_0$.
(2) *Conversely, if* $\mathbb{E}_0$ *is a lattice of* $\mathbb{M}$ *on* U_0, *with analytic germ* $\mathcal{E}$ *at* 0 *and if* $\mathbb{E}_\infty$ *is a solution to Problem 5.5 for* $\mathbb{E}_0$, *a basis of the trivial bundle defined by* $(\mathbb{E}_0, \mathbb{E}_\infty)$ *induces a basis of* $\mathcal{E}$ *in which the connection matrix takes the form* (5.2). □

If $d = 1$, there exists a solution to Problem 5.1 (cf. Remark III.2.13-(2)). On the other hand, Corollary 2.6 implies immediately:

5.7 Corollary (Solution to Birkhoff's problem in the irreducible case). *If* $(\mathscr{M}, \nabla)$ *is irreducible, or if the monodromy of* $(\mathscr{M}, \nabla)_{|\mathbb{C}^*}$ *is semisimple, the Problem 5.3 has a solution.* □

5.8 Remark (Birkhoff's problem with a sesquilinear form). Let $(\mathscr{E}, \nabla)$ be a lattice which is a solution to Problem 5.4. There exists a nondegenerate bilinear (resp. sesquilinear) form of weight $w \in \mathbb{Z}$ on $(\mathscr{E}, \nabla)$ if and only if, after conjugating by the same constant matrix $C \in \mathrm{GL}_d(\mathbb{C})$, the matrices B_{-k} are skewsymmetric (resp. $(-1)^k$-symmetric) for $k \geqslant 2$ and $B_{-1} + {}^t B_{-1} = w\,\mathrm{Id}$. In order that such a form exists, it is necessary that we have $(\mathscr{M}^*, \nabla^*) \simeq (\mathscr{M}, \nabla)$ (resp. $\simeq (\overline{\mathscr{M}}, \overline{\nabla})$).

5.b M. Saito's criterion

We will now give an existence criterion for Birkhoff's problem as stated in Problem 5.5, a criterion which is due to M. Saito (in a slightly different way however; cf. [Sai89] and also [Sab06]). Let us take up the situation and the notation of §III.2.b.

Let $\mathcal{V}'$ be a Deligne lattice of the germ $\mathcal{M}'$ of $\mathcal{M}$ at ∞, for instance that used in Chapter III, such that the eigenvalues of the residue of ∇ have their real part in $]-1,0]$. Let $\mathbb{V}'$ be the $\mathbb{C}[\tau']$-submodule of $\mathbb{M}$, the analytic germ of which at ∞ is equal to $\mathcal{V}'$. We have in an evident way $\mathcal{V}'/\tau'\mathcal{V}' = \mathbb{V}'/\tau'\mathbb{V}'$, a space that we will denote by H', equipped with its monodromy T'.

Let us consider the increasing filtration of $\mathbb{M}$ by the $\tau^{-k}\mathbb{E}_0 = \tau'^{k}\mathbb{E}_0$, for $k \in \mathbb{Z}$. It induces an increasing filtration on H' by the formula

$$G'_k = \mathbb{V}' \cap \tau'^{k}\mathbb{E}_0/(\tau'\mathbb{V}') \cap \tau'^{k}\mathbb{E}_0.$$

This filtration is still exhaustive (cf. Lemma III.2.2).

5.9 Theorem. *Let $G'_\bullet$ be the filtration of H' associated to $\mathbb{E}_0$. Then, any decomposition $H' = \bigoplus_{p\in\mathbb{Z}} H'_p$ with respect to which T' is lower triangular (i.e., $T'(H'_p) \subset H'_p \oplus H'_{p+1} \oplus \cdots$ for any p) and compatible with the filtration $G'_\bullet$ (i.e., $G'_k = \bigoplus_{p\leqslant k} H'_p$ for any k) defines a canonical solution to Birkhoff's Problem 5.5.*

It will be useful to formulate the condition in the theorem in terms of a pair of filtrations:

5.10 Lemma. *There exists a decomposition as in Theorem 5.9 if and only if there exists an exhaustive decreasing filtration $H'^\bullet$ of H' which is stable by the monodromy T' and which is opposite to the increasing filtration $G'_\bullet$, that is, such that*

$$\frac{H'^\ell \cap G'_k}{(H'^{\ell+1} \cap G'_k) + (H'^\ell \cap G'_{k-1})} = 0 \quad for\ k \neq \ell.$$

Proof. It is clear that the condition in the theorem implies the condition in the lemma, by setting $H'^\ell = \bigoplus_{p\geqslant \ell} H'_p$.

Conversely, if the condition in the lemma is fulfilled, we first show by decreasing induction on $\ell - k$ that $H'^\ell \cap G'_k = 0$ if $\ell > k$: this is clearly true for $\ell - k \gg 0$ and we then use that, for $\ell > k$, we have $H'^\ell \cap G'_k = (H'^{\ell+1} \cap G'_k) + (H'^\ell \cap G'_{k-1})$. We deduce that, if we put $H'_p = H'^p \cap G'_p$, we have $H'_p \cap H'_q = 0$ for $p \neq q$. Lastly, we show by increasing induction on $j \geqslant 0$ that the intersection $H'^{k-j} \cap G'_k$ can be decomposed as a sum (which is direct, after what precedes) $H'_k + \cdots + H'_{k-j}$. $\qquad\square$

5.11 Example. Let us assume that $\dim H' = 2$ and that T' is not semisimple. There exists thus a single eigen line L for T'. If a nontrivial decomposition

$H' = H'_p \oplus H'_q$ exists, with $p < q$, we must have $H'_q = L$ and, consequently, $G'_p \neq L$. In other words, the filtration $G'_\bullet$ defined by

$$G'_{-1} = 0, \quad G'_0 = L, \quad G'_1 = H'$$

does not fulfill the condition in the theorem. On the other hand, if $G'_0 \neq L$, the condition of the theorem is satisfied, as in the case where T' is semisimple.

5.12 Exercise. Give an explicit form to the condition of the theorem for $\dim H' = 3$.

Proof (of Theorem 5.9). Let us begin with a result concerning opposite filtrations. Let V be a finite dimensional $\mathbb{C}$-vector space equipped with an exhaustive decreasing filtration $V^\bullet$ (all filtrations are indexed by $\mathbb{Z}$). We denote by $\mathrm{gr}V = \bigoplus_\ell (V^\ell/V^{\ell+1})$ the graded vector space. We moreover assume that V is equipped with two filtrations $F_\bullet V$ (exhaustive increasing) and $F'^\bullet V$ (exhaustive decreasing). They induce on $\mathrm{gr}V$ the following filtrations

$$F_\bullet \mathrm{gr}V = \bigoplus_\ell \left[(F_\bullet \cap V_\ell)/(F_\bullet \cap V_{\ell+1}) \right]$$
$$F'^\bullet \mathrm{gr}V = \bigoplus_\ell \left[(F'^\bullet \cap V_\ell)/(F'^\bullet \cap V_{\ell+1}) \right].$$

Lemma. *Let us suppose that $F_\bullet \mathrm{gr}V$ and $F'^\bullet \mathrm{gr}V$ are opposite. Then $F_\bullet V$ and $F'^\bullet V$ are so and, for any k, we have $\mathrm{gr}(F_k V \cap F'^k V) = F_k \mathrm{gr}V \cap F'^k \mathrm{gr}V$.*

Proof. We will give a "geometric" proof of this result. In order to do so, let us introduce a new variable u. The $\mathbb{C}$-vector space $\mathbb{F} = \bigoplus_k u^k F_k V$ is naturally equipped with the structure of a $\mathbb{C}[u]$-module, as $F_\bullet$ is increasing. Similarly, $\mathbb{F}' = \bigoplus_k u^k F'^k V$ is naturally equipped with the structure of a $\mathbb{C}[u^{-1}]$-module. Both are contained in the free $\mathbb{C}[u, u^{-1}]$-module $\mathbb{C}[u, u^{-1}] \otimes_{\mathbb{C}} V$. One checks (by taking a basis adapted to the corresponding filtration) that $\mathbb{F}$ is $\mathbb{C}[u]$-free of rank $\dim V$, and $\mathbb{F}'$ is $\mathbb{C}[u^{-1}]$-free, of the same rank. Moreover,

$$\mathbb{C}[u, u^{-1}] \underset{\mathbb{C}[u]}{\otimes} \mathbb{F} = \mathbb{C}[u, u^{-1}] \underset{\mathbb{C}}{\otimes} V = \mathbb{C}[u, u^{-1}] \underset{\mathbb{C}[u^{-1}]}{\otimes} \mathbb{F}'.$$

These data enable us to define, by a gluing procedure, a vector bundle $\mathscr{F}(F_\bullet V, F'^\bullet V)$ on $\mathbb{P}^1$.

Exercise. The filtrations $F_\bullet V$ and $F'^\bullet V$ are opposite if and only if the vector bundle $\mathscr{F}(F_\bullet V, F'^\bullet V)$ is isomorphic to the trivial bundle (of rank $\dim V$).

The lemma can be shown by induction on the length of the filtration $V^\bullet$. One is reduced to the case of a filtration of length two, that is, to an exact sequence

$$0 \longrightarrow V^1 = \mathrm{gr}^1 V \longrightarrow V = V^0 \longrightarrow V/V^1 = \mathrm{gr}^0 V \longrightarrow 0.$$

By definition of the induced filtration on the graded pieces, one has an exact sequence of vector bundles on $\mathbb{P}^1$:

$$0 \to \mathscr{F}(F_\bullet \mathrm{gr}^1 V, F'^\bullet \mathrm{gr}^1 V) \longrightarrow \mathscr{F}(F_\bullet V, F'^\bullet V) \longrightarrow \mathscr{F}(F_\bullet \mathrm{gr}^0 V, F'^\bullet \mathrm{gr}^0 V) \to 0.$$

The assumption means that both extreme terms are isomorphic to trivial bundles. Using an argument analogous to that of Theorem I.2.8, the previous extension is split, hence the filtrations $F_\bullet V$ and $F'^\bullet V$ are opposite.

For any ℓ and any k, the map

$$(*) \quad \mathrm{gr}^\ell(F_k V \cap F'^k V) := (V^\ell \cap F_k V \cap F'^k V)/(V^{\ell+1} \cap F_k V \cap F'^k V)$$
$$\longrightarrow F_k \mathrm{gr}^\ell V \cap F'^k \mathrm{gr}^\ell V \subset \mathrm{gr}^\ell V$$

is injective. Therefore,

$$(**) \qquad \dim \mathrm{gr}^\ell(F_k V \cap F'^k V) \leqslant \dim(F_k \mathrm{gr}^\ell V \cap F'^k \mathrm{gr}^\ell V).$$

On the other hand, $\dim(F_k V \cap F'^k V) = \dim \mathrm{gr}(F_k V \cap F'^k V)$. Therefore

$$\begin{aligned}
\dim V &= \sum_k \dim(F_k V \cap F'^k V) && \text{(opposite filtrations)} \\
&= \sum_k \dim \mathrm{gr}(F_k V \cap F'^k V) \\
&= \sum_k \sum_\ell \dim \mathrm{gr}^\ell(F_k V \cap F'^k V) \\
&\leqslant \sum_k \sum_\ell \dim(F_k \mathrm{gr}^\ell V \cap F'^k \mathrm{gr}^\ell V) && \text{after } (**) \\
&= \sum_k \dim(F_k \mathrm{gr} V \cap F'^k \mathrm{gr} V) = \dim \mathrm{gr} V && \text{(opposite filtrations)} \\
&= \dim V,
\end{aligned}$$

whence the equality in $(**)$. Consequently, $(*)$ is also surjective. $\qquad \square$

Let us come back to the proof of Theorem 5.9. We will construct a $\mathbb{C}[\tau']$-module $\mathbb{E}_\infty$, which is a solution to Problem 5.5 for $\mathbb{E}_0$. Let us consider the $\mathbb{C}[\tau', \tau'^{-1}]$-module

$$\mathbb{M}^{\mathrm{reg}} := \bigoplus_{k \in \mathbb{Z}} (\tau'^k \mathbb{V}'/\tau'^{k+1} \mathbb{V}') \simeq \mathbb{C}[\tau', \tau'^{-1}] \underset{\mathbb{C}}{\otimes} H'.$$

It is equipped with a natural connection induced by that of $\mathbb{M}$ and, if e is a basis of H', hence a $\mathbb{C}[\tau', \tau'^{-1}]$-basis of $\mathbb{M}^{\mathrm{reg}}$, the connection matrix in this basis is nothing but $\mathrm{Res}_{\mathcal{V}'} \nabla \cdot d\tau'/\tau'$. Let us note that $\mathbb{M}^{\mathrm{reg}}$ has a regular singularity at ∞ and at 0. This module can be identified, as mentioned above, to the $\mathbb{C}[\tau, \tau^{-1}]$-module with connection $\mathrm{gr}_{\mathcal{V}'} \mathcal{M}$ that we have yet considered in II.2.22. When $\mathbb{M}$ does not have a regular singularity at 0, one cannot identify $\mathbb{M}$ with $\mathbb{M}^{\mathrm{reg}}$, as the latter module is regular at 0. On the other hand, if $\mathbb{M}$ has a regular singularity, both modules are isomorphic.

The filtration $H'^{\bullet}$ enables us to construct a logarithmic lattice $\mathbb{E}_{\infty}^{\mathrm{reg}}$ of $\mathbb{M}^{\mathrm{reg}}$: we set

$$\mathbb{E}_{\infty}^{\mathrm{reg}} = \bigoplus_{k \in \mathbb{Z}} \tau'^{k} H'^{-k}.$$

On the other hand, the lattice $\mathbb{E}_0$ of $\mathbb{M}$ enables us to construct a lattice $\mathbb{E}_0^{\mathrm{reg}}$ of $\mathbb{M}^{\mathrm{reg}}$: we set

$$\mathbb{E}_0^{\mathrm{reg}} = \bigoplus_{k \in \mathbb{Z}} (\mathbb{E}_0 \cap \tau'^{k} \mathbb{V})/(\mathbb{E}_0 \cap \tau'^{k+1} \mathbb{V}).$$

The condition of the theorem (or, more precisely, that of the lemma) exactly means that $\mathbb{E}_{\infty}^{\mathrm{reg}}$ is a solution to Problem 5.5 for the lattice $\mathbb{E}_0^{\mathrm{reg}}$ of $\mathbb{M}^{\mathrm{reg}}$.

As $\mathbb{M}$ has a regular singularity at infinity, we can regard $\mathbb{M}^{\mathrm{reg}}$ as a $\mathbb{C}[\tau, \tau^{-1}]$ submodule of $\mathcal{M}'$ (cf. Remark II.2.22). The lattice $\mathbb{E}_{\infty}^{\mathrm{reg}}$ defines a lattice $\mathcal{E}_{\infty} \subset \mathcal{M}'$ if we put $\mathcal{E}_{\infty} = \mathbb{C}\{\tau'\} \otimes_{\mathbb{C}[\tau']} \mathbb{E}_{\infty}^{\mathrm{reg}}$. There exists then a unique lattice $\mathbb{E}_{\infty}$ of $\mathbb{M}$ on the chart U_{∞}, the germ of which at ∞ is equal to $\mathcal{E}_{\infty}$. By construction, we have

$$\mathrm{gr}_{V'} \mathbb{E}_{\infty} = \mathrm{gr}_{V'} \mathcal{E}_{\infty} = \mathbb{E}_{\infty}^{\mathrm{reg}}.$$

Let us consider the filtrations $\mathbb{E}_k = \tau'^{k} \mathbb{E}_0$ and $\mathbb{E}_{\infty}^k = \tau'^{k} \mathbb{E}_{\infty}$. The assumption is then that the filtrations induced on $\mathbb{V}'^0/\mathbb{V}'^1$ are opposite. Multiplying by τ'^j, this implies that they induce opposite filtrations on $\mathbb{V}'^j/\mathbb{V}'^{j+1}$ for any j. It follows from the previous lemma that, for any $\ell \geqslant 1$, the filtrations they induce on $\mathbb{V}'^0/\mathbb{V}'^{\ell}$ are opposite and that, using the notation of Lemma 5.10,

$$\dim\left[(\mathbb{E}_k \cap \mathbb{V}'^0 + \mathbb{V}'^{\ell}) \cap (\mathbb{E}_{\infty}^k \cap \mathbb{V}'^0 + \mathbb{V}'^{\ell}) \bmod \mathbb{V}'^{\ell}\right] = \sum_{j=0}^{\ell-1} \dim(G'_{k-j} \cap H'^{k-j}).$$

But, as k is fixed, we have $\mathbb{E}_k \cap \mathbb{V}'^{\ell} = 0$ and $\mathbb{V}'^{\ell} \not\subset \mathbb{E}_{\infty}^k$ for $\ell \gg 0$, hence the left-hand term above is nothing but $\mathbb{E}_k \cap \mathbb{E}_{\infty}^k \cap \mathbb{V}'^0$ and, multiplying by τ'^{-k}, we get

$$\dim(\mathbb{E}_0 \cap \mathbb{E}_{\infty} \cap \mathbb{V}'^{-k}) = \dim(\mathbb{E}_k \cap \mathbb{E}_{\infty}^k \cap \mathbb{V}'^0) = \sum_{j \geqslant 0} \dim(G'_{k-j} \cap H'^{k-j}).$$

For $k \gg 0$, this gives

$$\dim(\mathbb{E}_0 \cap \mathbb{E}_{\infty}) = \dim(\mathbb{V}'^0/\mathbb{V}'^1) = \mathrm{rk}\,\mathbb{E}_0.$$

In order to conclude, it is enough to show (cf. Proposition I.4.15) that the natural map $\mathbb{E}_0 \cap \mathbb{E}_{\infty} \to \mathbb{E}_0/\tau \mathbb{E}_0$ is injective, as it will then be bijective. Using oppositeness once more, we get, for any k,

$$\mathbb{V}'^k \cap \tau \mathbb{E}_0 \cap \mathbb{E}_{\infty} \subset \mathbb{V}'^{k+1},$$

hence
$$\mathbb{V}'^{k} \cap \tau\mathbb{E}_0 \cap \mathbb{E}_\infty \subset \mathbb{V}'^{k+1} \cap \tau\mathbb{E}_0 \cap \mathbb{E}_\infty.$$

Iterating the assertion, we deduce that, for any k, $\mathbb{V}'^{k} \cap \tau\mathbb{E}_0 \cap \mathbb{E}_\infty$ is zero, hence (taking $k \ll 0$) $\tau\mathbb{E}_0 \cap \mathbb{E}_\infty$ is zero, whence the desired injectivity.   □

5.13 Remarks.

(1) One can show (see for instance [Sai89, Sab06]) that, if one fixes an order as in §III.1.c and if for any λ one can find a decomposition of H'_λ as in Theorem 5.9 compatible with the filtration naturally induced by $\mathbb{G}_\bullet$ on H'_λ, then the characteristic polynomial of B_{-1} for the canonical solution to Birkhoff's problem given by Theorem 5.9 is nothing but the characteristic polynomial $\chi^\infty_{\mathcal{E}_0}(s)$ associated to the lattice $\mathcal{E}_0$ as in §III.2.b.

(2) Proposition III.1.6 applied to the lattice $\mathbb{E}^{\mathrm{reg}}_\infty$ shows that, for the canonical solution to Birkhoff's problem given by Theorem 5.9, the matrix B_{-1} is *semisimple* if and only if the unipotent part T'_u of T' is lower triangular with respect to the decomposition $\oplus H'_p$, that is, $(T'_u - \mathrm{Id})(H'_p) \subset H'_{p+1} \oplus \cdots$.

(3) Assume that we are given a nondegenerate bilinear or sesquilinear form of weight w on $\mathbb{E}_0$. It amounts to giving a basis of $\mathbb{E}_0$ in which the matrix of ∇ takes the form

$$\left(\frac{1}{\tau}\left(A_{-r-1}\tau^{-r} + \cdots + A_{-1}\right) + \sum_{k \geqslant 0} A_l \tau^k \right) d\tau$$

where, for $k \neq -1$, A_k is skewsymmetric (resp. $(-1)^k$-symmetric) and A_{-1} satisfies $A_{-1} + {}^t A_{-1} = w\,\mathrm{Id}$.

We wish to give a condition in order to find a Birkhoff normal form with the same property. It is a matter of extending to $\mathbb{E}$ the nondegenerate bilinear (or sesquilinear) form defined on $\mathbb{E}_0$. Let us note that this form can be extended in an evident way as a form of the same kind on the meromorphic bundle $(\mathbb{M}, \nabla)$ and defines thus, by the Riemann-Hilbert correspondence (cf. Corollary III.1.12 and Exercise III.1.13), a nondegenerate bilinear form on H'.

It immediately follows from this Corollary that the form can be extended as a form of weight w on $\mathbb{E}$ if and only if, with respect to the bilinear form $\langle\,,\,\rangle$ on H', the filtration $H'^\bullet$ of Lemma 5.10 satisfies

$$(H'^{k}_{\neq 1})^{\perp} = H'^{-k-w}_{\neq 1} \quad \text{and} \quad (H'^{k}_{1})^{\perp} = H'^{-k-w+1}_{1}.$$

Indeed, these conditions are equivalent to the fact that the form on $(\mathbb{M}, \nabla)$ induces on $\mathbb{E}_\infty$ a nondegenerate form of weight w, whence the existence, by gluing, of the desired form on $\mathbb{E}$.

V

Fourier-Laplace duality

Introduction

The Fourier transform exchanges functions of two real variables x, y with rapid decay at infinity as well as all their derivatives and functions of the same kind in the variables η, ξ. If we put $t = x + iy$ and $\theta = \xi + i\eta$ (variable denoted by τ' from §2 on), it is expressed by the formula

$$\widehat{\varphi}(\eta, \xi) = \int_{\mathbb{R}^2} \varphi(x, y) e^{\overline{t\theta} - t\theta} \, dt \wedge d\overline{t}.$$

The inverse transform is expressed by an analogous formula by using the kernel $e^{t\theta - \overline{t\theta}}$. One extends these transforms, by the usual duality procedure, to temperate distributions.

If the temperate distribution φ is a solution to a linear holomorphic differential equation with polynomial coefficients

$$a_d(t) \left(\frac{\partial}{\partial t} \right)^d \varphi + \cdots + a_1(t) \frac{\partial \varphi}{\partial t} + a_0(t) \varphi = 0$$

then the distribution $\widehat{\varphi}$ is itself a solution to the differential equation

$$a_d(-\partial/\partial\theta) \left(\theta^d \widehat{\varphi} \right) + \cdots + a_1(-\partial/\partial\theta)(\theta\widehat{\varphi}) + a_0(-\partial/\partial\theta)\widehat{\varphi} = 0.$$

In this chapter, we will extend the algebraic aspect of this transform, as it appears in the differential equations above, to lattices of $(\mathbb{C}(t), \nabla)$-vector spaces. We will thus make a correspondence, under some conditions, between the solutions of Birkhoff's problem and the solutions of the partial Riemann-Hilbert problem.

The Fourier transform also exhibits a local variant, the *microlocalization*, which will be useful to analyze duality on lattices and its behaviour by Fourier transform.

The Fourier transform does not induce a correspondence between $(\mathbb{C}(t), \nabla)$-vector spaces and $(\mathbb{C}(\theta), \nabla)$-vector spaces: as a matter of fact, torsion phenomena may occur because, in the case of temperate distributions for instance, the Fourier transform of a constant function is a distribution with punctual support.

The Fourier transform naturally acts on modules over the one-variable Weyl algebra. One will find a detailed study of this transform in [Mal91]. We will only consider the properties which are useful for the study of the Fourier transform of lattices.

1 Modules over the Weyl algebra

We will not give complete proofs for the following results. The reader can refer for instance to [Bjö79, Bd83, Cou95, Ehl87, Kas95, Sab93, GM93, Mal91, Sch94].

1.a The one-variable Weyl algebra

This is the quotient algebra of the free algebra generated by the polynomial algebras $\mathbb{C}[t]$ and $\mathbb{C}[\partial_t]$ by the relation $[\partial_t, t] = 1$. We will denote it by $\mathbb{C}[t]\langle \partial_t \rangle$. It is a noncommutative algebra. Its elements are the *holomorphic differential operators* with polynomial coefficients. Such an operator can be written in a unique way as

$$P(t, \partial_t) = a_d(t)\partial_t^d + \cdots + a_1(t)\partial_t + a_0(t)$$

where the a_i are polynomials in t, and $a_d \neq 0$. We call d the *degree* of the operator. The product of two such operators can be reduced to this form by using the commutation relation

$$(1.1) \qquad\qquad \partial_t a(t) = a(t)\partial_t + a'(t).$$

Its degree is equal to the sum of the degrees of its factors. The *symbol* of an operator is the class of the operator modulo the operators of strictly lower degree. The set of symbols of differential operators, equipped with the induced operations, is identified to the algebra of polynomials of two variables with coefficients in $\mathbb{C}$; it is thus a Noetherian ring. This enables one to prove that the Weyl algebra is itself left and right Noetherian.

The Weyl algebra contains as subalgebras the algebras $\mathbb{C}[t]$ (operators of degree 0) and $\mathbb{C}[\partial_t]$ (operators with constant coefficients).

The *transposition* is the involution $P \mapsto {}^tP$ of $\mathbb{C}[t]\langle \partial_t \rangle$ defined by the following properties:

(1) ${}^t(P \cdot Q) = {}^tQ \cdot {}^tP$ for all $P, Q \in \mathbb{C}[t]\langle \partial_t \rangle$,
(2) ${}^tP = P$ for any $P \in \mathbb{C}[t]$,
(3) ${}^t\partial_t = -\partial_t$.

1.b Holonomic modules over the Weyl algebra

There is an equivalence between $\mathbb{C}[t]\langle\partial_t\rangle$-modules and $\mathbb{C}[t]$-modules equipped with a connection: the Leibniz rule for connections is translated to Relation (1.1) in the Weyl algebra, when acting on the left on the $\mathbb{C}[t]$-module M, extending thus the natural action of $\mathbb{C}[t]$. In the following, when saying "module" over the Weyl algebra, we will usually mean *left module of finite type* over $\mathbb{C}[t]\langle\partial_t\rangle$.

Left to right, right to left. If M^d is a right module over $\mathbb{C}[t]\langle\partial_t\rangle$, equip it canonically with a structure of left module M^l by setting, for any $m \in \mathrm{M}^d$,

$$P \cdot m := m^t P.$$

Conversely, to any left module is canonically associated a right module.

Any module has thus a presentation

$$(1.2) \qquad \mathbb{C}[t]\langle\partial_t\rangle^p \xrightarrow{\;\cdot A\;} \mathbb{C}[t]\langle\partial_t\rangle^q \longrightarrow \mathrm{M} \longrightarrow 0,$$

where A is a $p \times q$ matrix with entries in $\mathbb{C}[t]\langle\partial_t\rangle$. The vectors are here written as line vectors, multiplication by A is done on the right and commutes thus with the left action of $\mathbb{C}[t]\langle\partial_t\rangle$.

We will say that a module M over $\mathbb{C}[t]\langle\partial_t\rangle$ is *holonomic* if any element of M is killed by some nonzero operator of $\mathbb{C}[t]\langle\partial_t\rangle$, i.e., satisfies a nontrivial differential equation. This equation can have degree 0: we will then say that the element is *torsion*. A holonomic module contains by definition no submodule isomorphic to $\mathbb{C}[t]\langle\partial_t\rangle$. In an exact sequence of $\mathbb{C}[t]\langle\partial_t\rangle$-modules

$$0 \longrightarrow \mathrm{M}' \longrightarrow \mathrm{M} \longrightarrow \mathrm{M}'' \longrightarrow 0$$

the middle term is holonomic if and only if the extreme ones are so.

1.3 Exercise (Products of operators and exact sequences of modules).

(1) Prove that, in the Weyl algebra $\mathbb{C}[t]\langle\partial_t\rangle$, we have

$$PQ = 0 \implies P = 0 \text{ or } Q = 0$$

(consider the highest degree terms in ∂_t).

(2) Show that, if $Q \neq 0$, right multiplication by Q induces an isomorphism

$$\mathbb{C}[t]\langle\partial_t\rangle/(P) \xrightarrow[\sim]{\;\cdot Q\;} (Q)/(PQ),$$

if (R) denotes the left ideal $\mathbb{C}[t]\langle\partial_t\rangle \cdot R$.

(3) Prove that the sequence of left $\mathbb{C}[t]\langle\partial_t\rangle$-modules is exact:

$$0 \longrightarrow \mathbb{C}[t]\langle\partial_t\rangle/(P) \xrightarrow{\;\cdot Q\;} \mathbb{C}[t]\langle\partial_t\rangle/(PQ) \longrightarrow \mathbb{C}[t]\langle\partial_t\rangle/(Q) \longrightarrow 0.$$

Given $\delta = (\delta_1, \ldots, \delta_q) \in \mathbb{Z}^q$, let us call δ-degree of an element $(P_1, \ldots, P_q)$ of $\mathbb{C}[t]\langle \partial_t \rangle^q$ the integer $\max_{k=1,\ldots,q}(\deg P_k - \delta_k)$. Let us call *good filtration* any filtration of $\mathbb{M}$ obtained as the image, in some presentation like (1.2), of the filtration of $\mathbb{C}[t]\langle \partial_t \rangle^q$ by the δ-degree, for some $\delta \in \mathbb{Z}^q$.

The graded module of $\mathbb{M}$ with respect to some good filtration is a module of finite type over the ring of polynomials of two variables.

1.4 Proposition.

(1) *A module is holonomic if and only if its graded module with respect to some (or any) good filtration has a nontrivial annihilator.*

(2) *Any holonomic module can be generated by one element.*   □

We deduce that a holonomic module has a presentation (1.2) with $q = 1$.

1.5 Examples.

(1) If $P \notin \mathbb{C}$ is an operator in $\mathbb{C}[t]\langle \partial_t \rangle$, the quotient module $\mathbb{M}$ of $\mathbb{C}[t]\langle \partial_t \rangle$ by the left ideal over $\mathbb{C}[t]\langle \partial_t \rangle \cdot P$ is a holonomic module: it has finite type, as it admits the presentation

$$0 \longrightarrow \mathbb{C}[t]\langle \partial_t \rangle \xrightarrow{\cdot P} \mathbb{C}[t]\langle \partial_t \rangle \longrightarrow \mathbb{M} \longrightarrow 0.$$

The graded module with respect to the filtration induced by the degree is equal to the quotient of the ring of polynomials by the ideal generated by the symbol of P; its annihilator is nontrivial, hence the module is holonomic.

(2) If δ_0 denotes the Dirac distribution at 0 on $\mathbb{C}$, the submodule of temperate distributions generated by δ_0 is holonomic: it has finite type by definition; moreover we have $t\delta_0 = 0$, whence an isomorphism of $\mathbb{C}$-vector spaces

$$\mathbb{C}[\partial_t] \xrightarrow{\sim} \mathbb{C}[t]\langle \partial_t \rangle \cdot \delta_0;$$

lastly, it is easy to check that $t^\ell \partial_t^k \delta_0 = 0$ as soon as $\ell > k$. This module is thus a torsion module. It takes the form given in Example (1) by taking for P the degree 0 operator equal to t.

(3) More generally, any torsion module is a direct sum of modules isomorphic to modules of the kind $\mathbb{C}[t]\langle \partial_t \rangle/(t - t^o)$ (the reader should check this).

(4) Any holonomic module $\mathbb{M}$ can be included in a short exact sequence

$$0 \longrightarrow \mathbb{K} \longrightarrow \mathbb{C}[t]\langle \partial_t \rangle/(P) \longrightarrow \mathbb{M} \longrightarrow 0$$

where $\mathbb{K}$ is torsion: in order to do so, we choose a generator e of $\mathbb{M}$ (Proposition 1.4-(2)) and we take for P an operator of minimal degree (say d) which kills e; it is a matter of seeing that the kernel of the surjective morphism $\mathbb{C}[t]\langle \partial_t \rangle/(P) \to \mathbb{M}$ which sends the class of 1 to e is torsion; if we work in the ring of fractions $\mathbb{C}[t, a_d^{-1}]$, in which we can invert the

1.c Duality

If $\mathbb{M}$ is a left $\mathbb{C}[t]\langle\partial_t\rangle$-module, then $\mathrm{Hom}_{\mathbb{C}[t]\langle\partial_t\rangle}(\mathbb{M}, \mathbb{C}[t]\langle\partial_t\rangle)$ is equipped with a structure of right $\mathbb{C}[t]\langle\partial_t\rangle$-module (by $(\varphi \cdot P)(m) = \varphi(m)P$). More generally, the spaces $\mathrm{Ext}^i_{\mathbb{C}[t]\langle\partial_t\rangle}(\mathbb{M}, \mathbb{C}[t]\langle\partial_t\rangle)$ are right $\mathbb{C}[t]\langle\partial_t\rangle$-modules[4].

1.8 Proposition (Duality preserves holonomy). *If $\mathbb{M}$ is holonomic,*

(1) *the right modules $\mathrm{Ext}^i_{\mathbb{C}[t]\langle\partial_t\rangle}(\mathbb{M}, \mathbb{C}[t]\langle\partial_t\rangle)$ are zero for $i \neq 1$;*
(2) *the left module $D\mathbb{M}$ associated to the right module $\mathrm{Ext}^1_{\mathbb{C}[t]\langle\partial_t\rangle}(\mathbb{M}, \mathbb{C}[t]\langle\partial_t\rangle)$ is holonomic;*
(3) *we have $D(D\mathbb{M}) \simeq \mathbb{M}$.*

Proof. By using the long Ext exact sequence associated to the short exact sequence 1.5-(4), we are reduced to showing the proposition for $\mathbb{M} = \mathbb{C}[t]\langle\partial_t\rangle/(P)$. This module has the resolution

$$0 \longrightarrow \mathbb{C}[t]\langle\partial_t\rangle \xrightarrow{\ \cdot P\ } \mathbb{C}[t]\langle\partial_t\rangle \longrightarrow \mathbb{M} \longrightarrow 0$$

and, applying the functor $\mathrm{Hom}_{\mathbb{C}[t]\langle\partial_t\rangle}(\bullet, \mathbb{C}[t]\langle\partial_t\rangle)$ to this resolution, we get the exact sequence

$$0 \longrightarrow \mathrm{Hom}_{\mathbb{C}[t]\langle\partial_t\rangle}(\mathbb{M}, \mathbb{C}[t]\langle\partial_t\rangle) \longrightarrow \mathbb{C}[t]\langle\partial_t\rangle \xrightarrow{\ P\cdot\ } \mathbb{C}[t]\langle\partial_t\rangle$$
$$\longrightarrow \mathrm{Ext}^1_{\mathbb{C}[t]\langle\partial_t\rangle}(\mathbb{M}, \mathbb{C}[t]\langle\partial_t\rangle) \longrightarrow 0$$

in which we see that $\mathrm{Ext}^1_{\mathbb{C}[t]\langle\partial_t\rangle}(\mathbb{M}, \mathbb{C}[t]\langle\partial_t\rangle) = \mathbb{C}[t]\langle\partial_t\rangle/(P)$ (here, (P) denotes the *right* ideal generated by P) and that all other Ext vanish. We thus have here

$$D\mathbb{M} = \mathbb{C}[t]\langle\partial_t\rangle/({}^t P). \qquad \square$$

Let $\mathbb{M}$ be a meromorphic bundle with poles at Σ, equipped with a connection. If $p(t)$ is a polynomial whose roots are the $\sigma \in \Sigma$, the bundle $\mathbb{M}$ is a free $\mathbb{C}[t, 1/p]$-module of finite rank. It is also a $\mathbb{C}[t]\langle\partial_t\rangle$-holonomic module (cf. Proposition 1.6). Its dual $D\mathbb{M}$ as a $\mathbb{C}[t]\langle\partial_t\rangle$-module is holonomic, but is not necessarily a meromorphic bundle. We will see that the module $(D\mathbb{M})(*\Sigma)$ is a meromorphic bundle with connection, i.e., Σ contains the set singular of $D\mathbb{M}$.

On the other hand, the dual $\mathbb{M}^* = \mathrm{Hom}_{\mathbb{C}[t,1/p]}(\mathbb{M}, \mathbb{C}[t, 1/p])$ of the meromorphic bundle $\mathbb{M}$ is naturally equipped with a connection (cf. 0.11.b).

We will compare the meromorphic bundles with connection $\mathbb{M}^*$ and $(D\mathbb{M})(*\Sigma)$.

1.9 Proposition (Localization is compatible with duality). *We have a canonical isomorphism $\mathbb{M}^* \simeq (D\mathbb{M})(*\Sigma)$.*

---

[4] The reader can refer to [God64] for the elementary homological algebra used below.

Proof. Let us introduce the algebra $\mathbb{C}[t, 1/p]\langle\partial_t\rangle$ of differential operators with coefficients in $\mathbb{C}[t, 1/p]$. Then $\mathbb{M}$ is also a $\mathbb{C}[t, 1/p]\langle\partial_t\rangle$-module. By considering a resolution of $\mathbb{M}$ by free $\mathbb{C}[t]\langle\partial_t\rangle$-modules, we see that

$$DM(*\Sigma) = \mathbb{C}[t, 1/p] \underset{\mathbb{C}[t]}{\otimes} DM = \mathrm{Ext}^1_{\mathbb{C}[t,1/p]\langle\partial_t\rangle}(\mathbb{M}, \mathbb{C}[t,1/p]\langle\partial_t\rangle)^l.$$

We will now construct a free resolution of $\mathbb{M}$ as a $\mathbb{C}[t, 1/p]\langle\partial_t\rangle$-module. In order to do that, let us forget for a while the connection on the module $\mathbb{M}$ and let us consider it only as a free $\mathbb{C}[t, 1/p]$-module. Then $\mathbb{C}[t, 1/p]\langle\partial_t\rangle \otimes_{\mathbb{C}[t,1/p]} \mathbb{M}$ is a free $\mathbb{C}[t, 1/p]\langle\partial_t\rangle$-module. It can be identified with $\mathbb{C}[\partial_t] \otimes_{\mathbb{C}} \mathbb{M}$. Any element can be written in a unique way as $\sum_{k\geqslant 0} \partial_t^k \otimes m_k$, and the (left) action of $\mathbb{C}[t, 1/p]\langle\partial_t\rangle$ is given by the formulas

$$(1.10) \qquad \partial_t\Big(\sum_{k\geqslant 0} \partial_t^k \otimes m_k\Big) = \sum_{k\geqslant 0} \partial_t^{k+1} \otimes m_k$$

$$(1.11) \qquad f(t)(\partial_t^k \otimes m_k) = \partial_t^k \otimes f(t)m_k + \sum_{\ell=0}^{k-1} \partial_t^\ell \otimes g_\ell(t)m_k,$$

where we have set $[f(t), \partial_t^k] = \sum_{\ell=0}^{k-1} \partial_t^\ell g_\ell(t)$ in the Weyl algebra $\mathbb{C}[t]\langle\partial_t\rangle$.

Let us consider the homomorphism

$$(1.12)$$
$$\mathbb{C}[t, 1/p]\langle\partial_t\rangle \underset{\mathbb{C}[t,1/p]}{\otimes} \mathbb{M} \xrightarrow{\;\partial_t \otimes 1 - 1 \otimes \partial_t\;} \mathbb{C}[t, 1/p]\langle\partial_t\rangle \underset{\mathbb{C}[t,1/p]}{\otimes} \mathbb{M}$$
$$\sum_{k\geqslant 0} \partial_t^k \otimes m_k \longmapsto \sum_{k\geqslant 0} \partial_t^k \otimes (m_{k-1} - (\partial_t m_k)),$$

where we now use the action of ∂_t on $\mathbb{M}$. Let us show that the morphism (1.12) is $\mathbb{C}[t, 1/p]\langle\partial_t\rangle$-linear; we will for instance check that

$$f(t) \cdot (\partial_t \otimes 1 - 1 \otimes \partial_t)(1 \otimes m) = (\partial_t \otimes 1 - 1 \otimes \partial_t)(1 \otimes f(t)m),$$

letting the reader check the other properties as an exercise; we have

$$f(t) \cdot (\partial_t \otimes 1 - 1 \otimes \partial_t)(1 \otimes m) = \partial_t \otimes f(t)m - 1 \otimes f'(t)m - 1 \otimes f(t)\partial_t m$$
$$= \partial_t \otimes f(t)m - 1 \otimes \partial_t(f(t)m).$$

Let us show the injectivity of (1.12): let $\sum_{k\geqslant 0} \partial_t^k \otimes m_k$ be such that $m_{k-1} - (\partial_t m_k) = 0$ for any $k \geqslant 0$ (by setting $m_{-1} = 0$); as $m_k = 0$ for $k \gg 0$, we deduce that $m_k = 0$ for any k.

We then identify the cokernel of (1.12) to $\mathbb{M}$ (as a $\mathbb{C}[t, 1/p]\langle\partial_t\rangle$-module) by the mapping

$$\sum_{k\geqslant 0} \partial_t^k \otimes m_k \longmapsto \sum_{k\geqslant 0} \partial_t^k m_k.$$

Let us now come back to the computation of Ext^1 with this resolution. A left $\mathbb{C}[t, 1/p]\langle \partial_t \rangle$-linear morphism $\mathbb{C}[t, 1/p]\langle \partial_t \rangle \otimes M \to \mathbb{C}[t, 1/p]\langle \partial_t \rangle$ is determined by its restriction to $1 \otimes M$, which must be $\mathbb{C}[t, 1/p]$-linear. We deduce an identification

$$\mathrm{Hom}_{\mathbb{C}[t,1/p]\langle \partial_t \rangle}\left(\mathbb{C}[\partial_t] \otimes_{\mathbb{C}} M, \mathbb{C}[t, 1/p]\langle \partial_t \rangle\right) = M^* \underset{\mathbb{C}}{\otimes} \mathbb{C}[\partial_t],$$

where the structure of right $\mathbb{C}[t, 1/p]\langle \partial_t \rangle$-module on the right-hand term is given formulas analogous to (1.10) and (1.11). Therefore, applying the functor

$$\mathrm{Hom}_{\mathbb{C}[t,1/p]\langle \partial_t \rangle}(\bullet, \mathbb{C}[t, 1/p]\langle \partial_t \rangle)^l$$

to the complex (1.12), we get a complex of the same kind, where we have replaced M with M*. Taking up the argument used for (1.12), we deduce an identification

$$\mathrm{Ext}^1_{\mathbb{C}[t,1/p]\langle \partial_t \rangle}(M, \mathbb{C}[t, 1/p]\langle \partial_t \rangle)^l \simeq M^*.$$

In doing so, we also have proved that $\mathrm{Ext}^1_{\mathbb{C}[t,1/p]\langle \partial_t \rangle}(M, \mathbb{C}[t, 1/p]\langle \partial_t \rangle)^l$ is a meromorphic bundle.   □

1.13 Exercise (An explicit form of the isomorphism of Proposition 1.9). We assume that M has the presentation $\mathbb{C}[t, 1/p]\langle \partial_t \rangle/(P)$, where $P = \partial_t^d + \sum_{i=0}^{d-1} a_i \partial_t^i$ is a differential operator with coefficients a_i in $\mathbb{C}[t, 1/p]$ and (P) denotes the left ideal that it generates. We denote by e_k the class of ∂_t^k in M. Hence e_0 is a $\mathbb{C}[t, 1/p]\langle \partial_t \rangle$-generator of M and $(e_0, \ldots, e_{d-1})$ is a $\mathbb{C}[t, 1/p]$-basis of M. The matrix of the connection in this basis takes the form

$$\begin{pmatrix} 0 & 0 & \ldots & 0 & -a_0 \\ 1 & 0 & \ldots & 0 & -a_1 \\ \vdots & & \vdots & \vdots & \vdots \\ 0 & 0 & \ldots & 1 & -a_{d-1} \end{pmatrix}$$

The dual module $M' := DM(*\Sigma)$ is therefore isomorphic to $\mathbb{C}[t, 1/p]\langle \partial_t \rangle/({}^t P)$. The isomorphism of Proposition 1.9 amounts to the existence of a nondegenerate $\mathbb{C}[t, 1/p]$-bilinear pairing $S : M \otimes_{\mathbb{C}[t,1/p]} M' \to \mathbb{C}[t, 1/p]$ which is compatible with the connection (cf. Definition III.1.10).

(1) Prove that the matrix of the connection on M* in the dual basis $(e_0^*, \ldots, e_{d-1}^*)$ takes the form

$$\begin{pmatrix} 0 & -1 & 0 & \ldots & 0 \\ 0 & 0 & -1 & 0 & 0 \\ \vdots & \vdots & \ddots & \ddots & \vdots \\ 0 & \ldots & \ldots & 0 & -1 \\ a_0 & \ldots & \ldots & a_{d-2} & a_{d-1} \end{pmatrix}$$

(2) Prove that $\varepsilon_0 := e_{d-1}^*$ satisfies ${}^tP\varepsilon_0 = 0$. Conclude that the morphism $M' \to M^*$ sending the class of 1 to ε_0 is an isomorphism of $\mathbb{C}[t, 1/p]\langle\partial_t\rangle$-modules.

(3) Equip M^* with the basis $(\varepsilon_0, \ldots, \varepsilon_{d-1})$ with $\varepsilon_k = (-\partial_t)^k\varepsilon_0$. Prove that the natural pairing $S : M \otimes_{\mathbb{C}[t, 1/p]} M^* \to \mathbb{C}[t, 1/p]$ is given by the following formulas:

$$S(e_i, \varepsilon_j) = \begin{cases} 0 & \text{if } i + j < d - 1, \\ 1 & \text{if } i + j = d - 1, \\ -\left[(-\partial_t)^k a_{d-1} + \cdots + a_{d-k-1}\right] & \text{if } \begin{cases} i + j = d + k, \\ 1 \leqslant k \leqslant d - 1. \end{cases} \end{cases}$$

1.d Regularity

A differential operator with coefficients in $\mathbb{C}[t]$

$$P(t, \partial_t) = a_d(t)\partial_t^d + \cdots + a_1(t)\partial_t + a_0(t)$$

is said to have a *regular singularity* at one of its singularities t^o (i.e., some root of a_d) if it fulfills *Fuchs condition* at t^o (cf. Theorem II.4.3)

(1.14) $\qquad \forall i \in \{1, \ldots, d-1\}, \quad i - v_{t^o}(a_i) \leqslant d - v_{t^o}(a_d)$,

where v_{t^o} denotes the valuation at t^o (i.e., the vanishing order at t^o).

Let us set $t = 1/t'$ and $\partial_t = -t'^2\partial_{t'}$ and let $P_\infty(t', \partial_{t'}) = P(t, \partial_t)$ be the operator with coefficients in $\mathbb{C}[t, t^{-1}] = \mathbb{C}[t', t'^{-1}]$ obtained from P by this change of variables. We will say that P has *regular singularity at infinity* if P_∞ does so at $t' = 0$.

1.15 Exercise (Regular singularity at infinity).

(1) Prove that the operator P has a regular singularity at infinity if and only if its coefficients satisfy the inequalities $\deg a_i - i \leqslant \deg a_d - d$ for any $i \leqslant d$ (we define $\deg a_i = -\infty$ if $a_i \equiv 0$).

(2) Show that the regularity condition at t^o (for $t^o \in \mathbb{C}$ or $t^o = \infty$) is stable by transposition.

1.16 Proposition. *Let us fix $t^o \in \mathbb{C} \cup \{\infty\}$, and let us set*

$$\boldsymbol{k} = \begin{cases} \mathbb{C}\{t - t^o\}[(t - t^o)^{-1}] & \text{if } t^o \in \mathbb{C}, \\ \mathbb{C}\{t'\}[t'^{-1}] & \text{if } t^o = \infty. \end{cases}$$

Let M be a holonomic module and $\mathcal{M} = \boldsymbol{k} \otimes_{\mathbb{C}[t]} M$ be the associated $(\boldsymbol{k}, \nabla)$-vector space with connection. The following properties are equivalent:

(1) *any element of $\mathcal{M}$ is killed by some operator $P \in \mathbb{C}[t]\langle\partial_t\rangle$ with regular singularity at t^o,*

(2) *the $(\boldsymbol{k}, \nabla)$-vector space $(\mathcal{M}, \nabla)$ has a regular singularity (in the sense of §II.2.a).*   □

When the conditions of the proposition are fulfilled, we say that $\mathbb{M}$ has a *regular singularity* at t^o. Exercise 1.15-(2) shows that this property is stable by duality.

2 Fourier transform

2.a Fourier transform of a module over the Weyl algebra

The relation $[\partial_t, t] = 1$ defining the Weyl algebra can also be written as $[(-t), \partial_t] = 1$, so that we have an isomorphism of algebras

$$\mathbb{C}[t]\langle \partial_t \rangle \longrightarrow \mathbb{C}[\tau']\langle \partial_{\tau'} \rangle$$
$$t \longmapsto -\partial_{\tau'}$$
$$\partial_t \longmapsto \tau'$$

denoted by $P \mapsto \widehat{P}$.

Any module $\mathbb{M}$ over $\mathbb{C}[t]\langle \partial_t \rangle$ becomes *ipso facto* a module over $\mathbb{C}[\tau']\langle \partial_{\tau'} \rangle$; we denote it then[5] by $\widehat{\mathbb{M}}$: this is the *Fourier transform* of $\mathbb{M}$. It is immediate that $\mathbb{M}$ is holonomic if and only if its Fourier transform is so. Similarly, if $\mathbb{M}$ is holonomic, the module

$$\mathbb{M}[\partial_t^{-1}] := \mathbb{C}[\partial_t, \partial_t^{-1}] \otimes_{\mathbb{C}[\partial_t]} \mathbb{M}$$

is still holonomic: we have indeed $\widehat{\mathbb{M}[\partial_t^{-1}]} = \widehat{\mathbb{M}}[\tau'^{-1}]$, a module which is holonomic after Proposition 1.6. Moreover, the kernel and the cokernel of the natural morphism $\mathbb{M} \to \mathbb{M}[\partial_t^{-1}]$ are isomorphic to modules of the kind $\mathbb{C}[t]^p$ (with the usual action of ∂_t): as a matter of fact, the kernel and the cokernel of the localization morphism $\widehat{\mathbb{M}} \to \widehat{\mathbb{M}}[\tau'^{-1}]$ are holonomic modules with support in $\tau' = 0$, hence take the form $(\mathbb{C}[\tau']\langle \partial_{\tau'} \rangle/(\tau'))^p$; we deduce the assertion by the inverse Fourier transform. In particular, the singular points of $\mathbb{M}$ are also that of $\mathbb{M}[\partial_t^{-1}]$.

2.1 Example. The Fourier transform of $\mathbb{M} = \mathbb{C}[t]\langle \partial_t \rangle/(P)$ is $\mathbb{C}[\tau']\langle \partial_{\tau'} \rangle/(\widehat{P})$. If $P = t - t^o$ (we then identify $\mathbb{M}$ with $\mathbb{C}[\partial_t] \cdot \delta_{t^o}$), we have $\widehat{P} = -(\partial_{\tau'} + t^o)$ and $\widehat{\mathbb{M}}$ is identified to the space of entire functions $\mathbb{C}[\tau'] \cdot e^{-t^o \tau'}$.

2.2 Proposition (Relations between $\mathbb{M}$ and $\widehat{\mathbb{M}}$, cf. [Mal91, Chap. V, §1]). *Let us assume that the $\mathbb{C}[t]\langle \partial_t \rangle$-holonomic module $\mathbb{M}$ has a regular singularity at infinity. Then,*

---

[5] This notation is traditional and one should not confuse it with the—also traditional—notation that we have used for formalization.

(1) *its Fourier transform $\widehat{M}$ has a singularity only at $\tau' = 0$ and $\tau' = \infty$. The former is regular and the latter has Poincaré rank (cf. §0.14) less than or equal to 1;*
(2) *the kernel and the cokernel of the left action of $\partial_t : M \to M$ have finite dimension and we have*

$$\dim \operatorname{Coker} \partial_t - \dim \operatorname{Ker} \partial_t = \operatorname{rk} \widehat{M} - \operatorname{rk} M.$$

In particular, if moreover $M = M[\partial_t^{-1}]$, we have $\operatorname{rk} \widehat{M} = \operatorname{rk} M$.

Proof.

(1) We reduce, according to Proposition 1.16, to the case where $M = \mathbb{C}[t]\langle \partial_t \rangle/(P)$, where P has a regular singularity at infinity. It will be convenient to write $P = \sum_i \partial_t^i a_i(t)$ with $a_i \in \mathbb{C}[t]$ and $a_d \neq 0$. The regular singularity condition says that $\deg a_i \leqslant \deg a_d - (d - i)$ (Exercise 1.15). Let us set $a_i(t) = \sum_j a_{i,j} t^j$ and $\widehat{d} = \deg a_d$. We thus have

$$P = a_{d,\widehat{d}} \partial_t^d t^{\widehat{d}} + \sum_{\substack{j < \widehat{d} \\ i \leqslant d}} a_{i,j} \partial_t^i t^j$$

with $a_{i,j} \neq 0$ only if $j - i \leqslant \widehat{d} - d$. The degree of $\widehat{P}$ in $\partial_{\tau'} = -t$ is equal to $\widehat{d}$ and the coefficient of $\partial_{\tau'}^{\widehat{d}}$ is $(-1)^{\widehat{d}} a_{d,\widehat{d}} \tau'^d$, which shows the nonexistence of a singularity away from $\{0, \infty\}$. Moreover, the Fuchs condition (1.14) at $\tau' = 0$ is clearly satisfied. Lastly, in the coordinate $\tau = 1/\tau'$, one can write

$$(*) \qquad \tau^d \widehat{P} = a_{d,\widehat{d}} \big(\tau^2 \partial_\tau \big)^{\widehat{d}} + \sum_{\substack{j < \widehat{d} \\ i \leqslant d}} a_{i,j} \tau^{d-i} (\tau^2 \partial_\tau)^j.$$

This shows that $\widehat{M}[\tau'^{-1}] = \mathbb{C}[\tau, \tau^{-1}]\langle \partial_\tau \rangle/(\widehat{P})$ is a free $\mathbb{C}[\tau, \tau^{-1}]$-module having as a basis the classes of $(\tau^2 \partial_\tau)^j$ $(0 \leqslant j < \widehat{d})$. In this basis, the matrix of $\tau^2 \partial_\tau$ has polynomial coefficients in τ, which shows that the Poincaré rank is $\leqslant 1$.

(2) Let us first notice that, if M' is a submodule of M and M'' is the quotient module M/M', then M' and M'' also have a regular singularity at infinity. Moreover, if the assertion is true for M' and M'', it is also true for M (use the snake lemma). It is thus enough to show it for modules of the kind $\mathbb{C}[t]\langle \partial_t \rangle/(P)$. After the computation above, we have in this case $\operatorname{rk} M = d$ and $\operatorname{rk} \widehat{M} = \widehat{d}$. We can write

$$\tau'^{\widehat{d}-d} \widehat{P} = b(\tau' \partial_{\tau'}) + \tau' Q(\tau', \tau' \partial_{\tau'}) \quad \text{if } \widehat{d} \geqslant d,$$

$$\widehat{P} = \tau'^{d-\widehat{d}} \cdot [b(\tau' \partial_{\tau'}) + \tau' Q(\tau', \tau' \partial_{\tau'})] \quad \text{if } d \geqslant \widehat{d},$$

where b is a non identically zero polynomial of degree $\min(d, \widehat{d})$ and Q is an operator of degree $\leqslant d$ with respect to τ' and of degree $< \widehat{d}$ with respect to $\partial_{\tau'}$. When $\widehat{P} = \tau'^k$ or $\widehat{P} = \partial_{\tau'}^k$, $(k \in \mathbb{N})$, the assertion of the proposition is easy to check directly. After Exercise 1.3, it is enough to show this assertion when $\widehat{P} = b(\tau'\partial_{\tau'}) + \tau'Q(\tau', \tau'\partial_{\tau'})$, where $\deg_{\tau'} \widehat{P} = \deg_{\partial_{\tau'}} \widehat{P} = \deg b$. It is matter here of proving that the $\mathbb{C}$-vector spaces $\operatorname{Ker} \tau'$ and $\operatorname{Coker} \tau'$ have the same finite dimension.

In order to avoid any computation, we will use a notion similar to that of a "Deligne lattice". To this end, let us set, for $k \in \mathbb{Z}$,

$$V_k \mathbb{C}[\tau']\langle \partial_{\tau'} \rangle = \begin{cases} \tau'^{|k|} \mathbb{C}[\tau']\langle \tau'\partial_{\tau'} \rangle & \text{if } k \leqslant 0, \\ \sum_{j=0}^{k} \partial_{\tau'}^j \mathbb{C}[\tau']\langle \tau'\partial_{\tau'} \rangle & \text{if } k \geqslant 0. \end{cases}$$

We have $\tau' V_k \subset V_{k-1}$ with equality if $k \leqslant 0$, and $\partial_{\tau'} V_k \subset V_{k+1}$. Moreover, $V_0/V_{-1} = \mathbb{C}[\tau'\partial_{\tau'}]$. Let us also set

$$U_k \widehat{\mathbb{M}} = V_k \mathbb{C}[\tau']\langle \partial_{\tau'} \rangle \Big/ (\widehat{P}) \cap V_k \mathbb{C}[\tau']\langle \partial_{\tau'} \rangle.$$

According to the form of $\widehat{P}$,

$$U_k \widehat{\mathbb{M}}/U_{k-1}\widehat{\mathbb{M}} \simeq \mathbb{C}\langle \tau'\partial_{\tau'} \rangle/(b(\tau'\partial_{\tau'} + k))$$

for any $k \in \mathbb{Z}$. In particular, all the spaces U_k/U_{k-1} have the same finite dimension. Lastly, the mapping

$$(2.3) \qquad U_k/U_{k-1} \xrightarrow{\;\tau'\;} U_{k-1}/U_{k-2}$$

is bijective as soon as 0 is not a root of the polynomial $b(s + k + 1)$, as the composition (on the right and on the left) with $\partial_{\tau'} : U_{k-1}/U_{k-2} \rightarrow U_k/U_{k-1}$ is then bijective. There exists thus $k_0 \geqslant 0$ such that (2.3) is an isomorphism for any $k \geqslant k_0$ and $k \leqslant -k_0$.

One can show (cf. for instance [Sab93, §4.2.4]) that $\tau' : \widehat{\mathbb{M}}/U_{k_0} \rightarrow \widehat{\mathbb{M}}/U_{k_0-1}$ and $\tau' : U_{-k_0} \rightarrow U_{-k_0-1}$ are bijective (the only delicate point is the injectivity of the latter mapping).

We deduce that the kernel (resp. the cokernel) of $\tau' : \widehat{\mathbb{M}} \rightarrow \widehat{\mathbb{M}}$ is also that of $\tau' : U_{k_0}/U_{-k_0} \rightarrow U_{k_0-1}/U_{-k_0-1}$. As the source and the target of the latter linear mapping are spaces having the same finite dimension, so are the kernel and the cokernel. $\Box$

2.4 Remark. The Fourier transform depends on the choice of an origin on the affine line with coordinate t. Translating the coordinate amounts to tensoring the Fourier transform with an exponential factor, as in Example 2.1.

2.b Fourier transform and duality

The coordinate t being fixed, let us consider the involution

$$\mathbb{C}[t]\langle \partial_t \rangle \longrightarrow \mathbb{C}[t]\langle \partial_t \rangle$$
$$P(t, \partial_t) \longmapsto \overline{P}(t, \partial_t) := P(-t, -\partial_t).$$

The Fourier transform and the transposition are related by the formula

$$^t(\widehat{P}) = \widehat{{}^t\overline{P}}$$

By reading §1.c "with a hat", we get an isomorphism

$$(2.5) \qquad\qquad D(\widehat{\mathbb{M}}) \simeq \widehat{D\overline{\mathbb{M}}}.$$

2.6 Corollary. *Let $\mathbb{M}$ be a holonomic module with a regular singularity at infinity. We suppose that there exists a morphism $D\mathbb{M} \to \mathbb{M}$, the kernel and the cokernel of which are isomorphic to some power of the module $\mathbb{C}[t]$ equipped with its usual derivation. The meromorphic bundle $\widehat{\mathbb{M}}[\tau'^{-1}]$ is then equipped with a nondegenerate sesquilinear form.*

Proof. Let us apply Fourier transform: we get a morphism

$$\widehat{D\mathbb{M}} \longrightarrow \overline{\widehat{\mathbb{M}}},$$

the kernel and the cokernel of which are supported on $\tau' = 0$. After localizing, the morphism becomes an isomorphism

$$D(\widehat{\mathbb{M}})[\tau'^{-1}] \xrightarrow{\sim} \overline{\widehat{\mathbb{M}}[\tau'^{-1}]}.$$

According to Proposition 1.9, we get an isomorphism

$$(\widehat{\mathbb{M}}[\tau'^{-1}])^* \xrightarrow{\sim} \overline{\widehat{\mathbb{M}}[\tau'^{-1}]}. \qquad\qquad \square$$

2.c Fourier transform of a lattice

The Fourier transform does not induce a correspondence between lattices of $\mathbb{M}$ and lattices of $\widehat{\mathbb{M}}$ for the very reason that $\tau' = \partial_t$ does not act on a lattice of $\mathbb{M}$. We will see however that, under some conditions, it transforms a lattice of $\mathbb{M}$ to a lattice of $\widehat{\mathbb{M}}[\tau'^{-1}]$ in the variable $\tau = \tau'^{-1}$. In other words, at the level lattices, the natural kernel to consider is not $e^{-t\tau'}$, but $e^{-t/\tau}$.

Let $\mathbb{M}$ be a $\mathbb{C}[t]\langle \partial_t \rangle$-holonomic module and let $\widehat{\mathbb{M}}$ be its Fourier transform (cf. §2.a), which is a $\mathbb{C}[\tau']\langle \partial_{\tau'} \rangle$-holonomic module. We will assume in the following that $\mathbb{M}$ has *a regular singularity at infinity*. We then know that $\widehat{\mathbb{M}}$ has a singularity at $\tau' = 0$ (regular singularity) and at $\tau' = \infty$ (possibly irregular singularity) only. The localized module $\widehat{\mathbb{M}}[\tau'^{-1}]$ is still holonomic and it is a free $\mathbb{C}[\tau', \tau'^{-1}]$-module of finite rank. We can consider it as a meromorphic bundle on the Riemann sphere $\widehat{\mathbb{P}}^1$, covered by the charts with respective coordinates τ' and τ, related by $\tau = \tau'^{-1}$ on their intersection. It is therefore also a free $\mathbb{C}[\tau, \tau^{-1}]$-module.

2.7 Theorem. *With these conditions, let* $\mathbb{E}$ *be a* $\mathbb{C}[t]$ *submodule of finite type of* $\mathbb{M}$. *Let* $\widetilde{\mathbb{E}}$ *be the image of* $\mathbb{E}$ *by the natural localization morphism* $\mathbb{M} \rightarrow \mathbb{M}[\partial_t^{-1}]$. *Let* $\widehat{\mathbb{E}}$ *be the* $\mathbb{C}[\tau]$-*module generated by* $\widetilde{\mathbb{E}}$ *in* $\widehat{\mathbb{M}[\partial_t^{-1}]} = \widehat{\mathbb{M}}[\tau'^{-1}]$, *in other words,*

$$\widehat{\mathbb{E}} = \sum_{k \geqslant 0} \partial_t^{-k} \widetilde{\mathbb{E}} = \mathbb{C}[\tau] \cdot \widetilde{\mathbb{E}}.$$

Then

(1) $\widehat{\mathbb{E}}$ *is a free* $\mathbb{C}[\tau]$-*module of finite type;*

(2) *if* $\mathbb{E}$ *generates* $\mathbb{M}$ *over* $\mathbb{C}[t]\langle \partial_t \rangle$, $\widehat{\mathbb{E}}$ *is a lattice of the* $\mathbb{C}[\tau, \tau^{-1}]$-*module* $\widehat{\mathbb{M}}[\tau'^{-1}]$;

(3) *the meromorphic connection* $\widehat{\nabla}$ *on* $\widehat{\mathbb{E}}$ *has a pole of order at most 1 at* $\tau = 0$.

Proof.

(1) If we know that $\widehat{\mathbb{E}}$ has finite type, we deduces that it is free, as it has no $\mathbb{C}[\tau]$-torsion (see for instance [Lan65, Chap. XV, §2]), being contained in $\widehat{\mathbb{M}}[\tau'^{-1}]$. The finiteness is the essential point. Let us begin by showing it in a simple case: we suppose that $\mathbb{M} = \mathbb{C}[t]\langle \partial_t \rangle / (P)$, where P is an operator, that we write as in the proof of Proposition 2.2

$$P = \sum_{i=1}^{d} \partial_t^i a_i(t),$$

with $a_i \in \mathbb{C}[t]$ and $a_d \not\equiv 0$. Let us first consider the lattice $\mathbb{E}$ generated by the class e of 1, which satisfies

$$\partial_t^{-d} P \cdot e = 0,$$

an equality which shows (cf. (∗) in the proof of Proposition 2.2) that $e, \ldots, t^{\widehat{d}-1} e$ is a system of generators of the $\mathbb{C}[\tau]$-module $\widehat{\mathbb{E}}$, whence finiteness in this case.

Let us notice now that, for any $k \geqslant 1$, we have

$$(2.8) \qquad \mathbb{C}[\tau] \cdot \left(\widetilde{\mathbb{E}} + \cdots + \partial_t^k \widetilde{\mathbb{E}} \right) = \widehat{\mathbb{E}} + \cdots + \tau'^k \widehat{\mathbb{E}} = \tau'^k \widehat{\mathbb{E}},$$

as $\tau'^\ell \widehat{\mathbb{E}} \subset \tau'^k \widehat{\mathbb{E}}$ for $\ell \leqslant k$. Therefore, finiteness is satisfied for lattices $\mathbb{E}_k = \mathbb{E} + \cdots + \partial_t^k \mathbb{E}$.

Lastly, any lattice of $\mathbb{M}$ is contained in some lattice $\mathbb{E}_k$, for a sufficiently large integer k. We deduce that its Fourier transform also has finite type over $\mathbb{C}[\tau]$ (as $\mathbb{C}[\tau]$ is Noetherian).

After Proposition 1.4, any holonomic module $\mathbb{M}$ is the quotient of a module of the kind $\mathbb{M}' = \mathbb{C}[t]\langle \partial_t \rangle / (P)$ by a torsion module. It follows in particular that P has a regular singularity at infinity if $\mathbb{M}$ does so. Any lattice $\mathbb{E}$ of $\mathbb{M}$ is contained in the image of a lattice $\mathbb{E}'_k$ of $\mathbb{M}'$. Therefore, $\widehat{\mathbb{E}}$ is the quotient of $\widehat{\mathbb{E}'_k}$, hence also has finite type.

(2) We apply (2.8), true for any lattice $\mathbb{E}$.
(3) It is a matter of proving that $\widehat{\mathbb{E}}$ is stable under the action of $\tau^2\partial_\tau$ alias t. This follows from the fact that $\mathbb{E}$ is stable by t.   □

We will have to use Theorem 2.7 in the following form:

2.9 Corollary. *Let $\mathbb{E}$ be a free $\mathbb{C}[t]$-module of rank d, equipped with a meromorphic connection ∇ having poles in some finite set Σ. Let us suppose that*

(1) *the connection ∇ has a regular singularity at infinity;*
(2) *any element of $\mathbb{E}$ admits a unique primitive, i.e., ∂_t^{-1} acts in a natural way on $\mathbb{E}$.*

Then, if we denote by $\widehat{\mathbb{E}}$ the $\mathbb{C}$-vector space $\mathbb{E}$ equipped with its structure of $\mathbb{C}[\partial_t^{-1}]$-module and if we set $\tau = \partial_t^{-1}$,

(1) *the $\mathbb{C}[\tau]$-module $\widehat{\mathbb{E}}$ is free of rank $\widehat{d} = d$,*
(2) *the operator $\widehat{\nabla}_{\partial_\tau} := (\nabla_{\partial_t})^2 t$ is a meromorphic connection on $\widehat{\mathbb{E}}$ having a pole at $\tau = 0$ only, this one being of type 1, and having a regular singularity at infinity.*

Proof. We consider the $\mathbb{C}[t]\langle\partial_t\rangle$-module $M' = \mathbb{E}(*\Sigma)$. As $\mathbb{E}$ is free over $\mathbb{C}[t]$, the natural morphism $\mathbb{E} \to M'$ is injective, and $\mathbb{E}$ is a lattice of the meromorphic bundle M'. On the other hand, we know (cf. Proposition 1.6) that the connection ∇ on $\mathbb{E}$ makes M' a $\mathbb{C}[t]\langle\partial_t\rangle$-holonomic module, with a regular singularity at infinity and, by Proposition 1.7, $\mathbb{E}$ is a lattice as such. Let us set $M = M'[\partial_t^{-1}]$. This is also a $\mathbb{C}[t]\langle\partial_t\rangle$-holonomic module, with a regular singularity at infinity. Then the second hypothesis of the corollary implies that the natural mapping $\mathbb{E} \to M$ induced by the localization mapping is injective and thus $\mathbb{E} = \widehat{\mathbb{E}} = \widehat{\mathbb{E}}$. We now deduce Properties (1) and (2) of the corollary from Theorem 2.7.

The equality $\widehat{d} = d$ is a consequence of Proposition 2.2.   □

We will now consider a converse statement of this corollary. In other words, let $\widehat{\mathbb{E}}$ be a free $\mathbb{C}[\tau]$-module of rank $\widehat{d}$, equipped with a meromorphic connection $\widehat{\nabla}$. Let us assume that the connection $\widehat{\nabla}$ has a pole at $\tau = 0$, this being of type 1, a regular singularity at infinity, and no other pole. Let us define the action of t on $\widehat{\mathbb{E}}$ as that of $\tau^2\nabla_{\partial_\tau}$. Therefore, $\widehat{\mathbb{E}}$ becomes a $\mathbb{C}[t]$-module, that we then denote by $\mathbb{E}$.

Let us set $\widehat{M} = \widehat{\mathbb{E}}[\tau^{-1}]$ and $\tau' = \tau^{-1}$. Then $\widehat{M}$ is a $\mathbb{C}[\tau']\langle\partial_{\tau'}\rangle$-holonomic module, having a regular singularity at $\tau' = 0$ and possibly an irregular one at $\tau' = \infty$. As $\widehat{\mathbb{E}}$ is free over $\mathbb{C}[\tau]$, the natural morphism $\widehat{\mathbb{E}} \to \widehat{M}$ is injective. Let us denote by M the inverse Fourier transform of $\widehat{M}$. We have then $M = M[\partial_t^{-1}]$, and $\mathbb{E}$ is a $\mathbb{C}[t]$-submodule, stable by ∂_t^{-1}, which generates $\widehat{M}$ over $\mathbb{C}[t]\langle\partial_t\rangle$. It is not clear, in general, whether it has finite type over $\mathbb{C}[t]$. We will give a criterion in order that such is indeed the case.

2.10 Proposition. *Let us assume that $\widehat{\mathbb{E}}$ has a basis ε in which the matrix of $\widehat{\nabla}$ takes the form $\left(\dfrac{B_0}{\tau} + B_\infty\right)\dfrac{d\tau}{\tau}$, where $B_\infty + k\,\mathrm{Id}$ is invertible for any $k \in \mathbb{N}$ (in particular $(\widehat{\mathbb{E}}, \widehat{\nabla})$ has type 1 at $\tau = 0$, the only singularities are $\tau = 0$ and $\tau = \infty$ and the connection can be extended with a logarithmic pole having residue $-B_\infty$ on the trivial bundle of rank $\widehat{d}$ on $\widehat{\mathbb{P}}^1$). Then*

(1) *the inverse Fourier transform $\mathbb{E}$ is a free $\mathbb{C}[t]$-module of rank $d = \widehat{d}$, the action of τ^{-1} defines a meromorphic connection ∇ on this bundle, all the singularities of which (even at $t = \infty$) are regular;*

(2) *The poles of the connection are located at the eigenvalues of B_0 and the connection can be extended with a logarithmic pole at $t = \infty$ on the trivial bundle on $\mathbb{P}^1$;*

(3) *the bundle with meromorphic connection $(\mathbb{E}, \nabla)$ is logarithmic if and only if B_0 is semisimple with distinct eigenvalues.*

The inverse Fourier transform gives thus a one-to-one correspondence between (trivial) bundles of rank $\widehat{d}$ on the affine space, equipped with a connection, the matrix of which can take the form $\left(\dfrac{B_0}{\tau} + B_\infty\right)\dfrac{d\tau}{\tau}$ with $B_\infty + k\,\mathrm{Id}$ invertible for any $k \in \mathbb{N}$, and that for which the matrix can take the form $(B_\infty - \mathrm{Id})(t\,\mathrm{Id} - B_0)^{-1}\,dt$. Let us notice however that we do not obtain in this way all the trivial bundles on $\mathbb{P}^1$ with logarithmic connection. On the other hand, if needed, we can add $m\dfrac{d\tau}{\tau}$ to the connection, for a suitable m, in order that $B_\infty + k\,\mathrm{Id}$ is invertible for any $k \in \mathbb{N}$.

Therefore, a solution to Birkhoff's Problem IV.5.1 for $r = 1$ gives a solution to a partial Riemann problem analogous to Problem IV.1.3 and, when the eigenvalues of B_0 are simple, to the Riemann-Hilbert Problem IV.1.3.

Proof. It will be convenient to write the basis ε in a column. By definition,

$$t \cdot \varepsilon = \left({}^t B_0 + \tau \cdot {}^t B_\infty\right) \cdot \varepsilon,$$

whence, for any $k \geqslant 1$, using the assumption on B_0,

$$\tau^k \cdot \varepsilon = \prod_{\ell=0}^{k-1} \left[({}^t B_\infty + \ell\,\mathrm{Id})^{-1}(t\,\mathrm{Id} - {}^t B_0) \right] \cdot \varepsilon.$$

This shows that $\mathbb{E}$ is generated by ε over $\mathbb{C}[t]$. We will now show that $M \subset \mathbb{C}(t) \otimes_{\mathbb{C}[t]} \mathbb{E}$. For that purpose, it is now enough to check that, for any $k \geqslant 1$, $\tau^{-k}\varepsilon \subset \mathbb{C}(t) \otimes_{\mathbb{C}[t]} \mathbb{E}$. We can write the relation defining t as

$$\tau^{-1} t \cdot \varepsilon = \left(\tau^{-1}\,{}^t B_0 + {}^t B_\infty\right) \cdot \varepsilon$$

(if the basis ε is written as a column); from the commutation relation $\tau^{-1} t = t\tau^{-1} + 1$, we deduce

$$(t\,\mathrm{Id} - {}^t B_0)\tau^{-1} \cdot \varepsilon = \left({}^t B_\infty - \mathrm{Id}\right) \cdot \varepsilon,$$

and thus

$$(*) \qquad \tau^{-1} \cdot \varepsilon = (t \, \mathrm{Id} - {}^t B_0)^{-1} ({}^t B_\infty - \mathrm{Id}) \cdot \varepsilon.$$

Iterating the process gives the assertion, and therefore

$$\mathbb{C}(t) \otimes_{\mathbb{C}[t]} \mathbb{M} = \mathbb{C}(t) \otimes_{\mathbb{C}[t]} \mathbb{E}.$$

As the rank of $\mathbb{M}$ is equal to $\widehat{d}$ (Proposition 2.2), the rank of $\mathbb{E}$ (i.e., the dimension of $\mathbb{C}(t) \otimes_{\mathbb{C}[t]} \mathbb{E}$ over $\mathbb{C}(t)$) is equal to $\widehat{d}$. It follows that $\widehat{\mathbb{E}}$ is free of rank $\widehat{d}$: as a matter of fact, the generating set ε gives a surjective morphism $\mathbb{C}[t]^{\widehat{d}} \to \mathbb{E}$, which induces an isomorphism after tensoring with $\mathbb{C}(t)$, as the $\mathbb{C}(t)$-vector spaces have the same rank; its kernel is then torsion; the latter must be zero, as $\mathbb{C}[t]^{\widehat{d}}$ has no torsion submodule. This proves the first point.

The matrix of ∇ in the basis ε is $(B_\infty - \mathrm{Id})(t \, \mathrm{Id} - B_0)^{-1} \, dt$: indeed, this is exactly what gives the relation $(*)$. The poles of this matrix are located at the eigenvalues of B_0. If these ones are simple, the poles are simple.

Lastly, if we extend $\mathbb{E}$ as a trivial bundle on $\mathbb{P}^1$, the connection matrix of ∇ in the coordinate $t' = 1/t$ is equal to $-(B_\infty - \mathrm{Id})(\mathrm{Id} - t' B_0)^{-1} dt'/t'$. It has thus a simple pole at $t' = 0$. $\qquad \square$

3 Fourier transform and microlocalization

In this section, we assume that $\mathbb{M}$ is a $\mathbb{C}[t]\langle \partial_t \rangle$-holonomic module with regular singularities including at infinity, and that $\mathbb{F}$ is a lattice of it[6]. Let $\mathbb{G}$ be the localized Fourier transform of $\mathbb{M}$ and let $\widehat{\mathbb{F}}$ be that of $\mathbb{F}$, as in Theorem 2.7. The connection ∇ on $\mathbb{G}$ has a pole of order at most 1 at $\tau = 0$. We will analyze the formal structure of $(\mathbb{G}, \widehat{\mathbb{F}})$ at $\tau = 0$. First, it is easy (by considering the example of $\mathbb{C}[t]\langle \partial_t \rangle/(P)$) to check that the eigenvalues of the most polar part of the connection matrix are exactly the singular points of $\mathbb{M}$. We will give a more precise statement by considering the formal decomposition of $\mathbb{G}$— a decomposition that can be obtained with the method of Turrittin (Exercise II.5.9)—to the germs of $\mathbb{M}$ at each of its singularities. This is nothing but an algebraic version of the classic *stationary phase method*.

3.a Formal microlocalization

Let $\mathscr{D}$ be the sheaf of holomorphic differential operators on $\mathbb{C}$ (coordinate t): a section of $\mathscr{D}$ on an open set U of $\mathbb{C}$ is a polynomial $\sum_{i \geqslant 0} a_i(t) \partial_t^i$, where the a_i are holomorphic on U. It is a sheaf of noncommutative rings, the commutation relation (1.1) determining the product.

Let us also introduce the sheaf of *formal microdifferential operators* $\mathscr{E}$: by definition, this is the sheaf on $\mathbb{C}$ (coordinate t), the sections of which on

---

[6] The need of a change of notation will soon be clear.

an open set $U \subset \mathbb{C}$ are the formal Laurent series $\sum_{i \geqslant i_0} a_i(t) \tau^i$, where the coefficients a_i are holomorphic on U. This is a sheaf of *noncommutative* rings. If we write the factors τ^i on the right, the product is determined by the relation[7]

$$(3.1) \qquad \tau \cdot f(t) = \sum_{k=0}^{\infty} (-1)^k f^{(k)}(t) \tau^{k+1}.$$

We deduce in particular that $[\tau, f(t)] = -\tau \cdot f'(t) \cdot \tau$, that is,

$$(3.2) \qquad [\tau^{-1}, f(t)] = f'(t).$$

The sheaf $\mathscr{E}$ is a sheaf of left and right $\mathscr{D}$-modules: the right action of ∂_t is right multiplication by τ^{-1} and the left action of $f(t)$ is the multiplication in $\mathscr{E}$; the left action of ∂_t is left multiplication by τ^{-1}, that one computes using Relation (3.2); lastly, the right action of $f(t)$ is right multiplication by $f(t)$ in $\mathscr{E}$, that one computes using (3.1).

Any element of $\mathscr{E}(U)$ can also be written in a unique way as $\sum_{i \geqslant i_0} \tau^i a_i(t)$ with $a_i \in \mathscr{O}(U)$.

The subsheaf of rings consisting of operators without pole at $\tau = 0$ is denoted by $\mathscr{E}(0)$. This is a sheaf of left and right $\mathscr{O}$-modules, if $\mathscr{O}$ denotes the sheaf of holomorphic functions on $\mathbb{C}$.

Let M be as above and let $\mathscr{M} = \mathscr{O} \otimes_{\mathbb{C}[t]} \mathrm{M}$ be the associated left $\mathscr{D}$-module. The *microlocalized* module of $\mathscr{M}$ is the left $\mathscr{E}$-module

$$\mathscr{M}^\mu := \mathscr{E} \otimes_{\mathscr{D}} \mathscr{M}.$$

3.3 Lemma. *The microlocalized module $\mathscr{M}^\mu$ has support in the set of singular points of $\mathscr{M}$.*

Proof. It is enough to check that the microlocalized module of $\mathscr{O}$ is zero, as, in the neighbourhood of any nonsingular point, $\mathscr{M}$ is isomorphic to $\mathscr{O}^d$. We have $\mathscr{O} = \mathscr{D}/\mathscr{D} \cdot \partial_t$, and therefore $\mathscr{O}^\mu = \mathscr{E}/\mathscr{E} \cdot \tau^{-1} = 0$. $\qquad\square$

Let $\mathscr{F}^\mu = \mathrm{image}\,[\mathscr{E}(0) \otimes_{\mathscr{O}} \mathscr{F} \to \mathscr{M}^\mu]$ be the microlocalized module of the lattice $\mathscr{F} = \mathscr{O} \otimes_{\mathbb{C}[t]} \mathrm{F}$ of $\mathscr{M}$. This is a lattice of $\mathscr{M}^\mu$ (a coherent $\mathscr{E}(0)$-module which generates $\mathscr{M}^\mu$ as a left $\mathscr{E}$-module). We will denote by $\widehat{k}$ the field $\mathbb{C}[\![\tau]\!][\tau^{-1}]$ and by $\widehat{\nabla}$ the connection induced by $\partial/\partial\tau$ on $\widehat{k}$ (here, the "hat" has a double meaning: it indicates that we work with the coordinate τ of the "Fourier" line, and with formal series in τ). For any $c \in \mathbb{C}$, the ring $\mathscr{E}_c$ is a left module over $\widehat{k}\langle \partial_\tau \rangle$ if $\tau^2 \partial_\tau$ acts by left multiplication by t. Similarly, $\mathscr{M}_c^\mu$ is a $(\widehat{k}, \widehat{\nabla})$-vector space with connection. We will denote (according to the notation of §II.5.b) by $\mathscr{E}^{c/\tau} \otimes \mathscr{M}^\mu$ the $\widehat{k}$-vector space $\mathscr{M}^\mu$ equipped with the connection for which $\tau^2 \partial_\tau$ acts as left multiplication by $t - c$ (we thus have "translated c to the origin").

---

[7] This relation is best understood if one regards τ as the "primitivation" operator ∂_t^{-1}.

3.4 Proposition. *At any singular point c of $\mathscr{M}$, the germ $\mathscr{E}^{c/\tau} \otimes \mathscr{M}_c^\mu$ is a $(\widehat{\boldsymbol{k}}, \widehat{\nabla})$-vector space with a regular singularity and $\mathscr{E}^{c/\tau} \otimes \mathscr{F}_c^\mu$ is a lattice of it, stable by $\tau^2 \nabla_{\partial_\tau}$.*

Before beginning the proof, let us establish a result which will also be useful later on:

3.5 Lemma. *For any $c \in \mathbb{C}$,*

(1) *$\mathscr{E}_c$ (resp. $\mathscr{E}_c(0)$) is flat as a left and right module over $\mathscr{D}_c$ (resp. over $\mathscr{O}_c$);*
(2) *$\mathscr{E}_c$ is flat as a left and right module over $\widehat{\boldsymbol{k}}\langle \partial_\tau \rangle$.*

The first assertion of this statement, for instance, means that, for any short exact sequence

$$0 \longrightarrow \mathscr{M}' \longrightarrow \mathscr{M} \longrightarrow \mathscr{M}'' \longrightarrow 0$$

of $\mathscr{D}_c$-modules of finite type, the corresponding sequence

$$0 \longrightarrow \mathscr{E}_c \underset{\mathscr{D}_c}{\otimes} \mathscr{M}' \longrightarrow \mathscr{E}_c \underset{\mathscr{D}_c}{\otimes} \mathscr{M} \longrightarrow \mathscr{E}_c \underset{\mathscr{D}_c}{\otimes} \mathscr{M}'' \longrightarrow 0$$

is also exact.

Proof. Let us notice first that, for any open set U, the $\mathscr{O}(U)$-module $\mathscr{E}(0)(U)$ is flat, as it is flat over $\mathscr{O}(U)[\tau]$ and as the latter ring is flat (being free) over $\mathscr{O}(U)$.

Let us show that $\mathscr{E}_c(0)$ is flat over $\mathscr{O}_c$: it is a matter of checking (cf. [AM69]) that, if $f_1, \ldots, f_p \in \mathscr{O}_c$ satisfy a relation $\sum_{i=1}^p P_i f_i = 0$ with $P_i \in \mathscr{E}_c(0)$, then the vector $(P_1, \ldots, P_p)$ is a combination with coefficients in $\mathscr{E}_c(0)$ of vectors of relations in $\mathscr{O}_c$ between $f_1, \ldots, f_p$; but the f_i and the P_i are defined on some open neighbourhood U of c, on which the desired assertion is true, as $\mathscr{E}(0)(U)$ is flat over $\mathscr{O}(U)$; by taking the germs at c, we get the desired assertion for $\mathscr{E}_c(0)$.

Once this result is proved, the flatness of $\mathscr{E}_c$ over $\mathscr{D}_c$ is a consequence of general (and easy) results concerning flatness over filtered rings (see for instance [Sch85, Prop. II.1.2.4]): we consider the τ-adic filtration on $\mathscr{E}_c$; this is a *Zariskian* filtration, that is, if $P_{-1} \in \tau\mathscr{E}(0)$, then $1 - P_{-1}$ is invertible in $\mathscr{E}_c(0)$ (checking this is easy here, as we do not ask any convergence condition for the series in powers of τ); similarly, we equip $\mathscr{D}_c$ with the filtration by the degree in ∂_t; let us work as above on some open set U, so that the graded rings are respectively the commutative rings $\mathscr{O}(U)[\tau][\tau^{-1}]$ and $\mathscr{O}(U)[\tau^{-1}]$; the former is flat over the latter (cf. for instance [AM69]); we can apply the statement of [Sch85] and then come back to germs at c.

The ring $\widehat{\boldsymbol{k}}\langle \partial_\tau \rangle$ can be identified with the subring of $\mathscr{E}_c$ consisting of series $\sum_{i \geqslant i_0} a_i(t)\tau^i$, where the a_i are polynomials of bounded degree. In order to show flatness, we apply the same technique as above: we equip the ring $\mathscr{E}_c$ with the $(t - c)$-adic filtration; one similarly checks that it is a Zariskian filtration, whose associated graded ring can be identified with $\widehat{\boldsymbol{k}}\langle \partial_\tau \rangle$; we conclude by using once more [Sch85]. $\qquad\square$

Proof (of Proposition 3.4). This is the microlocal analogue of Theorem 2.7. The question is local at the singular point c, that we assume to be 0 in order to simplify notation. Let us begin with the regularity of $\mathcal{M}_0^\mu$. The kernel and the cokernel of the localization homomorphism $\mathcal{M} = \mathcal{M}_0 \to \mathcal{M}[t^{-1}]$ are $\mathcal{D}_0$-holonomic modules having support at the origin and $\mathcal{M}_0[t^{-1}]$ is a $(\boldsymbol{k}, \nabla)$-connection with regular singularity. By an extension argument and according to the flatness lemma above, it is enough to consider the case where the germ $\mathcal{M}$ has support at the origin or the case where $\mathcal{M}$ is a $(\boldsymbol{k}, \nabla)$-connection with regular singularity.

In the first case, it is also enough to consider $\mathcal{M} = \mathcal{D}_0/\mathcal{D}_0 \cdot t$ (cf. Example 1.5-(3)) and the lattice $\mathcal{F}$ generated by the class of 1. Then $\mathcal{M}^\mu = \mathcal{E}_0/\mathcal{E}_0 \cdot t = \mathbb{C}[\![\tau]\!][\tau^{-1}] = \widehat{\boldsymbol{k}}$ and $\mathcal{F}^\mu = \mathcal{E}_0(0)/\mathcal{E}_0(0) \cdot t = \mathbb{C}[\![\tau]\!]$, the connection being the connection $\widehat{\nabla}$ on $\widehat{\boldsymbol{k}}$.

In the second case, one can assume, according to Theorem II.2.25 and by the same extension argument as above, that $\mathcal{M}$ has rank one, hence is isomorphic to $\mathcal{D}_0/\mathcal{D}_0 \cdot (\partial_t t - \alpha)$ with $\operatorname{Re}\alpha \in [0, 1[$. We have $\mathcal{M}^\mu = \mathcal{E}_0/\mathcal{E}_0 \cdot (t - \alpha\tau)$ and we conclude as in the first case.

For what concerns $\mathscr{F}_0^\mu$, the only point to check is the finiteness as a $\mathbb{C}[\![\tau]\!]$-module, because the stability under the action of $\tau^2 \nabla_{\partial_\tau}$ is true by definition of the connection. Moreover, if the finiteness is true for some lattice $\mathscr{F}_0$, it is also true for any other lattice of $\mathcal{M}_0$. We can then use once more the extension argument and reduce to checking the assertion for the lattice generated by the class of 1 in each of the previous examples, which is easy and left to the reader. $\qquad\square$

3.b Formal decomposition of the Fourier transform

3.6 Proposition. *The composed $\mathbb{C}[\![\tau]\!]$-linear mapping*

$$\widehat{\mathbb{G}} := \widehat{\boldsymbol{k}} \otimes_{\mathbb{C}[\tau^{-1}]} \widehat{\mathbb{M}} \longrightarrow \Gamma(\mathbb{C}, \widehat{\boldsymbol{k}} \otimes_{\mathbb{C}[\partial_t]} \mathcal{M}) \longrightarrow \Gamma(\mathbb{C}, \mathcal{M}^\mu)$$

is an isomorphism compatible with the action of t, which identifies the formalized module $\widehat{\mathbb{F}}$ of the lattice $\widehat{\mathbb{F}}$ of $\mathbb{G}$ to $\Gamma(\mathbb{C}, \mathscr{F}^\mu)$.

We thus get the formal decomposition[8] of the meromorphic bundle with connection $(\mathbb{G}, \nabla)$ at 0: as a matter of fact, as $\mathcal{M}^\mu$ has support at the singular points c of $\mathcal{M}$, we can write $\Gamma(\mathbb{C}, \mathcal{M}^\mu) = \bigoplus_c \mathcal{M}_c^\mu$. On any summand, the left multiplication by t is identified with the action of $\tau^2 \nabla_{\partial_\tau}$; moreover, $\mathcal{E}^{c/\tau} \otimes \mathcal{M}_c^\mu$ is a $(\widehat{\boldsymbol{k}}, \widehat{\nabla})$-connection with regular singularity (Proposition 3.4).

---

[8] A more precise statement relating the Stokes structure of $\mathbb{G}$ at $\tau = 0$ with properties of $\mathcal{M}$ also exists. One can write both Stokes matrices introduced in Example II.6.11 in terms of the *variation* operators acting on the solutions of $\mathcal{M}$. We will not use this result and refer the interested reader to [Mal91, Chap. XII] and to the references given therein.

Proof (of Proposition 3.6). Let us first notice that

$$\mathbb{G}^\wedge := \widehat{\boldsymbol{k}} \underset{\mathbb{C}[\tau^{-1}]}{\otimes} \widehat{\mathbb{M}} = \widehat{\boldsymbol{k}} \underset{\mathbb{C}[\tau,\tau^{-1}]}{\otimes} \left(\mathbb{C}[\tau,\tau^{-1}] \underset{\mathbb{C}[\tau^{-1}]}{\otimes} \widehat{\mathbb{M}} \right)$$

$$= \widehat{\boldsymbol{k}} \underset{\mathbb{C}[\tau,\tau^{-1}]}{\otimes} \mathbb{G} = \mathbb{C}[\![\tau]\!] \underset{\mathbb{C}[\tau]}{\otimes} \mathbb{G},$$

so that $\mathbb{G}^\wedge$ is indeed the formalized module of $\mathbb{G}$ at $\tau = 0$.

Assertion. *The $\widehat{\boldsymbol{k}}$-vector spaces $\mathbb{G}^\wedge$ and $\bigoplus_c \mathscr{M}_c^\mu$ have the same dimension.*

Let us take this statement for granted and end the proof. First, it is clear that the mapping is compatible with the action of t and that it sends $\widehat{\mathbb{F}}$ in $\bigoplus_c \mathscr{F}_c^\mu$.

Let us now show that the mapping $\widehat{\mathbb{F}} \to \mathscr{F}_c^\mu$, composition with the projection to the summand of index c, is *surjective*. After Nakayama's lemma, it is enough to show that the composed mapping $\widehat{\mathbb{F}} \to \mathscr{F}_c^\mu/\tau\mathscr{F}_c^\mu$ is so. As $\mathscr{F}_c = \mathscr{O}_c \otimes_{\mathbb{C}[t]} \mathbb{F}$, there exists $m_1, \ldots, m_p \in \mathbb{F}$ such that any germ $m \in \mathscr{F}_c$ has a decomposition $m = \sum_i \varphi_i m_i$ with $\varphi_i \in \mathscr{O}_c$. On the other hand, there exists an operator $P \in \mathbb{C}[t]\langle \partial_t \rangle$ annihilating all the m_i; in other words, there exists $d \in \mathbb{N}$ and $a_d(t) \in \mathbb{C}[t]$ such that

$$\partial_t^d a_d(t) \cdot m_i \in \sum_{k=1}^d \partial_t^{d-k} \mathbb{F} \quad (i = 1, \ldots, p).$$

If n denotes the order of vanishing of a_d at c, we deduce that the element $(t-c)^n (1 \otimes m_i)$ belongs to $\tau\mathscr{F}_c^\mu$ for any i. Taking the Taylor expansion of the φ_i up to order n, we can write $m = m' + (t-c)^n \sum_i \psi_i m_i$ with $m' \in \mathbb{F}$ and $\psi_i \in \mathscr{O}_c$. We deduce that $1 \otimes m \equiv 1 \otimes m' \mod \tau\mathscr{F}_c^\mu$ in $\mathscr{F}_c^\mu$, whence the desired surjectivity.

We also deduce that $\mathbb{G}^\wedge \to \mathscr{M}_c^\mu$ is surjective for any c. Therefore, so is $\mathcal{E}^{c/\tau} \otimes \mathbb{G}^\wedge \to \mathcal{E}^{c/\tau} \otimes \mathscr{M}_c^\mu$, which shows (according to Exercise II.5.9) that, for any c, $\mathcal{E}^{c/\tau} \otimes \mathbb{G}^\wedge$ can be decomposed as the direct sum of a $(\widehat{\boldsymbol{k}}, \widehat{\nabla})$-vector space with a purely irregular singularity and of a $(\widehat{\boldsymbol{k}}, \widehat{\nabla})$-vector space with regular singularity, the latter being mapped onto $\mathcal{E}^{c/\tau} \otimes \mathscr{M}_c^\mu$: indeed, any homomorphism from a $(\widehat{\boldsymbol{k}}, \widehat{\nabla})$-vector space with purely irregular singularity to a $(\widehat{\boldsymbol{k}}, \widehat{\nabla})$-vector space with regular singularity is zero (cf. Exercise II.5.5). Coming back to $\mathbb{G}^\wedge$, we thus see that the map in the proposition can be decomposed as the direct sum of its projections on each of the summands $\mathscr{M}_c^\mu$. It is thus surjective and, after the assertion, it is bijective. We also deduce that it induces an isomorphism $\widehat{\mathbb{F}}^\wedge \to \bigoplus_c \mathscr{F}_c^\mu$. $\qquad\square$

Proof (of the assertion). According to an argument that has yet been used, it is enough to show it for modules $\mathbb{M}$ of the kind $\mathbb{C}[t]\langle \partial_t \rangle/(P)$, where $P = a_d(t)\partial_t^d + \cdots$ has regular singularities even at infinity. In this case, we have seen that the rank of $\mathbb{G}$ is equal to the degree of a_d, while that of $\mathscr{M}_c^\mu$ is equal to the valuation of a_d at c, that is, the multiplicity of c as a root of a_d. Therefore, the assertion is clear. $\qquad\square$

3.c A microdifferential criterion for the symmetry of the characteristic polynomial

The lattice $\widehat{\mathbb{F}}$ of $\mathbb{G}$ admits a characteristic polynomial at infinity, constructed in §III.2.b and denoted by $\chi_{\widehat{\mathbb{F}}}^{\infty}(s)$. This is a polynomial of degree $\widehat{d} = \mathrm{rk}\,\widehat{\mathbb{F}}$. We will give a condition on the microlocalized module $\mathscr{F}^{\mu}$ in order to get a symmetry relation

$$\chi_{\widehat{\mathbb{F}}}^{\infty}(s) = (-1)^{\widehat{d}}\chi_{\widehat{\mathbb{F}}}^{\infty}(w - s)$$

for some suitable $w \in \mathbb{Z}$.

Let us first notice that, after Corollary III.2.8, it is enough that $\mathbb{G}$ comes equipped with a nondegenerate sesquilinear form of weight w with respect to $\widehat{\mathbb{F}}$.

In order to obtain a sesquilinear form on $\mathbb{G}$, it is enough to be given a homomorphism $D\mathbb{M} \to \mathbb{M}$ the kernel and the cokernel of which are free $\mathbb{C}[t]$-modules of finite rank: as a matter of fact, such a homomorphism induces, by Fourier transform and according to (2.5), a homomorphism $D\widehat{\mathbb{M}} \to \widehat{\mathbb{M}}$ the kernel and the cokernel of which have support at $\tau' = 0$, so that, by localizing with respect to τ, we get an isomorphism $(D\widehat{\mathbb{M}})[\tau'^{-1}] \xrightarrow{\sim} \mathbb{G}$; lastly, according to Proposition 1.9, we can identify the left-hand term to $\mathbb{G}^{*}$.

Controlling the weight with respect to $\widehat{\mathbb{F}}$ can be done in a microdifferential way. Let us first note that, according to the flatness of $\mathscr{E}$ over $\mathscr{D}$, we have

$$\mathscr{E}xt_{\mathscr{E}}^{i}(\mathscr{M}^{\mu}, \mathscr{E}) = \mathscr{E} \underset{\mathscr{D}}{\otimes} \mathscr{E}xt_{\mathscr{D}}(\mathscr{M}, \mathscr{D})$$

(analytic and sheaf-theoretic versions of duality in §1.c), in other words, $D(\mathscr{M}^{\mu}) = (D\mathscr{M})^{\mu}$. In a way analogous to Proposition 1.9, we have:

3.7 Proposition. *For any $c \in \mathbb{C}$, we have an isomorphism*

$$D(\mathscr{M}_{c}^{\mu}) \simeq \overline{\mathrm{Hom}_{\widehat{\boldsymbol{k}}}(\mathscr{M}_{c}^{\mu}, \widehat{\boldsymbol{k}})} := (\overline{\mathscr{M}_{c}^{\mu}})^{*}.$$

Proof (Sketch). We proceed in a way analogous to that of Proposition 1.9. We first show that the sequence

$$0 \longrightarrow \widehat{\boldsymbol{k}}\langle\partial_{\tau}\rangle \underset{\widehat{\boldsymbol{k}}}{\otimes} \mathscr{M}_{c}^{\mu} \xrightarrow{t \otimes 1 - 1 \otimes t} \widehat{\boldsymbol{k}}\langle\partial_{\tau}\rangle \underset{\widehat{\boldsymbol{k}}}{\otimes} \mathscr{M}_{c}^{\mu} \longrightarrow \mathscr{M}_{c}^{\mu}$$

is exact, and we deduce, by flatness 3.5-(2), that the sequence

$$0 \longrightarrow \mathscr{E}_{c} \underset{\widehat{\boldsymbol{k}}}{\otimes} \mathscr{M}_{c}^{\mu} \xrightarrow{t \otimes 1 - 1 \otimes t} \mathscr{E}_{c} \underset{\widehat{\boldsymbol{k}}}{\otimes} \mathscr{M}_{c}^{\mu} \longrightarrow \mathscr{M}_{c}^{\mu}$$

is so, then we end as in the proof of Proposition 1.9. $\square$

3.8 Proposition. *Let $D\mathbb{M} \to \mathbb{M}$ be a homomorphism of $\mathbb{C}[t]\langle\partial_t\rangle$-modules, whose kernel and cokernel are $\mathbb{C}[t]$-free of finite rank. If there exists $w \in \mathbb{Z}$ such that, for any singular point c of $\mathcal{M}$, the induced microlocal isomorphism $D(\mathcal{M}_c^\mu) \simeq (\overline{\mathcal{M}_c^\mu})^* \to \mathcal{M}_c^\mu$ sends $(\overline{\mathscr{F}_c^\mu})^*$ in $\tau^{-w}\mathscr{F}_c^\mu$, then the induced homomorphism $\overline{\mathbb{G}}^* \to \mathbb{G}$ defines a nondegenerate sesquilinear form on $\mathbb{G}$ of weight w with respect to $\widehat{\mathbb{F}}$.*

Proof. We have yet indicated that the sesquilinear form is nondegenerate. According to Proposition 3.6, the formalized module at $\tau = 0$ of $\widehat{\overline{\mathbb{F}}}^*$ is sent bijectively to the formalized module at 0 of $\tau^{-w}\widehat{\mathbb{F}}$. This shows that the kernel and the cokernel of $\widehat{\overline{\mathbb{F}}}^* \to \tau^{-w}\widehat{\mathbb{F}}$ have support in $\tau \neq 0$. But as, on the other hand, the morphism that we deduce after localization with respect to τ is an isomorphism (it is nothing but $\overline{\mathbb{G}}^* \to \mathbb{G}$), this kernel and this cokernel are zero. □

VI

Integrable deformations of bundles with connection on the Riemann sphere

Introduction

When considering deformations of one-variable meromorphic differential systems, one is lead to ask questions that we treat in this chapter:

(1) Given a meromorphic differential system analytically depending on parameters, let us assume that we have solved, for some particular value of the parameter, the Riemann-Hilbert Problem IV.1.1 or one of its variants. Can we find a solution to this problem for nearby values of the parameter, a solution which would depend analytically on the parameter?

We can moreover ask whether, the initial solution being fixed, the family of solutions thus found is unique, as in the Cauchy Theorem for instance. In order to get existence and uniqueness results, it is essential to impose the *integrability* condition: it must be possible to complete the given differential system to an integrable system of linear partial differential equations.

(2) When the Riemann-Hilbert problem or Birkhoff's problem is solved in a family, the integrability condition takes the form of a nonlinear matrix differential system which is often remarkable. The resolution of this system is thus equivalent to that of the Riemann-Hilbert problem or of Birkhoff's problem, whence a "geometric" method of resolution of such a system.

(3) Lastly, some integrable deformations of a one-variable meromorphic differential system are *universal*. The properties of the system are hidden in the parameter space of the deformation, which underlies then a rich structure. We will exploit this structure in the study of Frobenius manifolds done in Chapter VII.

We will illustrate these questions for the Riemann-Hilbert problem, as well as for Birkhoff's problem for a meromorphic differential system with pole of order one.

It is impossible not to mention here the Painlevé equations, which are one of the main sources of examples of isomonodromic deformations, since

R. Fuchs' discovery relating the sixth Painlevé equation to a universal isomonodromic deformation of a system of rank two with four regular singular points on the Riemann sphere (cf. §1.b). The reader can refer to the books [IKSY91], [IN86] and [Con99] for more details, as well as to the article [JMU81].

The contents of this chapter are inspired by the articles [JMU81, JM81, Mal83c, Mal83a, Mal86].

We take up the notation of §I.5. In particular, if X is a manifold and x^o is a base point of X, we denote by A^o the restriction to $\mathbb{P}^1 \times \{x^o\}$ of the object A defined on $\mathbb{P}^1 \times X$.

1 The Riemann-Hilbert problem in a family

1.a Integrable deformations of solutions to the Riemann-Hilbert problem

We take up in this section the situation described in Remark III.1.22. In other words, let X be a connected complex analytic manifold of dimension n and, for any $i \in \{0, \ldots, p\}$, let $\psi_i : X \to \mathbb{P}^1$ be a holomorphic mapping. We assume that, for any $x \in X$, the values $\psi_i(x)$ $(i = 0, \ldots, p)$ are pairwise distinct. Therefore, the graphs $\Sigma_i \subset \mathbb{P}^1 \times X$ of the mappings ψ_i do not intersect each other.

There exists a family of automorphisms of $\mathbb{P}^1$ parametrized by X sending $\{\infty\} \times X$ to Σ_0. We have even a better result:

1.1 Lemma. *There exists an automorphism φ of $\mathbb{P}^1 \times X$ such that the following diagram commutes:*

$$\mathbb{P}^1 \times X \xrightarrow{\;\;\varphi\;\;} \mathbb{P}^1 \times X$$
$$\searrow \qquad \swarrow$$
$$X$$

satisfying $\varphi(\{\infty\} \times X) = \Sigma_0$, $\varphi(\{0\} \times X) = \Sigma_1$ and $\varphi(\{1\} \times X) = \Sigma_2$.

Proof. Let t be a coordinate on $\mathbb{P}^1 \setminus \{\infty\}$. Then φ is given by

$$\varphi(t, x) = \frac{\psi_0(x)t - \psi_1(x)\left(\dfrac{\psi_0(x) - \psi_2(x)}{\psi_1(x) - \psi_2(x)}\right)}{t - \left(\dfrac{\psi_0(x) - \psi_2(x)}{\psi_1(x) - \psi_2(x)}\right)}. \qquad \square$$

We will then suppose in the following that $\psi_0 \equiv \infty$ and we will fix a coordinate t on $U_0 = \mathbb{P}^1 \setminus \{\infty\}$.

1.2 Theorem ([Miw81, Mal83c]). *Let us suppose that the manifold X is 2-connected (i.e., 1-connected and having its second homotopy group equal to zero) and let us fix some point x^o of X. Let us pick $A_1^o, \ldots, A_p^o \in M_d(\mathbb{C})$. There exist then matrices $A_1, \ldots, A_p$, which are holomorphic in the neighbourhood of x^o and meromorphic on X, such that the matrix 1-form*

$$\Omega = \sum_{i=1}^p A_i(x) \frac{d(t - \psi_i(x))}{t - \psi_i(x)}$$

fulfills the integrability condition $d\Omega + \Omega \wedge \Omega = 0$. Moreover, the matrices A_i are determined in a unique way by the initial condition $A_i(x^o) = A_i^o$ $(i = 1, \ldots, p)$.

Let us first be more explicit on the integrability condition.

1.3 Proposition (Schlesinger equations [Sch12]). *Let $A_1, \ldots, A_p$ be matrices of holomorphic functions on some connected open set X' of X. Then the matrix 1-form*

$$\Omega = \sum_{i=1}^p A_i(x) \frac{d(t - \psi_i(x))}{t - \psi_i(x)}$$

is integrable if and only if the matrices A_i satisfy on X' the differential system

$$(1.4) \qquad dA_i - \sum_{j \neq i} [A_i, A_j] \frac{d(\psi_i - \psi_j)}{\psi_i - \psi_j} = 0.$$

1.5 Remarks.

(1) We will see below that the *local* existence in Theorem 1.2 is an easy result, being a straightforward consequence of integrability, in the sense of Frobenius, of Schlesinger equations (Proposition 1.8). The *geometric* method that we will use in the proof gives the *global* existence (in the meromorphic sense) of the solutions to these equations. We say that the system of Schlesinger equations fulfills *Painlevé's property*.

(2) If the matrices A_i are solutions to Schlesinger equations on the connected open set X', their diagonal parts Δ_i satisfy $d\Delta_i = 0$, hence the Δ_i are constant matrices. The proposition also enables one to show that the sum $\sum_{i=1}^p A_i$ of the matrices A_i is independent of $x \in X'$.

(3) Moreover, the characteristic polynomial of each A_i does not depend on x: indeed, it is a matter of verifying that it is locally constant on the connected open set X'; in order to do that, it is enough, as the determinations of the logarithm are indexed by the discrete set $\mathbb{Z}$ and according to Exercise II.4.5, to show that the characteristic polynomial of each monodromy matrix T_i does not depend on x; this is a consequence of the isomonodromy of the family of connections on the trivial bundle $\mathscr{O}_{\mathbb{P}^1}^d$ of matrix $\sum_i \frac{A_i(x)}{t - \psi_i(x)} dt$, which is a consequence of the fact that Ω satisfies the integrability condition (cf. Proposition 0.16.6).

(4) If we have for any i the equality $A_i^o + {}^tA_i^o = w_i \operatorname{Id}$, then we also have $A_i + {}^tA_i \equiv w_i \operatorname{Id}$. Indeed, one verifies that $A_i - w_i \operatorname{Id}$ and $-{}^tA_i$ are solutions to Schlesinger equations with the same initial conditions. According to the uniqueness assertion in Theorem 1.2, these solutions coincide everywhere.

Proof (of Proposition 1.3). We have

$$d\Omega = \sum_{i=1}^{p} \frac{dA_i}{t - \psi_i} \wedge dt - \sum_{i=1}^{p} \frac{dA_i \wedge d\psi_i}{t - \psi_i}$$

and

$$\Omega \wedge \Omega = \sum_{i,j} A_i A_j \frac{d(t - \psi_i) \wedge d(t - \psi_j)}{(t - \psi_i)(t - \psi_j)}$$

$$= \sum_{i,j} A_i A_j \frac{[dt \wedge d(\psi_i - \psi_j)] + [d\psi_i \wedge d\psi_j]}{(t - \psi_i)(t - \psi_j)}$$

$$= \sum_{i,j} \frac{[A_i, A_j]}{(t - \psi_i)} \cdot \frac{[dt \wedge d(\psi_i - \psi_j)] + [d\psi_i \wedge d\psi_j]}{(\psi_i - \psi_j)}.$$

The integrability condition $d\Omega + \Omega \wedge \Omega = 0$ is then equivalent to

(1.6)
$$\sum_i \frac{dA_i}{t - \psi_i} = \sum_{i,j} \frac{[A_i, A_j]}{(t - \psi_i)} \frac{d(\psi_i - \psi_j)}{(\psi_i - \psi_j)}$$

and

(1.7)
$$\sum_i \frac{dA_i \wedge d\psi_i}{t - \psi_i} = \sum_{i,j} \frac{[A_i, A_j]}{(t - \psi_i)} \frac{d\psi_i \wedge d\psi_j}{(\psi_i - \psi_j)}.$$

By taking the residue of (1.6) along the hypersurfaces $\Sigma_k = \{t - \psi_k(x) = 0\}$, we find the Schlesinger equations (1.4). Conversely, Equations (1.4) imply (1.6). They also imply

$$\sum_i \frac{dA_i \wedge d\psi_i}{t - \psi_i} = \sum_{i,j} \frac{[A_i, A_j]}{(t - \psi_i)} \frac{d(\psi_i - \psi_j) \wedge d\psi_i}{(\psi_i - \psi_j)},$$

whence (1.7).   □

Proof (of Theorem 1.2: existence). Let (E^o, ∇^o) be the trivial bundle $\mathscr{O}_{\mathbb{P}^1}^d$ equipped with the connection ∇^o having matrix $\sum_{i=1}^{p} \dfrac{A_i^o}{t - \psi_i(x^o)}$ in the canonical basis ε^o. It also has a logarithmic pole at infinity. As X is 2-connected we can construct on $\mathbb{P}^1 \times X$, according to Remark III.1.22, a bundle (E, ∇) with connection having logarithmic poles along the Σ_i $(i = 0, \ldots, p)$ such that $(E, \nabla)_{|\mathbb{P}^1 \times \{x^o\}} = (E^o, \nabla^o)$.

Let ∇ be the connection induced by ∇ on $\Sigma_0 = \{\infty\} \times X$, in the sense of §0.14.b. This is a *flat* connection on $E_\infty := i_\infty^* E = E_{|\{\infty\} \times X}$. Therefore, as X is 1-connected, the bundle E_∞ is trivializable; more precisely, there exists a unique basis ε' of E_∞ which extends the canonical basis of $E_\infty^o = \mathbb{C}^d$ and which is ∇-horizontal.

As E^o is trivial, there exists, after Corollary I.5.8, a hypersurface Θ of X, not containing x^o, such that the restriction morphism $\mathscr{E}(*\pi^{-1}\Theta) \to \pi^* i_\infty^* \mathscr{E}(*\pi^{-1}\Theta)$ is an isomorphism. It follows that the meromorphic bundle $\mathscr{E}(*\pi^{-1}\Theta)$ is trivializable and that it is equipped with a basis ε, obtained by lifting the basis ε', which restricts to the canonical basis ε^o at x^o.

Let Ω be the matrix of ∇ in the basis ε. It is a matrix of meromorphic differential 1-forms on $\mathbb{P}^1 \times X$, with logarithmic poles along the Σ_i ($i = 0, \ldots, p$) and also with poles along $\mathbb{P}^1 \times \Theta$.

Let A_i be the residue of Ω along Σ_i. It is a matrix of meromorphic functions on $\Sigma_i \simeq X$ with poles along Θ. Let us consider then the matrix

$$\Omega' = \Omega - \sum_{i=1}^p A_i(x) \frac{d(t - \psi_i(x))}{t - \psi_i(x)}.$$

For x fixed in $X \smallsetminus \Theta$, its entries are meromorphic differential forms on $\mathbb{P}^1$, with at most logarithmic poles at infinity. As $H^0(\mathbb{P}^1, \Omega_{\mathbb{P}^1}^1(\infty)) = 0$ (because $\Omega_{\mathbb{P}^1}^1(\infty) \simeq \mathscr{O}_{\mathbb{P}^1}(-1)$), we conclude that the coefficient of dt in Ω' is zero. We can thus write, in local coordinates, $\Omega' = \sum_{j=1}^n B_k(t, x)\, dx_k$, where the B_k are holomorphic on $\mathbb{C} \times (X \smallsetminus \Theta)$. As the matrix of ∇ in the basis ε' is zero, we have

$$\lim_{|t| \to \infty} B_k(t, x) = 0$$

and thus $B_k \equiv 0$. We deduce that $\Omega = \sum_{i=1}^p A_i(x) \dfrac{d(t - \psi_i(x))}{t - \psi_i(x)}$. Therefore, the matrices $A_1, \ldots, A_p$ are a solution to the problem. $\qquad\square$

Proof (of Theorem 1.2: uniqueness). The question of uniqueness in Theorem 1.2 is a local problem, as the open set where two systems $A_1, \ldots, A_n$ and $A_1', \ldots, A_n'$ of solutions are holomorphic is connected, being the complement of an analytic hypersurface in X (cf. Lemma 0.2.1). We can consider the system (1.4) of Schlesinger equations as a Pfaff system on the manifold $X \times M_d(\mathbb{C})^p$ in the neighbourhood of the point $(x^o, A_1^o, \ldots, A_p^o)$. According to §0.13.b, in order to obtain the local uniqueness, it is enough to show:

1.8 Proposition (The Schlesinger equations are integrable). *The Pfaff system of Schlesinger equations is integrable and defines a foliation of dimension $n = \dim X$ in the neighbourhood of $(x^o, A_1^o, \ldots, A_p^o)$.*

The tangent space to the leaf passing through $(x^o, A_1^o, \ldots, A_p^o)$ is transversal to the subspace $\{0\} \times M_d(\mathbb{C})^p$ of $T_{(x^o, A_1^o, \ldots, A_p^o)}(X \times M_d(\mathbb{C})^p)$.

Indeed, once this result is obtained, the inverse function theorem gives a local parametrization of the leaf going through $(x^o, A_1^o, \ldots, A_p^o)$ by a neighbourhood of x^o in X, whence the uniqueness.   □

Proof (of Proposition 1.8). Let us begin by showing that when $n = p$, $X = X_p$ is the open set of $\mathbb{C}^p$ consisting of the $(x_1, \ldots, x_p)$ such that $x_i \neq x_j$ if $i \neq j$ and $\psi_i(x) = x_i$. Let $\mathscr{J}$ be the subsheaf of $\Omega_{X_p \times M_d(\mathbb{C})^p}^1$ generated by the entries of the matrices

$$\Omega_i = dA_i - \sum_{j \neq i} [A_i, A_j] \frac{d(x_i - x_j)}{x_i - x_j}.$$

It is a matter of verifying (cf. Lemma 0.13.2) that $\mathscr{J}$ is a subbundle of rank p of $\Omega_{X_p \times M_d(\mathbb{C})^p}^1$ and that the entries of $d\Omega_i$ vanish modulo $\mathscr{J} \wedge \Omega_{X_p \times M_d(\mathbb{C})^p}^1$. But one can write, modulo $\mathscr{J} \wedge \Omega_{X_p \times M_d(\mathbb{C})^p}^1$,

$$-d\Omega_i = \sum_{j \neq i} ([dA_i, A_j] + [A_i, dA_j]) \frac{d(x_i - x_j)}{x_i - x_j}$$

$$\equiv \sum_{j \neq i} \sum_{k \neq i} [[A_i, A_k], A_j] \frac{d(x_i - x_k)}{x_i - x_k} \wedge \frac{d(x_i - x_j)}{x_i - x_j}$$

$$+ \sum_{j \neq i} \sum_{k \neq j} [A_i, [A_j, A_k]] \frac{d(x_j - x_k)}{x_j - x_k} \wedge \frac{d(x_i - x_j)}{x_i - x_j}.$$

For $j, k \neq i$ and $j < k$, the coefficient of $dx_j \wedge dx_k$ in the right-hand term is equal to

$$\frac{[[A_i, A_j], A_k] - [[A_i, A_k], A_j]}{(x_i - x_j)(x_i - x_k)} - \frac{1}{(x_j - x_k)} \left(\frac{[A_i, [A_k, A_j]]}{(x_i - x_k)} + \frac{[A_i, [A_j, A_k]]}{(x_i - x_j)} \right)$$

and, according to Jacobi's identity, it is zero. The coefficients of $dx_i \wedge dx_j$ are treated similarly.

Let us denote by $\xi_1, \ldots, \xi_p$ the coordinates on $T_{x^o}^* X_p$ and by $b_{k\ell}^{(i)}$ that on the ith factor $M_d(\mathbb{C})$, $k, \ell = 1, \ldots, d$. Therefore, an element of the cotangent bundle $T^*(X_p \times M_d(\mathbb{C})^p)$ to $X_p \times M_d(\mathbb{C})^p$ has coordinates

$$\left(x_1, \ldots, x_p, (a_{k\ell}^{(i)})_{\substack{k,\ell=1,\ldots,d \\ i=1,\ldots,p}}, \xi_1, \ldots, \xi_p, (b_{k\ell}^{(i)})_{\substack{k,\ell=1,\ldots,d \\ i=1,\ldots,p}} \right).$$

Equations (1.4) define a subbundle F of $T^*(X_p \times M_d(\mathbb{C})^p)$ isomorphic to the bundle $T^*(X_p) \times M_d(\mathbb{C})^p$ by

$$\left(x_1, \ldots, x_p, (a_{k\ell}^{(i)})_{\substack{k,\ell=1,\ldots,d \\ i=1,\ldots,p}}, \xi_1, \ldots, \xi_p \right)$$

$$\longmapsto \left(x_1, \ldots, x_p, (a_{k\ell}^{(i)})_{\substack{k,\ell=1,\ldots,d \\ i=1,\ldots,p}}, \xi_1, \ldots, \xi_p, (b_{k\ell}^{(i)})_{\substack{k,\ell=1,\ldots,d \\ i=1,\ldots,p}} \right)$$

with

$$b_{k\ell}^{(i)} = \sum_j c_{k\ell}^{(i,j)} \frac{\xi_i - \xi_j}{x_i - x_j},$$

where the coefficients $c_{k\ell}^{(i,j)}$ are computed from the entries of A_i and A_j.

For any $(x, A_1, \ldots, A_p) \in X_p \times M_d(\mathbb{C})^p$ the fibre $F_x \subset \mathbb{C}^p \times M_d(\mathbb{C})^p$ is thus a subspace of dimension p, transversal to the subspace having equations $\xi_1 = \cdots = \xi_p = 0$. The proposition is thus proved in this particular case.

For X arbitrary, the Pfaff system (1.4) is the pullback by the mapping

$$(\psi_1, \ldots, \psi_p, \mathrm{Id}) : X \times M_d(\mathbb{C})^p \longrightarrow X_p \times M_d(\mathbb{C})^p$$

of the Pfaff system considered above. It is then integrable (cf. Remark 0.13.3). One can see as above that it defines a subbundle of rank n of $T^*(X \times M_d(\mathbb{C})^p)$ isomorphic to $T^*(X) \times M_d(\mathbb{C})^p$ and the fibres of which are transversal to $X \times T^*(M_d(\mathbb{C})^p)$. □

1.b An example : the sixth Painlevé equation as an isomonodromy equation

We will give details for the particular case of Schlesinger's system (1.4) with four singular points in $\mathbb{P}^1$ (we fix the three first ones at $0, 1, \infty$, according to Lemma 1.1, and only the last one is variable, but distinct from $0, 1, \infty$), and where the size d of the matrices is equal to 2.

We thus assume that $X = \mathbb{P}^1 \smallsetminus \{0, 1, \infty\}$, and that the three first functions $\psi_i(x)$ are constant, equal respectively to $0, 1, \infty$. Lastly, the fourth one is the identity $x \mapsto x$. With a slight change of notation, the differential form Ω can be written as

$$\Omega = A_0(x) \frac{dt}{t} + A_1(x) \frac{dt}{t-1} + A_x(x) \frac{d(t-x)}{(t-x)},$$

and Schlesinger equations take the form

$$dA_0 = [A_0, A_x] \frac{dx}{x}, \; dA_1 = [A_1, A_x] \frac{dx}{x-1}, \; dA_x = [A_x, A_0] \frac{dx}{x} + [A_x, A_1] \frac{dx}{x-1}.$$

If we define A_∞ by the relation $A_0 + A_1 + A_x + A_\infty = 0$, we conclude that A_∞ is constant, and that the system reduces to the system

$$A_0'(x) = -\frac{1}{x}[A_0, A_1 + A_\infty], \quad A_1'(x) = -\frac{1}{x-1}[A_1, A_0 + A_\infty],$$

if we denote by $A'(x)$ the derivative of A with respect to x.

Remark. The differential equation for A_0 (resp. A_1) as given above is said to have the *Lax form*.

Recall (cf. Remark 1.5) that the trace and the determinant of these matrices are constant. By adding to each A_i the matrix $-\frac{1}{2}\operatorname{tr} A_i\operatorname{Id}$, we can assume that the A_i have trace zero. Let us assume for instance that both (opposite) eigenvalues of A_∞, denoted by $\pm\lambda_\infty/2$, are nonzero. We can choose a basis in which A_∞ is diagonal.

The diagonal parts and the determinant of A_0 and A_1 being constant, it remains for one entry to be determined in each matrix. One can show (see for instance [IN86] or [Con99, p. 48]) that each one can be obtained from a solution $y(x)$ of the sixth Painlevé equation

$$
\text{(P VI)}\quad y'' = \frac{1}{2}\Big(\frac{1}{y} + \frac{1}{y-1} + \frac{1}{y-x}\Big)(y')^2 - \Big(\frac{1}{x} + \frac{1}{x-1} + \frac{1}{y-x}\Big)y'
$$
$$
+ \frac{y(y-1)(y-x)}{x^2(x-1)^2}\Big(\alpha + \beta\frac{x}{y^2} + \gamma\frac{x-1}{(y-1)^2} + \delta\frac{x(x-1)}{(y-x)^2}\Big)
$$

where the parameters $\alpha, \beta, \gamma, \delta$ are determined by the eigenvalues (independent of x) of the matrices A_i:

$$
\alpha = \frac{(1-\lambda_\infty)^2}{2}, \quad \beta = -\frac{\lambda_0^2}{2}, \quad \gamma = \frac{\lambda_1^2}{2}, \quad \delta = \frac{1-\lambda_x^2}{2}.
$$

1.9 Remark. We have followed here in an opposite way the path which led to the discovery of Painlevé equations. The problem that Painlevé wanted to solve concerned differential equations of the kind

$$
y''(x) = R(x, y, y'),
$$

where R is a rational fraction. The solutions $y(x)$ of such a nonlinear equation may have poles (behaviour in $1/(x-c)^k$), ramification points (behaviour in $(x-c)^\alpha$ or in $\log(x-c)$, etc.) as singularities, and also essential singularities (for instance $y(x) = \exp(1/(x-c))$). The parameter c may depend on initial conditions (this possible dependence on the integration constant is due to the nonlinearity of the equation). The solutions are said to have no *moving singularity* if the ramification points or the essential singularities of the solutions do not depend on these integration constants.

Therefore, the goal of Painlevé was the determination of a minimal class of such differential equations, in such a way that the resolution of any differential equation of this kind, all the solutions of which have no "moving singularity", can be reduced to the resolution of some equation of the class. In particular, it was a matter of finding "new transcendental functions".

This nonexistence of moving singularities is called "Painlevé's property". It is easy to give examples of differential equations with moving singularities: take for instance the differential equation satisfied by $y(x) = \exp(1/(x-c))$.

The solutions of Schlesinger equations fulfill this property as they only have *poles* (along a divisor Θ which *a priori* depends on initial conditions).

1.c Universality

1.10 Definition (of universal integrable deformation). An *integrable* deformation (E, ∇) of (E^o, ∇^o) is a bundle with a flat meromorphic connection on $\mathbb{P}^1 \times X$, having logarithmic poles along a smooth hypersurface $\Sigma \cup (\{\infty\} \times X)$ of $\mathbb{P}^1 \times X$ which is a covering of degree $p + 1$ of X, inducing (E^o, ∇^o) when restricted to $x^o \in X$.

We say that the integrable deformation (E, ∇) of (E^o, ∇^o) is *complete* at x^o if, for any other integrable deformation (E', ∇', x'^o) of (E^o, ∇^o) parametrized by (X', x'^o), there exist neighbourhoods V, V' of x^o, x'^o in X, X' and an analytic mapping

$$f : (V', x'^o) \longrightarrow (V, x^o)$$

such that (E', ∇') is isomorphic (by an isomorphism inducing the identity when restricted to (E^o, ∇^o)) to the pullback of (E, ∇) by the mapping

$$\mathrm{Id} \times f : \mathbb{P}^1 \times V' \longrightarrow \mathbb{P}^1 \times V.$$

We say that the deformation is *universal* at x^o if, moreover, the germ of f at x'^o is unique.

Let X_d be the complementary set of the diagonal $x_i = x_j$ in $\mathbb{C}^d$, equipped with coordinates $x_1, \ldots, x_d$, having $x^o = (x_1^o, \ldots, x_d^o)$ as a base point. This space is not simply connected: its fundamental group is the *colored braid group with n strands* (see [Bri73] or also [AGZV88, vol. 2, p. 72]). One can show[1] that its higher homotopy groups are zero. As X_d is an open set of $\mathbb{C}^d$, its tangent bundle is trivial and equipped with the basis $\partial_{x_1}, \ldots, \partial_{x_d}$ associated to the coordinate system $x_1, \ldots, x_d$.

Let us denote by $\varpi : (\widetilde{X}_d, \widetilde{x}^o) \to (X_d, x^o)$ the universal covering with base point $\widetilde{x}^o$ of the space X_d equipped with its base point x^o. Its second homotopy group is zero, as the same property holds for X_d, and $\widetilde{X}_d$ is thus 2-connected. As $\widetilde{X}_d$ is a covering space of X_d, its tangent bundle $T\widetilde{X}_d$ is the pullback bundle of that of X_d by the covering map ϖ. It is therefore also trivialized and equipped with the basis $\partial_{x_1}, \ldots, \partial_{x_d}$. We will say that this trivialization is *canonical*.

Lastly, let us notice that the symmetric group $\mathfrak{S}_d$ acts freely and transitively on X_d. The quotient space Y_d is the space of monic polynomials without multiple root and the quotient mapping associates to $(x_1, \ldots, x_d) \in X_d$ the polynomial $\prod_{i=1}^{d}(s - x_i)$. Let us identify the space $X'_d = X_d \cap \{x_1 + \cdots + x_d = 0\}$ to $(\mathbb{C}^*)^{d-1}$ with the coordinates $x_i - x_{i+1}$ $(i = 1, \ldots, d-1)$. This space is stable by $\mathfrak{S}_d$ and the quotient space is the subspace Y'_d of Y_d consisting of polynomials for which the sum of roots is zero.

---

[1] By using the homotopy exact sequence [Ste51, p. 90] of the fibration $\pi_d : X_d \to X_{d-1}$ (forgetting the last component) and arguing by induction on $d \geqslant 3$; the vanishing of the higher homotopy groups follows then from the fact that the universal covering space of the complex line minus $d - 1$ points is contractible.

Fig. VI.1. The space Y_4' is the complement of the swallow tail.

We can now apply Theorem 1.2 by taking $d = p$, $X = \widetilde{X}_p$. Let (E, ∇) be the bundle on $\mathbb{P}^1 \times \widetilde{X}_p$ that we have constructed in the proof of this theorem and let Θ be the associated hypersurface.

1.11 Corollary. *The bundle with meromorphic connection (E, ∇) induces a universal deformation of its restriction $(E, \nabla)_{|\mathbb{P}^1 \times \{\widetilde{x}^o\}}$.*

Proof. Let (E', ∇') be an integrable deformation of (E^o, ∇^o) parametrized by a germ of manifold (X', x'^o). The manifold $\Sigma \subset \mathbb{C} \times X'$ is equal, in the neighbourhood of x'^o, to the union of the graphs of p functions $\psi_1, \ldots, \psi_p$. The mappings $\psi_1, \ldots, \psi_p$ define a unique mapping $f : X' \to X_p$. According to the uniqueness property in Theorem 1.2, the bundle (E', ∇') is isomorphic to the pullback of (E, ∇) by this mapping.   □

2 Birkhoff's problem in a family

2.a Integrable deformations of solutions to Birkhoff's problem

Let D be a disc in the complex line U_0 centered at the origin and let X be a connected complex analytic manifold of dimension n, that we also assume to be *simply connected* (up to changing X with a simply connected open set or, better, with its universal covering). Let us now denote by τ the coordinate on $\mathbb{P}^1$ in the chart U_0 centered at 0.

Let $(\widetilde{E}, \widetilde{\nabla})$ be a holomorphic bundle on $U := D \times X$, equipped with an *integrable* meromorphic connection having a pole of order one along $\{0\} \times X$ (cf. §0.14.c)[2].

Let us pick $x^o \in X$. We regard $(\widetilde{E}, \widetilde{\nabla})$ as an *integrable deformation* of the bundle with connection $(\widetilde{E}^o, \widetilde{\nabla}^o)$, restriction of $(\widetilde{E}, \widetilde{\nabla})$ to $U^o = D \times \{x^o\}$, in the sense of §0.14.a.

As we have fixed a coordinate τ on D, we can associate to $(\widetilde{E}, \widetilde{\nabla})$ a "residue" endomorphism R_0 of the bundle $\widetilde{E}_{|\{0\} \times X}$ as well as a Higgs field Φ on this bundle (cf. §0.14.c).

---

[2] As the connection is assumed to be integrable, the size of the disc is not significant and one could also start from a germ, along $\{0\} \times X$, of a holomorphic bundle with meromorphic connection.

Let us assume that we can solve Birkhoff's Problem IV.5.1 for $(\widetilde{E}^o, \widetilde{\nabla}^o)$, for the value x^o of the parameter. Let thus ε^o be a basis of $\widetilde{E}^o$ in which the connection matrix $\widetilde{\nabla}^o$ can be written as $\Omega^o = \left(\dfrac{B_0^o}{\tau} + B_\infty\right)\dfrac{d\tau}{\tau}$, where B_0^o and B_∞ are two matrices in $M_d(\mathbb{C})$.

2.1 Theorem. *With these conditions, there exist a hypersurface Θ of X not containing x^o and a unique basis ε of $\widetilde{\mathscr{E}}(*(D \times \Theta))$ which coincides with ε^o at x^o and in which the connection matrix $\widetilde{\nabla}$ takes the form*

$$(2.2) \qquad \Omega = \left(\frac{B_0(x)}{\tau} + B_\infty\right)\frac{d\tau}{\tau} + \frac{C(x)}{\tau},$$

with $B_0(x^o) = B_0^o$, where $B_0(x)$ (resp. $C(x)$) is a matrix of holomorphic functions (resp. of 1-forms) on $X \smallsetminus \Theta$ and meromorphic along Θ.

Therefore, there exists a canonical solution to Birkhoff's problem for almost all the values of the parameter, if we are given a solution for a particular value. Let us notice that the theorem implies in particular that the bundle $\widetilde{\mathscr{E}}(*(D \times \Theta))$ is trivializable.

Proof (of the existence).

(1) Let D' be an open disc of $\mathbb{P}^1$ centered at ∞ such that the annulus $A = D \cap D'$ is nonempty. Let us denote by τ' the coordinate on D', so that $\tau' = 1/\tau$ in A. On $A \times X$, the bundle $(\widetilde{E}, \widetilde{\nabla})_{|A \times X}$ is determined by its monodromy. As X is 1-connected, we have $\pi_1(A \times X) = \pi_1(A) = \mathbb{Z}$, so that $(\widetilde{E}, \widetilde{\nabla})_{|A \times X}$ is isomorphic to the pullback $p^+(\widetilde{E}^o, \widetilde{\nabla}^o)_{|A}$ by the projection $p : A \times X \to A$.

When restricted to $A \times X$, the trivial bundle $\mathscr{O}_{D' \times X}^d$, equipped with the connection having matrix $-(\tau' B_0^o + B_\infty)d\tau'/\tau'$, is thus isomorphic to $(\widetilde{E}, \widetilde{\nabla})_{|A \times X}$. Let us note that this connection has logarithmic poles along $\{\infty\} \times X$.

We can thus glue both bundles to get a bundle (E, ∇) on $\mathbb{P}^1 \times X$, equipped with a flat meromorphic connection having poles along $\{0, \infty\} \times X$, logarithmic at ∞ and of order one at 0. By assumption, the restriction E^o of E to $\mathbb{P}^1 \times \{x^o\}$ is trivializable.

(2) The bundle $i_\infty^* \mathscr{E}$ is equipped with a holomorphic flat connection ∇ (cf. §0.14.b). Giving such a connection is equivalent to giving the locally constant sheaf of its horizontal sections. As X is 1-connected, this locally constant sheaf is trivializable; so is thus the bundle $i_\infty^* \mathscr{E}$ and, more precisely, any basis of the fibre at x^o of this bundle can be extended in a unique way as a horizontal basis of the bundle (cf. Theorem 0.12.8).

(3) Corollary I.5.8 gives an isomorphism

$$\mathscr{E}(*\pi^{-1}\Theta) \simeq \pi^* i_\infty^* \mathscr{E}(*\pi^{-1}\Theta),$$

where Θ is the hypersurface associated, by Theorem I.5.3, to the bundle E constructed in (1).

(4) Let us apply this to the basis ε^o, to obtain a basis ε. It defines thus, after (3), a basis of the bundle $\mathscr{E}(*\pi^{-1}\Theta)$ as a $\mathscr{O}_{\mathbb{P}^1 \times X}(*\pi^{-1}\Theta)$-module; in other words, the elements of the basis can be extended as global sections of $\mathscr{E}$, at least on the complementary set of $\pi^{-1}\Theta$.

2.3 Lemma. *In such a basis ε and in the chart of $\mathbb{P}^1$ centered at 0, the connection matrix of ∇ takes the form (2.2).*

Proof. The matrix Ω in the basis ε has order one. It can be thus written as

$$\left(\frac{B_0(x)}{\tau} + B_\infty(x,\tau)\right)\frac{d\tau}{\tau} + C_0(x,\tau) + \frac{C(x)}{\tau},$$

where B_∞ and C_0 are holomorphic. The logarithmic behaviour at infinity shows that B_∞ and C_0 are independent of τ. The horizontality of the basis ε with respect to the restriction to infinity of the connection shows that $C_0 = 0$. The horizontality of the residue with respect to the connection ∇ (cf. Exercise 0.14.6-(2)) shows that the matrix B_∞ is constant. $\qquad\square$

Proof of the uniqueness. Let ε' and ε'' be two such bases, meromorphic along Θ' and Θ'' respectively. Let us set $\Theta = \Theta' \cup \Theta''$. There exists thus an isomorphism

$$(\mathscr{O}^d_{D\times X}(*(D\times\Theta)),\nabla') \xrightarrow[\sim]{P} (\mathscr{O}^d_{D\times X}(*(D\times\Theta)),\nabla''),$$

where ∇' and ∇'' take the form (2.2), whose restriction to $D \times \{x^o\}$ is the identity. Let us consider then its restriction to $D \times X^o$, with $X^o = X \setminus \Theta$. We will show that it is equal to the identity, which will give the desired result.

Let us note that the connections ∇' and ∇'', by their very definition, exist on the bundle $\mathscr{O}^d_{\mathbb{P}^1 \times X^o}$ and have a logarithmic pole along $\{\infty\} \times X^o$. As P preserves connections, it can be extended as an isomorphism of bundles with connection on $\mathbb{C} \times X^o$ (as the inclusion $D \times X^o \hookrightarrow \mathbb{C} \times X^o$ induces an isomorphism of fundamental groups).

Therefore, P is a solution of a differential system with regular singularity along the hypersurface $\{\infty\} \times X^o$, namely that associated to

$$\mathscr{H}om((\mathscr{O}^d_{\mathbb{P}^1 \times X^o}, \nabla'), (\mathscr{O}^d_{\mathbb{P}^1 \times X^o}, \nabla'')).$$

Theorem II.2.25 and an argument analogous to that of Exercise II.2.6 show that the entries of the matrix P are meromorphic along $\{\infty\} \times X^o$.

Let us also notice that the system satisfied by P is still in the Birkhoff normal form. If we write P as a vector, the system can thus be written, in the coordinate τ' at infinity, as

$$dP = \left[(A_\infty + \tau'A(x))\frac{d\tau'}{\tau'} + D(x)\tau'\right] \cdot P.$$

If we set $P = \sum_{\ell \geqslant \ell_0} \tau'^\ell P_\ell(x)$, we deduce in particular

$$(*) \qquad\qquad dP_\ell = D(x)P_{\ell-1}.$$

We have $P_\ell \equiv 0$ for $\ell < \ell_0$. If $\ell_0 < 0$ we also have $P_{\ell_0}(x^o) = 0$. According to $(*)$ we have $dP_{\ell_0} = 0$, hence, as $X \smallsetminus \Theta$ is connected (cf. 0.2.1), we have $P_{\ell_0} \equiv P_{\ell_0}(x^o) = 0$. We deduce that the entries of P are holomorphic on $\mathbb{P}^1 \times X^o$: they are thus independent of τ'. Therefore, $P(\tau', x) = P_0(x)$. We also have, after $(*)$, the identity $dP_0 = 0$, hence $P(\tau', x) \equiv P_0(x^o) = \mathrm{Id}$. $\qquad\square$

2.4 Exercise. Use Remark III.1.20 to show that P, defined on $D \times X^o$, can be extended holomorphically to $\mathbb{P}^1 \times X^o$.

Remark. The matrix $-B_\infty$ is the matrix of the residue at infinity of the connection in the horizontal basis ε. The integrability condition for ∇ can then be written, in this basis, as

$$(2.5) \qquad \begin{cases} dC = 0 \\ C \wedge C = 0 \\ [B_0, C] = 0 \\ dB_0 + C = [B_\infty, C]. \end{cases}$$

In local coordinates $x_1, \dots, x_n$ on X, we can write $C = \sum_{i=1}^n C^{(i)}(x)\, dx_i$, where the $C^{(i)}$ are matrices of meromorphic functions on X with poles along Θ. The system (2.5) takes the form

$$(2.6) \qquad \begin{cases} \dfrac{\partial C^{(i)}}{\partial x_j} = \dfrac{\partial C^{(j)}}{\partial x_i} & i, j = 1, \dots, n \\[2mm] [C^{(i)}, C^{(j)}] = 0 & i, j = 1, \dots, n \\[2mm] [B_0, C^{(i)}] = 0 & i = 1, \dots, n \\[2mm] \dfrac{\partial B_0}{\partial x_i} + C^{(i)} = [B_\infty, C^{(i)}] & i = 1, \dots, n. \end{cases}$$

If for instance the matrix $B_0(x)$ is *regular*[3] for any x, that is, if its minimal polynomial is equal to its characteristic polynomial, any matrix which commutes with $B_0(x)$ is a polynomial in $B_0(x)$. If the equations of the third line are satisfied, the $C^{(i)}(x)$ are polynomials in $B_0(x)$ and, in this case, the equations of the second line are also satisfied.

2.b Constructions with a "metric"

We keep the previous situation and we analyze the consequences of the existence of a duality on the bundle $(\widetilde{E}, \widetilde{\nabla})$. Let us consider the case of a sesquilinear duality, the coordinate τ being fixed on the disc D. The reader will check as an exercise that analogous arguments can be applied in the case of a bilinear duality.

---

[3] Recall that a *regular* matrix B is a matrix which has only one Jordan block for each eigenvalue. The matrices which commute to B are then polynomials in B and form a vector space of dimension d. The set of regular matrices is an open dense set of $M_d(\mathbb{C})$.

Let us thus assume that there exists a sesquilinear form $\widetilde{G}$ on $(\widetilde{E}, \widetilde{\nabla})$, of weight $w \in \mathbb{Z}$: if, as in §III.1.13, we denote by $(\overline{\widetilde{E}}, \overline{\widetilde{\nabla}})$ the conjugate of $(\widetilde{E}, \widetilde{\nabla})$ by the involution $\tau \mapsto -\tau$, we regard the form $\widetilde{G}$ as an isomorphism

$$(\widetilde{E}, \widetilde{\nabla}) \xrightarrow{\sim} (\overline{\widetilde{E}^*}, \overline{\widetilde{\nabla}^*} + w\frac{d\tau}{\tau}).$$

We will say that $\widetilde{G}$ is *Hermitian* of weight w if moreover its transpose conjugate ${}^t\overline{\widetilde{G}}$ is equal to $(-1)^w \widetilde{G}$.

If we regard $\widetilde{G}$ as a sesquilinear pairing

$$(\widetilde{\mathscr{E}}, \widetilde{\nabla}) \otimes (\widetilde{\mathscr{E}}, \widetilde{\nabla}) \longrightarrow (\tau^w \mathcal{O}_{D \times X}, d),$$

we have, for any two local sections e and e' of $\widetilde{\mathscr{E}}$, the identity

$$\overline{{}^t\widetilde{G}(e, e')} = \widetilde{G}(e', e).$$

Moreover, the coefficient of τ^w in $\widetilde{G}(e, e')$ only depends on the classes of e and e' in $\widetilde{\mathscr{E}}/\tau\widetilde{\mathscr{E}} = i_0^* \widetilde{\mathscr{E}}$. Therefore, given $\widetilde{G}$, we can define a bilinear form g_0 on the restriction $i_0^* \widetilde{E}$ of $\widetilde{E}$ to $\{0\} \times X$. We have

$$\widetilde{G}(e, e') = \tau^w g_0([e], [e']) + \tau^{w+1} g_0^{(1)}([e], [e']) + \cdots$$

As a consequence, if $\widetilde{G}$ is Hermitian of weight w, the form g_0 is *symmetric bilinear* on $i_0^* \widetilde{E}$. Similarly, we see that the "residue" R_0 of $\widetilde{\nabla}$ along $\{0\} \times X$, which is an endomorphism of the bundle $i_0^* \widetilde{E}$, is *self-adjoint* for g_0. The Higgs field Φ also satisfies $\Phi^* = \Phi$, that is, for any germ of vector field ξ on X, the endomorphism Φ_ξ of the bundle $i_0^* \widetilde{E}$ is self-adjoint with respect to g_0.

We can now extend the statement of Theorem 2.1 in the presence of $\widetilde{G}$.

2.7 Proposition. *With the conditions of Theorem 2.1, assume moreover that*

(1) *there exists a sesquilinear (resp. Hermitian) form $\widetilde{G}$ of weight w on $(\widetilde{E}, \widetilde{\nabla})$;*

(2) *when restricted to $\mathbb{P}^1 \times \{x^o\}$, the form $\widetilde{G}^o$ can be extended as a sesquilinear (resp. Hermitian) form G^o of weight w on (E^o, ∇^o).*

There exists then a unique sesquilinear (resp. Hermitian) form G of weight w on the meromorphic bundle $(\mathscr{E}(*\pi^{-1}\Theta), \nabla)$ which extends $\widetilde{G}$ and G^o.

Proof. It is a matter of seeing that the construction used in Theorem 2.1, taking $[(\widetilde{E}, \widetilde{\nabla}), (E^o, \nabla^o)]$ to (E, ∇), is functorial. If

$$(\widetilde{\varphi}, \varphi^o) : [(\widetilde{E}, \widetilde{\nabla}), (E^o, \nabla^o)] \longrightarrow [(\widetilde{E}', \widetilde{\nabla}'), (E'^o, \nabla'^o)]$$

is a homomorphism, one shows as in the "uniqueness" part in the proof of Theorem 2.1, that $\widetilde{\varphi}$ can be extended to $\mathbb{C} \times (X \smallsetminus \Theta)$, then that the homomorphism so defined is meromorphic along $\{\infty\} \times (X \smallsetminus \Theta)$ and, lastly, that its matrix in the bases ε and ε' is constant. This implies the existence and uniqueness of an extension $\varphi : (E, \nabla) \to (E', \nabla')$ of $(\widetilde{\varphi}, \varphi^o)$. Functoriality immediately follows.   □

2.8 Remark. If $\widetilde{G}$ and G^o are nondegenerate, so is G: in the proof above, it is a matter of seeing that, if $(\widetilde{\varphi}, \varphi^o)$ is an isomorphism, then so is its extension φ; this follows from uniqueness.

Let us now give an interpretation of the presence of a sesquilinear or Hermitian form on the bundle $(\mathscr{E}, \nabla)$ on $\mathbb{P}^1 \times X$, equipped with the meromorphic basis ε given by Theorem 2.1. Let us denote by $\mathbb{E}_0$ the free $\mathscr{O}_X[\tau]$-module of sections (polynomial in τ) of E on the open set $U_0 \times X$ centered at $\{0\} \times X$. Then $\overline{\mathbb{E}}_0$ is the $\mathscr{O}_X$-module $\mathbb{E}_0$ on which $\mathbb{C}[\tau]$ acts by $h(\tau) \cdot e = h(-\tau)e$ and, if $A(\tau, x)$ is the matrix of ∇_{∂_τ} in some $\mathscr{O}_X[\tau]$-base, that of $\overline{\nabla}_{\partial_\tau}$ in the same basis is $-A(-\tau, x)$.

On $U_0 \times X$, the form G defines a sesquilinear mapping

$$\mathbb{G}_0 : \mathbb{E}_0 \otimes \overline{\mathbb{E}}_0 \longrightarrow \tau^w \mathscr{O}_X[\tau]$$

which is compatible with connections, that is, such that

$$h(\tau)\mathbb{G}_0(e, e') = \mathbb{G}_0(h(\tau)e, e') = \mathbb{G}_0(e, h(-\tau)e')$$
$$\mathscr{L}_\xi \mathbb{G}_0(e, e') = \mathbb{G}_0(\nabla_\xi e, e') + \mathbb{G}_0(e, \overline{\nabla}_\xi e'),$$

where $\mathscr{L}_\xi \mathbb{G}_0(e, e')$ denotes the derivation of the function $\mathbb{G}_0(e, e')$ along the vector field ξ (Lie derivative). The coefficient of τ^w in $\mathbb{G}_0(e, e')$ only depends on the classes of e and e' in $\mathbb{E}_0/\tau\mathbb{E}_0$: this is $g_0([e], [e'])$. If we write as above

$$\mathbb{G}_0(e, e') = \tau^w g_0([e], [e']) + \tau^{w+1} g_0^{(1)}([e], [e']) + \cdots$$

we see that, if $e, e' \in \mathbb{E}_0$ are sections which can be extended as sections of E on $\mathbb{P}^1 \times X$, we have $\mathbb{G}_0(e, e') = \tau^w g_0([e], [e'])$.

2.9 Exercise. Working in the chart U_∞, define the form g_∞ on $i_\infty^* E$. Prove similarly that g_∞ is symmetric when G is Hermitian of weight w.

2.10 Proposition. *The form g_∞ is ∇-horizontal. If moreover G is Hermitian and nondegenerate of weight w, the forms g_0 and g_∞ are symmetric and nondegenerate. In such a case, if ε is a horizontal basis for ∇ (assuming X simply connected), so that, in this basis, the matrix of ∇ takes the form (2.2), we have on $X \smallsetminus \Theta$*

$$B_0^* = B_0, \quad B_\infty^* + B_\infty = w\,\mathrm{Id}, \quad C^* = C,$$

if B^* denotes the adjoint of B with respect to g.

Proof. The form G is a horizontal section of the bundle with connection $\mathscr{H}om_\mathscr{O}(E \otimes \overline{E}, \mathscr{O}[w])$; the natural connection on this bundle has a logarithmic pole at infinity. When restricted to $\tau = \infty$, the component on Ω_X^1 of the equation $\nabla G = 0$ is the equation $\nabla G(\infty, x) = 0$. This gives the first point.

On the other hand, we have an isomorphism of locally free $\mathcal{O}_X$-modules $i_\infty^*(E^*) \simeq (i_\infty^* E)^*$, so that, if G is nondegenerate, so is g_∞. The case of g_0 is similar and the symmetry has been seen above.

The matrix of ∇^* in the basis dual to ε is minus the transpose of that of ∇. That of $\overline{\nabla}^*[w]$ can thus be written as

$$\left(\frac{{}^t B_0(x)}{\tau} - {}^t B_\infty + w\,\mathrm{Id} \right) \frac{d\tau}{\tau} + \frac{{}^t C(x)}{\tau},$$

which gives the last point.   □

2.11 Remarks.

(1) Let $\mathcal{E}$ be the d-dimensional $\mathbb{C}$-vector space consisting of horizontal multivalued sections of ∇ on $\mathbb{C}^* \times X$. It is equipped with a monodromy endomorphism T. The form G induces a bilinear (resp. nondegenerate and symmetric) form $\mathcal{G}$ on this space and the monodromy is an automorphism of this bilinear form.

(2) In the following, we will commit the (usual) abuse to call "metric" the Hermitian form G, or the forms g_0 and g_∞. The latter are only complex symmetric nondegenerate bilinear forms and no positivity property is required.

2.c Summary of §§2.a and 2.b

Let E be a bundle of rank d on $\mathbb{P}^1 \times X$ equipped with a flat meromorphic connection ∇, having poles along $\{0, \infty\} \times X$, logarithmic along $\{\infty\} \times X$ and of order one along $\{0\} \times X$. Let us assume that the restriction E^o of E to $\mathbb{P}^1 \times \{x^o\}$ is trivial.

2.12 The hypersurface Θ. The closed set Θ of points $x \in X$ at which the restriction E_x to $\mathbb{P}^1 \times \{x\}$ is not the trivial bundle is empty or a hypersurface of X (hence if it has codimension $\geqslant 2$ in X, it is empty).

2.13 Meromorphic identification. Both bundles $E_0 := i_0^* E$ and $E_\infty := i_\infty^* E$ of rank d on X are identified in a meromorphic way along Θ through the isomorphisms of $\mathcal{O}_X(*\Theta)$-modules induced by the restrictions

$$\mathcal{E}_0(*\Theta) \xleftarrow{\sim} \pi_* \mathcal{E}(*\Theta) \xrightarrow{\sim} \mathcal{E}_\infty(*\Theta).$$

We will denote by $\mathcal{M}$ the meromorphic bundle $\pi_* \mathcal{E}(*\Theta)$ on X. It thus contains two locally free lattices, namely $\mathcal{E}_0$ and $\mathcal{E}_\infty$, and an intermediate "lattice" $\mathcal{E}_1$, namely $\pi_* \mathcal{E}$ (note that $\pi_* \mathcal{E}$ is a torsion free $\mathcal{O}_X$-coherent sheaf, hence is naturally contained in $\mathcal{M}$, but is not necessarily locally free).

2.14 Data at infinity. The lattice $\mathcal{E}_\infty$ is equipped with a flat connection $\triangledown$ and with a $\triangledown$-horizontal endomorphism R_∞ (residue of the connection ∇, with matrix $-B_\infty$ in a horizontal basis). We deduce a flat meromorphic connection and a horizontal meromorphic endomorphism on $\mathcal{M}$ and on the other lattices.

2.15 Data at 0. The lattice $\mathscr{E}_0$ is equipped with an endomorphism R_0 ("residue" of ∇) depending on the choice of the coordinate on the chart U_0 of $\mathbb{P}^1$ up to a multiplicative constant. It induces a meromorphic endomorphism on $\mathscr{M}$ and on the other lattices.

The lattice $\mathscr{E}_0$ is moreover equipped with a 1-form Φ with values in the endomorphisms of $\mathscr{E}_0$, which satisfies $\Phi \wedge \Phi = 0$; in other words $(\mathscr{E}_0, \Phi)$ is a *Higgs bundle*.

2.16 Behaviour with respect to parameters. The formation of Θ, $\mathscr{M}$, $\mathscr{E}_0$, $\mathscr{E}_\infty$, Φ, R_0, ∇ and R_∞ is compatible with base change: if $f : X' \to X$ is an analytic mapping and if E' denotes the pullback of E by $\mathrm{Id} \times f$ on $\mathbb{P}^1 \times X'$, equipped with the pullback connection ∇', the previous objects relative to E' are obtained from that relative to E by the pullback f^*.

2.17 Relations. In any ∇-horizontal basis, the matrix of ∇ takes the form (2.2) and the matrix of R_0 is B_0, that of R_∞ is $-B_\infty$, that of Φ is C. The integrability conditions (2.5) are then equivalent to the relations

$$\nabla^2 = 0, \quad \nabla(R_\infty) = 0, \quad \Phi \wedge \Phi = 0, \quad [R_0, \Phi] = 0$$
$$\nabla(\Phi) = 0, \quad \nabla(R_0) + \Phi = [\Phi, R_\infty].$$

Let us be more precise on the meaning of these relations:

- The connection ∇ on $\mathscr{E}$ induces in a natural way a connection on the bundle $\mathscr{H}om_{\mathscr{O}_X}(\mathscr{E}, \mathscr{E})$ (cf. §0.11.b), which is also denoted by ∇. We hence get sections $\nabla(R_\infty)$ and $\nabla(R_0)$ of $\Omega^1_X \otimes_{\mathscr{O}_X} \mathscr{H}om_{\mathscr{O}_X}(\mathscr{E}, \mathscr{E})$.
- We have seen in §0.12.a the definition of the curvature ∇^2 and, in §0.12.b, the definition of the expression $\Phi \wedge \Phi$ as sections of the sheaf $\Omega^2_X \otimes_{\mathscr{O}_X} \mathscr{H}om_{\mathscr{O}_X}(\mathscr{E}, \mathscr{E})$.
- We denote by $[R_0, \Phi] : \mathscr{E} \to \Omega^1_X \otimes_{\mathscr{O}_X} \mathscr{E}$ the homomorphism defined, for any vector field ξ, by

$$[R_0, \Phi]_\xi = [R_0, \Phi_\xi] : \mathscr{E} \longrightarrow \mathscr{E}.$$

The definition of $[\Phi, R_\infty]$ is similar.
- According to §§0.12.a-11.b we have an operator

$$\nabla : \Omega^1_X \otimes_{\mathscr{O}_X} \mathscr{H}om_{\mathscr{O}_X}(\mathscr{E}, \mathscr{E}) \longrightarrow \Omega^2_X \otimes_{\mathscr{O}_X} \mathscr{H}om_{\mathscr{O}_X}(\mathscr{E}, \mathscr{E}).$$

The expression $\nabla(\Phi) \in \Omega^2_X \otimes_{\mathscr{O}_X} \mathscr{H}om_{\mathscr{O}_X}(\mathscr{E}, \mathscr{E})$ is thus meaningful, as Φ can be regarded as a section of $\Omega^1_X \otimes_{\mathscr{O}_X} \mathscr{H}om_{\mathscr{O}_X}(\mathscr{E}, \mathscr{E})$.

2.18 Behaviour with respect to a "metric". Moreover, let us be given a nondegenerate Hermitian form G on E, compatible with the connection ∇ and which has weight $w \in \mathbb{Z}$. We deduce symmetric nondegenerate bilinear forms g_0 and g_∞ on $\mathscr{E}_0$ and $\mathscr{E}_\infty$, which coincide on $\mathscr{M}$. They satisfy, on $X \smallsetminus \Theta$,

$$\nabla(g) = 0, \quad R_\infty^* + R_\infty = -w \, \mathrm{Id}$$
$$\Phi^* = \Phi, \quad R_0^* = R_0.$$

The first (resp. the second) line can be extended to X by using $\mathscr{E}_\infty$ and g_∞ (resp. $\mathscr{E}_0$ and g_0).

Let us indicate that the equality $\Phi^* = \Phi$ means that, for any vector field ξ, we have $(\Phi_\xi)^* = \Phi_\xi$. Lastly, $\bigtriangledown(g) = 0$ means that, for any vector field ξ and all local sections e, e' of $\mathscr{E}$, we have

$$\mathscr{L}_\xi g(e, e') = g(\bigtriangledown_\xi(e), e') + g(e, \bigtriangledown_\xi(e')).$$

2.19 Converse statement. Let us be given a locally free $\mathcal{O}_X(*\Theta)$-module $\mathscr{M}$, equipped with a flat connection $\bigtriangledown$, with an endomorphism Φ taking values in $\Omega^1_X(*\Theta)$, with endomorphisms R_0 and R_∞, all being meromorphic along Θ and satisfying Relations 2.17. Then we can equip the bundle $\pi^*\mathscr{M}$, which is a bundle on $\mathbb{P}^1 \times (X \smallsetminus \Theta)$, meromorphic along Θ, with a flat connection ∇ having logarithmic poles along $\{\infty\} \times X$ and of order one along $\{0\} \times X$: we set

$$\nabla = \bigtriangledown + \left(\frac{R_0}{\tau} - R_\infty\right)\frac{d\tau}{\tau} + \frac{\Phi}{\tau}.$$

One could also choose

$$\nabla = \bigtriangledown - \left[\left(\frac{R_0}{\tau} + R_\infty\right)\frac{d\tau}{\tau} + \frac{\Phi}{\tau}\right].$$

We do not give here any precision concerning a possible holomorphic extension along $\mathbb{P}^1 \times \Theta$. If the relations 2.18 are also satisfied, we can lift the bilinear form g as a nondegenerate Hermitian form on $\pi^*\mathscr{M}$.

3 Universal integrable deformation for Birkhoff's problem

We adapt to Birkhoff's problem the notion of universal deformation introduced in §1.c.

3.1 Definition (of a universal integrable deformation). An *integrable deformation* (E, ∇) of (E^o, ∇^o) is a bundle with a flat meromorphic connection on $\mathbb{P}^1 \times X$, having poles of order one along $\{0\} \times X$ and logarithmic along $\{\infty\} \times X$, inducing (E^o, ∇^o) when restricted to $x^o \in X$.

We say that the integrable deformation (E, ∇) of (E^o, ∇^o) is *complete* at x^o if, for any other integrable deformation (E', ∇', x'^o) of (E^o, ∇^o) parametrized by (X', x'^o), there exist neighbourhoods V, V' of x^o, x'^o in X, X' and an analytic mapping

$$f : (V', x'^o) \longrightarrow (V, x^o)$$

such that (E', ∇') is isomorphic (by an isomorphism inducing the identity when restricted to (E^o, ∇^o)) to the pullback de (E, ∇) by the mapping

$$\mathrm{Id} \times f : \mathbb{P}^1 \times V' \longrightarrow \mathbb{P}^1 \times V.$$

We say that the deformation is *universal* at x^o if, moreover, the germ of f at x'^o is unique.

3.a Existence of a local universal deformation

Given two matrices B_0^o and B_∞ in $M_d(\mathbb{C})$, let (E^o, ∇^o) be the trivial bundle of rank d on $\mathbb{P}^1$ equipped with the connection having matrix $\Omega^o = (B_0^o/\tau + B_\infty)d\tau/\tau$.

3.2 Theorem ([Mal83a, Mal86]). *If the matrix B_0^o is regular, there exists a germ of universal deformation of (E^o, ∇^o). It has dimension d.*

One should notice that, here and in §3.b, unlike the general case of Theorem 2.1, we do not need to know $(\widetilde{E}, \widetilde{\nabla})$ to construct the matrix $B_0(x)$ and the matrix of 1-forms $C(x)$.

Proof (of Theorem 3.2). Let us consider the system (2.5) on a manifold X in the neighbourhood of a point x^o. Locally, the first equation can be solved in a unique way in the form $C(x) = d\Gamma(x)$, where Γ is a matrix of holomorphic functions satisfying the initial condition $\Gamma(x^o) = 0$. Moreover, as B_0^o is regular, so is $B_0(x)$, for x nearby x^o, for any holomorphic matrix $B_0(x)$ satisfying $B_0(x^o) = B_0^o$.

The system is therefore equivalent to the system for pairs $(B_0(x), \Gamma(x))$

$$[B_0, d\Gamma] = 0$$
$$d(B_0 + \Gamma) = [B_\infty, d\Gamma]$$

and thus equivalent to a system for Γ only, as the second line is equivalent to $B_0 = B_0^o - \Gamma + [B_\infty, \Gamma]$, because $\Gamma^o = 0$.

In conclusion, solving (2.5) amounts to solving the system

(3.3) $$[(B_0^o - \Gamma + [B_\infty, \Gamma]), d\Gamma] = 0$$

with initial condition $\Gamma^o = 0$. We can consider the system (3.3) as a Pfaff system on the space $M_d(\mathbb{C})$ of matrices Γ.

3.4 Lemma. *If B_0^o is regular, the system (3.3) is integrable on $M_d(\mathbb{C})$ and defines a foliation, the leaves of which have dimension d in the neighbourhood of the origin.*

Proof. Let us first compute the 2-form $d[(B_0^o - \Gamma + [B_\infty, \Gamma]), d\Gamma]$ with matrix values. We will use that

$$d[A, d\Gamma] = [dA, d\Gamma]_+ := dA \wedge d\Gamma + d\Gamma \wedge dA.$$

Then we have

$$d[(B_0^o - \Gamma + [B_\infty, \Gamma]), d\Gamma] = -2d\Gamma \wedge d\Gamma + 2[B_\infty, d\Gamma \wedge d\Gamma].$$

Let (γ_{ij}) be the canonical coordinates on $M_d(\mathbb{C})$. Therefore, $d\Gamma = (d\gamma_{ij})$. Let

$$\xi^{(1)} = \sum_{i,j} \xi_{ij}^{(1)} \frac{\partial}{\partial \gamma_{ij}} \quad \text{and} \quad \xi^{(2)} = \sum_{i,j} \xi_{ij}^{(2)} \frac{\partial}{\partial \gamma_{ij}}$$

be two vector fields on $M_d(\mathbb{C})$ in the neighbourhood of the origin. In order to show integrability, it is enough to verify (cf. Lemma 0.13.2) that, if $\xi^{(1)}$ and $\xi^{(2)}$ are annihilated by all the entries (1-forms) of $[(B_0^o-\varGamma+[B_\infty,\varGamma]),d\varGamma]$, then $\xi^{(1)} \wedge \xi^{(2)}$ is annihilated by all the entries (2-forms) of the matrix $d[(B_0^o-\varGamma+[B_\infty,\varGamma]),d\varGamma]$.

By assumption, the matrix $B_0 = B_0^o - \varGamma + [B_\infty,\varGamma] = (b_{ij})$ is regular for any $\varGamma$ near 0. We have

$$[B_0,d\varGamma]_{mn} = \sum_k (b_{mk}d\gamma_{kn} - b_{kn}d\gamma_{mk}),$$

so that, if we denote by $\varXi$ the matrix (ξ_{ij}), the vector field ξ belongs to the kernel of all 1-forms $[B_0,d\varGamma]_{mn}$ if and only if

(3.5) $$[B_0,\varXi] = 0.$$

One verifies in the same way that $\xi^{(1)} \wedge \xi^{(2)}$ belongs to the kernel of all 2-forms $(d[B_0,d\varGamma])_{mn}$ if and only if

(3.6) $$-2[\varXi^{(1)},\varXi^{(2)}] + 2[B_\infty,[\varXi^{(1)},\varXi^{(2)}]] = 0.$$

As B_0 is regular, (3.5) for $\varXi^{(1)}$ and $\varXi^{(2)}$ implies that $\varXi^{(1)}$ and $\varXi^{(2)}$ are polynomials in B_0, hence pairwise commute and, as a consequence, (3.6) is also satisfied.

Let us now compute the dimension of the leaves of the foliation we have obtained, in the neighbourhood of the origin. For any $\varGamma^o \in M_d(\mathbb{C})$ near 0, the matrix $A^o = B_0^o - \varGamma^o + [B_\infty,\varGamma^o]$ is regular. The vectors $\xi^o = \sum_{ij} \xi_{ij}^o \partial_{\gamma_{ij}}$ tangent to this leaf at A^o are the vectors such that $[A^o,\varXi^o] = 0$. As A^o is regular, this space has dimension d. $\qquad\square$

Let us now come back to Theorem 3.2. Let $Y \subset M_d(\mathbb{C})$ be the integral manifold of the system (3.3) going through 0, in the neighbourhood of the origin (we only consider its germ at 0). Let $\mathscr{F} = \mathscr{O}_Y^d$ be the trivial bundle of rank d equipped with the connection ∇ whose matrix in the canonical basis is

$$\left(\frac{B_0}{\tau} + B_\infty\right)\frac{d\tau}{\tau} + \frac{d\varGamma}{\tau},$$

where we have set $B_0 = B_0^o - \varGamma + [B_\infty,\varGamma]$, and where we restrict functions and 1-forms to the submanifold Y of $M_d(\mathbb{C})$. We will check that (F,∇) is a germ of universal deformation of (E^o,∇^o).

Let (E,∇) be an integrable deformation of (E^o,∇^o) parametrized by a manifold (X,x^o) of dimension n. As the problem is local, we may assume that X is a small open ball in $\mathbb{C}^n$ and, in particular, that X is 1-connected and, according to Corollary I.5.4, that the bundle E is trivial. Using the same argument as in Theorem 2.1, we see that there exists a unique basis ε of E in which the matrix of ∇ takes the form

$$\left(\frac{B_0(x)}{\tau} + B_\infty\right)\frac{d\tau}{\tau} + \frac{C(x)}{\tau},$$

with $B_0(x^o) = B_0^o$. According to (2.5), $dC(x) = 0$. There exists thus a holo-morphic mapping $\Gamma : X \to M_d(\mathbb{C})$ such that $d\Gamma(x) = C(x)$ and $\Gamma(x^o) = 0$. As the system (2.5) is satisfied by $(B(x), C(x))$, we deduce that Γ takes val-ues in Y and, by construction, we have $(E, \nabla) = \Gamma^+(F, \nabla)$. This shows that (F, ∇) is a complete integrable deformation of (E^o, Γ^o).

Let f be a holomorphic mapping $(X, x^o) \to (Y, 0)$ and let us set $(E, \nabla) = f^+(F, \nabla)$. The unique basis ε of (E, ∇) considered above is then the pullback of the canonical basis of F and the mapping f is nothing but the mapping Γ considered above.

If $f, f' : (X, x^o) \to (M_d(\mathbb{C}), 0)$ are such that the meromorphic bundles $f^+(F, \nabla)$ and $f'^+(F, \nabla)$ are isomorphic by an isomorphism inducing the iden-tity on (E^o, ∇^o), the "uniqueness" part in the proof of Theorem 2.1 shows that this isomorphism is the identity. The computation above proves then that $f = f'$, as this mapping is nothing but Γ. We deduce the universality property of the deformation.   □

3.b Existence and construction of a global universal deformation

Given two matrices $B_0^o, B_\infty \in M_d(\mathbb{C})$, let us assume now that B_0^o is *diagonal regular* (i.e., with distinct eigenvalues); we set $B_0^o = \mathrm{diag}(x_1^o, \ldots, x_d^o)$. In such a situation, we will construct an integrable deformation of the differential system with matrix $(B_0^o/\tau + B_\infty)d\tau/\tau$. We will see that this deformation is *universal*. This deformation takes two distinct forms (associated to two particular bases of the deformed bundle). We will analyze with details both forms and their relations. The semisimplicity of B_0^o will enable us to use Theorem III.2.10 and to give a more global version of Theorem 3.2.

We regard here the space X_d of §1.c as that of diagonal matrices

$$\begin{pmatrix} x_1 & & \\ & \ddots & \\ & & x_d \end{pmatrix}$$

having pairwise distinct eigenvalues.

3.7 Theorem ([JMU81, Mal83c]). *There exists a hypersurface Θ of $\widetilde{X}_d$ and a unique pair (B_0, C), meromorphic on $\widetilde{X}_d$ with poles along Θ, solution of Equations (2.5), such that $B_0(\widetilde{x}^o) = B_0^o = \mathrm{diag}(x_1^o, \ldots, x_d^o)$ and that, for any $\widetilde{x} \in X \smallsetminus \Theta$, the matrix $B_0(\widetilde{x})$ is conjugate to $\mathrm{diag}(x_1, \ldots, x_d)$.*

Let us begin with a statement in terms of bundles.

3.8 Proposition. *There exists on $\mathbb{P}^1 \times \widetilde{X}_d$ a bundle (E, ∇) such that*

(1) *the connection ∇ is flat, with pole of order one along $\{0\} \times \widetilde{X}_d$ and loga-rithmic along $\{\infty\} \times \widetilde{X}_d$,*

(2) *the restriction* (E^o, ∇^o) *of the bundle* (E, ∇) *at* $\widetilde{x}^o$ *has a basis in which the matrix of* ∇^o *is* $(B_0^o/\tau + B_\infty) \, d\tau/\tau$,

(3) *for any* $\widetilde{x} \in \widetilde{X}_d$, *the d-uple of eigenvalues of the "residue"* R_0 *(cf. §0.14.c) is equal to* $\varpi(\widetilde{x})$ *up to a permutation.*

Moreover, such a (E, ∇) *is unique up to isomorphism.*

Proof. Let D be a disc centered at the origin of $\mathbb{P}^1$. According to Theorem III.2.10, we can construct on $U = D \times \widetilde{X}_d$ a bundle $(\widetilde{E}, \widetilde{\nabla})$ with a flat connection having pole of order one along $\{0\} \times \widetilde{X}_d$ whose "residue" $R_0(\widetilde{x})$ has $\varpi(\widetilde{x}) = x = (x_1, \ldots, x_d)$ as its set of eigenvalues at any $\widetilde{x} \in \widetilde{X}_d$, unique up to isomorphism. We get (E, ∇) as in the first point of the proof of Theorem 2.1. The uniqueness is obtained as in Theorem 2.1. $\qquad\square$

As in Theorem 2.1, we thus get a hypersurface $\Theta(\widetilde{x}^o)$ of $\widetilde{X}_d$ and a flat connection ∇ on $i_\infty^* \mathscr{E}$, hence a flat connection, that we also denote by ∇, on $\pi_* \mathscr{E}(*\Theta)$, having poles along Θ.

Proof (of Theorem 3.7). The existence is obtained by using the same arguments as in Theorem 2.1, according to Corollary I.5.8 and Lemma 2.3. Let us consider the "uniqueness" part, which is local on $\widetilde{X}_d$.

Let $(B_0(\widetilde{x}), C(\widetilde{x}))$ be a meromorphic solution of the system (2.5) (for B_∞ fixed) which satisfies the initial condition $B_0(\widetilde{x}^o) = B_0^o$ and such that, for any $\widetilde{x} \in \widetilde{X}_d$ away from some hypersurface Θ, the matrix $B_0(\widetilde{x})$ is conjugate to $\mathrm{diag}(x_1, \ldots, x_d)$. Let V be a sufficiently small open neighbourhood of $\widetilde{x}^o$ on which exists a solution of $C(\widetilde{x}) = d\Gamma(\widetilde{x})$ with $\Gamma(\widetilde{x}^o) = 0$. Then Γ defines a mapping $f : (V, \widetilde{x}^o) \to (\mathrm{M}_d(\mathbb{C}), 0)$ with image contained in the local universal deformation $(Y, 0)$ of (E^o, ∇^o) given by Theorem 3.2. Let (F, ∇) be the universal bundle on Y. Then the spectrum of $B_0 = B_0^o - \Gamma + [B_\infty, \Gamma]$ defines a holomorphic mapping $g : (Y, 0) \to (X_d, x^o)$, that we can lift in a unique way (in the neighbourhood of 0 in Y) as a mapping $(Y, 0) \to (\widetilde{X}_d, \widetilde{x}^o)$.

By assumption, we have $g \circ f = \mathrm{Id}$. We deduce that $T_0 g \circ T_{\widetilde{x}^o} f = \mathrm{Id}$. As the tangent spaces $T_0 Y$ and $T_{\widetilde{x}^o} \widetilde{X}_d$ have the same dimension, the mappings f and g are inverse one to the other in the neighbourhood of $\widetilde{x}^o$. This shows that $\Gamma(\widetilde{x})$, hence $C(\widetilde{x})$ and $B_0(\widetilde{x})$, are determined in a unique way by the spectral mapping g in the neighbourhood of $\widetilde{x}^o$. $\qquad\square$

3.9 Proposition. *The bundle with meromorphic connection* (E, ∇) *given by Proposition 3.8 induces, for any* $\widetilde{x} \in \widetilde{X}_d \smallsetminus \Theta$, *a universal deformation of its restriction* $(E_{\widetilde{x}}, \nabla)$ *to the fibre* $\mathbb{P}^1 \times \{\widetilde{x}\}$.

Proof. Let us fix $\widetilde{x} \in \widetilde{X}_d \smallsetminus \Theta$. Then the bundle (E, ∇) constructed with the initial condition (E^o, ∇^o) at $\widetilde{x}^o$ is equal to the bundle constructed in the same way with the initial condition $(E, \nabla)_{|\mathbb{P}^1 \times \{\widetilde{x}\}}$ at $\widetilde{x}$, due to uniqueness. It is thus enough to showing universality at $\widetilde{x}^o$.

Let (E', ∇', x'^o) be an integrable deformation of (E^o, ∇^o). The "residue" R'_0 remains regular semisimple in some neighbourhood of x'^o and its eigenvalues define in a unique way a mapping $f : (V', x'^o) \to (\widetilde{X}_d, \widetilde{x}^o)$. That the bundle (E', ∇') is isomorphic to $f^+(E, \nabla)$ comes from the uniqueness, up to isomorphism, of (E', ∇') with the initial condition (E^o, ∇^o), which can be proved as in Proposition 3.8.   □

3.c Universal deformation with metric

In the setting of §3.b, we moreover assume that the matrix B_∞ satisfies ${}^t B_\infty + B_\infty = w \,\mathrm{Id}$ for some $w \in \mathbb{Z}$, i.e., the matrix $B_\infty - w/2 \,\mathrm{Id}$ is skewsymmetric. In other words, the form G^o on E^o for which $G^o(\varepsilon_i^o, \varepsilon_j^o) = \delta_{ij}$ defines an isomorphism $(E^o, \nabla^o) \xrightarrow{\sim} (\overline{E}^{o*}, \overline{\nabla}^{o*})[w]$. *Then this isomorphism can be extended in a unique way to an isomorphism which is meromorphic along the hypersurface* $\Theta(\widetilde{x}^o)$:

$$(\mathscr{E}(*\pi^{-1}\Theta), \nabla) \xrightarrow{\sim} (\overline{\mathscr{E}}^*(*\pi^{-1}\Theta), \overline{\nabla}^*)[w].$$

Indeed, if G^o can be extended, then the associated form g_∞ is horizontal for ∇ and thus $G(\varepsilon_i, \varepsilon_j) = \delta_{ij}$. Conversely, if we define G in this way, it is enough to check that the matrix B_0 and the matrices $C^{(i)}$ are symmetric. The skewsymmetry assumption for B_∞ shows that the system (2.5) (or its local expression (2.6)) is stable under transposition, that is, if $(B_0, C^{(1)}, \ldots, C^{(n)})$ is a solution, then $({}^t B_0, {}^t C^{(1)}, \ldots, {}^t C^{(n)})$ also. As B_0^o is diagonal, the uniqueness property in Theorem 3.7 shows that B_0 and the $C^{(i)}$ are symmetric.   □

3.d The basis ε

According to Theorem 2.1, there exists a unique basis ε of $E_0 := E_{|\{0\} \times \widetilde{X}_d}$, meromorphic along Θ, which coincides with ε^o at $\widetilde{x}^o$ and such that the connection matrix of ∇ in this basis lifted to E takes the form

$$\Omega = \left(\frac{B_0(\widetilde{x})}{\tau} + B_\infty \right) \frac{d\tau}{\tau} + \sum_{i=1}^{d} \frac{C^{(i)}(\widetilde{x})}{\tau} \, dx_i,$$

where B_0 and the $C^{(i)}$ are holomorphic on $\widetilde{X}_d \smallsetminus \Theta$ and meromorphic along Θ. The integrability condition for ∇ is equivalent to Conditions (2.5). Moreover, $B_0(\widetilde{x})$ is conjugate to $\mathrm{diag}(x_1, \ldots, x_d)$ for any $\widetilde{x}$.

3.e The basis e

As the "residue" R_0 of ∇ is regular semisimple and as $\widetilde{X}_d$ is 1-connected, the restriction E_0 of the bundle E to $\{0\} \times \widetilde{X}_d$ can be decomposed as a direct sum of eigen bundles of rank one. According to Theorem III.2.15 and Proposition

III.2.12, each of these bundles can be equipped with a flat connection and, as $\widetilde{X}_d$ is 1-connected, giving a nonzero section of each bundle at $\widetilde{x}^o$ also gives a global trivialization of it. As, by assumption, B_0^o is diagonal in the basis ε^o, this basis is adapted to the decomposition when restricted to $\widetilde{x}^o$. We thus have a trivialization of E_0, and more precisely a unique basis e which coincides with ε^o at $\widetilde{x}^o$ and which is compatible with the decomposition.

In the following, we will analyze the connection matrix of ∇ expressed in the basis e.

3.f Comparison of the bases ε and e

Theorem III.2.15 implies that there exists a base change, formal in τ, with holomorphic coefficients on $\widetilde{X}_d \smallsetminus \Theta$ and meromorphic along Θ, which transforms the matrix Ω of ∇ in the basis ε to a matrix $\widehat{\Omega}'$ which takes the form

$$\widehat{\Omega}' = -d\left(\frac{\Delta_0(\widetilde{x})}{\tau}\right) + \Delta_\infty \frac{d\tau}{\tau}, \quad`$$

where $\Delta_0(\widetilde{x}) = \mathrm{diag}(x_1, \ldots, x_d)$ and Δ_∞ is a constant diagonal matrix (which gives the "formal monodromy").

3.10 Exercise. Show, by adapting the proof of Theorem II.5.7, that Δ_∞ is nothing but the diagonal part of the matrix B_∞.

Let $\widehat{P}(\widetilde{x}, \tau) = \sum_{k=0}^\infty P_k(\widetilde{x})\tau^k$ be the matrix of this base change. We then have

$$\Omega = \widehat{P}\widehat{\Omega}'\widehat{P}^{-1} - d\widehat{P}\widehat{P}^{-1}$$

and thus $d\widehat{P} = \widehat{P}\widehat{\Omega}' - \Omega'\widehat{P}$. We deduce, by considering the coefficient of τ^k with $k = -2, -1, 0$, that we must have the relations

(3.11)
$$\begin{aligned} P_0 \Delta_0 &= B_0 P_0 \\ P_1 \Delta_0 - B_0 P_1 &= B_\infty P_0 - P_0 \Delta_\infty \\ C^{(i)} P_0 &= -P_0 D_i \\ \frac{\partial P_0}{\partial x_i} &= -C^{(i)} P_1 - P_1 D_i, \end{aligned}$$

where $D_i = \partial \Delta_0 / \partial x_i$ is the diagonal matrix whose only nonzero term is equal to 1 and is located at (i, i). Therefore, we have $\Delta_0 = \sum_i x_i D_i$.

Once this formal base change is known, the basis e introduced in §3.e is obtained by considering only the order 0 part of the base change, that is, the matrix P_0: as a matter of fact, we restrict the formal bundle associated to E to $\tau = 0$ to find the basis e. As a consequence, the matrix Ω' of ∇ in the basis e satisfies

$$\Omega' = P_0^{-1}\Omega P_0 + P_0^{-1} dP_0$$

with the normalization $P_0(\widetilde{x}^o) = \mathrm{Id}$, as e^o and ε^o coincide. If we use the relations above and if we set $T(\widetilde{x}) = -P_0(\widetilde{x})^{-1}P_1(\widetilde{x})$, we obtain

$$(3.12) \qquad \Omega' = -d\left(\frac{\Delta_0}{\tau}\right) + (\Delta_\infty + [\Delta_0, T])\frac{d\tau}{\tau} + [T, d\Delta_0].$$

One should notice that the perturbation $\widehat{\Omega}' - \Omega'$ of the diagonal matrix $\widehat{\Omega}'$ has no diagonal terms.

Lastly, the matrix of $\triangledown$ in the basis e is the restriction to $\tau = \infty$ of the part of Ω' which is independent of $d\tau$, that is here, $[T, d\Delta_0]$ that we can also write as $\sum_{i=1}^d A_i(\widetilde{x})\,dx_i$ with $A_i = -[D_i, T]$. We have

$$[T, d\Delta_0]_{ij} = T_{ij}(dx_j - dx_i).$$

Let us note that the last relation (3.11) can be replaced with the relation

$$\frac{\partial P_0}{\partial x_i} = P_0 A_i.$$

Let us set $M(\widetilde{x}) = [\Delta_0, T]$, so that $M_{ij} = (x_i - x_j)T_{ij}$. Giving M is equivalent to giving T, as the functions $x_i - x_j$ $(i \neq j)$ are invertible on $\widetilde{X}_d$. We have $M^o := M(\widetilde{x}^o) = B_\infty - \Delta_\infty$. Then the integrability condition on ∇ expressed in the basis e shows that M satisfies the differential system

$$(3.13) \qquad dM = [[d\Delta_0, T], M + \Delta_\infty].$$

3.14 Remarks.

(1) We have $\sum_i A_i = -[\sum_i D_i, T] = -[\mathrm{Id}, T] = 0$. Consequently, the matrix of $\nabla_{\Sigma_i \partial_{x_i}}$ in the basis e is zero. On the other hand, the matrix of Φ in the basis e is given by

$$\Phi_{\partial_{x_i}} = -D_i.$$

(2) The matrix T only shows up its nondiagonal entries in Equation (3.13). We can then replace, in the expression of M and in (3.13), the matrix T with the matrix $\widetilde{T}$ obtained by killing all diagonal coefficients of T. Let us also notice that $\widetilde{T}$ is known as soon as M is so.

3.g Case where B_∞ is skewsymmetric

The previous results become simpler if we suppose that B_∞ is skewsymmetric or, more generally, when $B_\infty - (w/2)\,\mathrm{Id}$ is so, with $w \in \mathbb{Z}$ (in this case $\Delta_\infty = (w/2)\,\mathrm{Id}$). Then:

3.15 Proposition. *If $B_\infty - (w/2)\,\mathrm{Id}$ is skewsymmetric, the matrix B_0 and the matrices $C^{(i)}$ are symmetric.*

Proof. It is enough to use that the system (2.5) is invariant by transposition (cf. also §3.c). $\qquad \square$

This property can also be read in the basis e:

3.16 Proposition. *The matrix $B_\infty - (w/2)\,\mathrm{Id}$ is skewsymmetric if and only if the matrix $M = [\Delta_0, T]$ is so. If this property holds, the basis e is g-orthonormal and the base change matrix P_0 between the bases ε and e satisfies $P_0{}^t P_0 = \mathrm{Id}$.*

Proof. By construction, we have $M^o = B_\infty - (w/2)\,\mathrm{Id}$, hence one of the ways is clear. If $B_\infty - (w/2)\,\mathrm{Id}$ is skewsymmetric, we deduce that the matrix $\widetilde{T}^o = (\mathrm{ad}\,\Delta_0)^{-1}(B_\infty)$ is symmetric. By considering the integrable system satisfied by $\widetilde{T}$ which is deduced from the system (3.13), we conclude that $\widetilde{T}$ is symmetric and thus M is skewsymmetric.

For the second point, let us consider the restriction to $\mathbb{P}^1 \times (\widetilde{X}_d \smallsetminus \Theta)$ of the nondegenerate Hermitian form G on (E, ∇) constructed in §3.c. The isomorphism

$$(3.17) \qquad\qquad (E, \nabla) \xrightarrow{\;\sim\;} (\overline{E}^*, \overline{\nabla}^*)[w]$$

that we deduce from it induces an isomorphism of the formalized modules along $\{0\} \times (\widetilde{X}_d \smallsetminus \Theta)$. As the rank one factors in the formal decomposition are pairwise nonequivalent, this isomorphism preserves the formal decomposition. By restricting the isomorphism (3.17) to $\{0\} \times (\widetilde{X}_d \smallsetminus \Theta)$, we thus get a nondegenerate bilinear form g_0 for which the basis e is orthogonal.

On the other hand, the form G induces a bilinear form g on $\pi_* \mathcal{E}_{|\widetilde{X}_d \smallsetminus \Theta}$ (by taking global sections). In the basis ε, it coincides with g_∞, and in the basis e with g_0. We deduce that $P_0{}^t P_0$ is equal to the diagonal matrix having the $g_0(e_i, e_i)$ as diagonal entries.

Lastly, let us show that $P_0{}^t P_0 = \mathrm{Id}$. It is enough to show that, for any $i = 1, \ldots, d$, we have $\dfrac{\partial (P_0{}^t P_0)}{\partial x_i} = 0$, as the equality is satisfied at $\widetilde{x}^o$. Let us first notice that, as $T = P_0^{-1} P_1$ is symmetric, we have the identity

$$P_0{}^t P_1 = P_1{}^t P_0.$$

On the other hand, using the two previous relations (3.11), we get

$$\frac{\partial P_0}{\partial x_i} \cdot {}^t P_0 = -(C^{(i)} P_1 + P_1 D_i) \cdot {}^t P_0 = [P_1{}^t P_0, C^{(i)}]$$
$$P_0 \cdot \frac{\partial\, {}^t P_0}{\partial x_i} = -P_0 \cdot ({}^t P_1 C^{(i)} + D_i{}^t P_1) = [C^{(i)}, P_0{}^t P_1].$$

The identity above enables us to conclude. $\qquad\qquad\square$

3.18 Exercise. We keep the skewsymmetry assumption for $B_\infty - (w/2)\,\mathrm{Id}$. In particular, the matrix T is symmetric. Let $v = \sum_i v_i(\widetilde{x}) e_i$ be a section of $\mathcal{E}$ on $\widetilde{X}_d \smallsetminus \Theta$. Prove the following properties:

(1) The section v is horizontal for $\triangledown$ if and only if the v_i satisfy the following equations $(i, j = 1, \ldots, d)$:

$$\frac{\partial v_i(\widetilde{x})}{\partial x_j} = -v_j(\widetilde{x})T_{ij}(\widetilde{x}) \quad \text{if } i \neq j \quad \text{and} \quad \frac{\partial v_i(\widetilde{x})}{\partial x_i} = \sum_{k \neq i} v_k(\widetilde{x})T_{ik}(\widetilde{x}).$$

(2) In such a case, the form $\sum_i v_i^2(\widetilde{x})\, dx_i$ is closed.

(3) Set $\mathfrak{E}_v = \sum_i x_i v_i(\widetilde{x})e_i$. If moreover v is an eigenvector of M with eigenvalue $\alpha \in \mathbb{C}$, we have

$$\triangledown \mathfrak{E}_v = \sum_{j=1}^{d}(\alpha + 1)e_j v_j \, dx_j + \sum_{i,j} M_{ij}e_j v_i \, dx_i$$

and for any i the function v_i^2 is homogeneous of degree 2α (i.e., $\sum_j x_j \partial_{x_j} v_i^2 = 2\alpha v_i^2$).

3.19 Exercise (After A. Givental [Giv98]).

(1) Prove that the integrability condition on the matrix $[T, d\Delta_0]$ of the connection $\triangledown$ can be expressed by the following relations, for any pair (i, j) with $i \neq j$:

$$\frac{\partial T_{ij}}{\partial x_\ell} = -T_{i\ell}T_{\ell j} \quad (\ell \neq i, j) \quad \text{and} \quad \sum_\ell \frac{\partial T_{ij}}{\partial x_\ell} = 0.$$

(2) We set $Q = P_0^{-1}P = (1 - \tau T + \tau^2 S + \cdots)$, so that $Q^{-1}\Omega'Q + Q^{-1}dQ = \widehat{\Omega}'$. By considering the coefficient of τ in this equality, prove that the diagonal entries of T also satisfy

$$\frac{\partial T_{ii}}{\partial x_\ell} = -T_{i\ell}T_{\ell i} \quad (\ell \neq i) \quad \text{and} \quad \sum_\ell \frac{\partial T_{ii}}{\partial x_\ell} = 0,$$

that is, $dT_{ii} = \sum_\ell T_{i\ell}T_{\ell i}(dx_i - dx_\ell)$.

(3) Deduce that, if $B_\infty - (w/2)\,\mathrm{Id}$ is skewsymmetric (hence if T is symmetric), the 1-form $\sum_i T_{ii}\, dx_i$ is closed.

3.20 Examples.

(1) If $d = 2$, one can easily check that M and $[d\Delta_0, T]$ are proportional, hence $dM = 0$ and M is constant. Let us set $u = x_1 - x_2$ and $v = x_1 + x_2$. Let $b = B_{\infty,12}$. We thus have $T_{12} = b/u$. Exercise 3.19-(2) shows then that T_{11} and T_{22} are independent of v and satisfy

$$\frac{\partial T_{11}}{\partial u} = -\frac{\partial T_{22}}{\partial u} = 2\frac{b^2}{u^2}$$

hence $T_{11} = T_{11}^o e^{-2b^2/u}$ and $T_{22} = T_{22}^o e^{2b^2/u}$.

(2) If $d = 3$, we set $u = x_1 - x_2$, $v = x_2 - x_3$ and $w = x_1 + x_2 + x_3$. We also set $T_{12} = a$, $T_{23} = b$ and $T_{13} = c$, so that $M_{12} = au$, $M_{23} = bv$ and $M_{13} = c(u + v)$. The system (3.13) can now be written as

$$d(au) = -bc(vdu - udv)$$
$$d(bv) = -ac(vdu - udv)$$
$$d(c(u + v)) = ab(vdu - udv).$$

It is convenient to consider the coordinates $\zeta = u/v \in \mathbb{C} \smallsetminus \{0, -1\}$ and $t = u + v \in \mathbb{C}^*$. We then see that the entries $\alpha = M_{12}$, $\beta = M_{23}$ and $\gamma = M_{13}$ of M only depend on ζ and satisfy the system

$$\alpha'(\zeta) = -\frac{1}{1+\zeta}\beta(\zeta)\gamma(\zeta)$$
$$\beta'(\zeta) = -\frac{1}{\zeta(1+\zeta)}\alpha(\zeta)\gamma(\zeta)$$
$$\gamma'(\zeta) = \frac{1}{\zeta}\alpha(\zeta)\beta(\zeta).$$

These equations clearly emphasize the invariance of the quantity $-R^2 = \alpha^2 + \beta^2 + \gamma^2$. Moreover, the eigenvector of M

$$\begin{pmatrix} \beta \\ \gamma \\ \alpha \end{pmatrix}$$

with respect to the eigenvalue 0 is horizontal for ∇.

Theorems 3.7 and III.2.15 show that there exists a solution (α, β, γ) to the previous system, which is a meromorphic function on the universal covering space of $\mathbb{C} \smallsetminus \{0, -1\}$. This solution is unique once the initial condition $M^o = B_\infty - \Delta_\infty$ fixed. The divisor Θ of poles a priori depends on this initial condition.

3.21 Remark (A Hamiltonian system). The notion of integrability for nonlinear differential systems, like those coming from physics, is not easy to define[4]. A way to consider it consists in giving these systems a Hamiltonian form and in looking for the maximal number of first integrals. For instance, K. Okamoto found a Hamiltonian form for Painlevé equations (cf. [IKSY91]). Let us see what happens to the system (2.5), written as (3.13).

---

[4] The introduction of [GR97] begins as follows: "It would seem fit for a paper entitled 'Integrability' to start with the definition of this notion. Alas, this is not possible. There exists a profusion of integrability definitions and where you have two scientists you have (at least) three definitions of integrability." N. Hitchin [Hit99] adopts the repartee of Louis Armstrong as his own: "If you gotta ask, you'll never know!"

We still assume that $B_\infty - w/2\,\mathrm{Id}$ is skewsymmetric. In the basis ε, the matrix of the endomorphism R_0 remains conjugate to the matrix $\mathrm{diag}(x_1, \ldots, x_d)$, although that of R_∞ remains constant. On the other hand, in the basis e, the matrix of R_0 is equal to $\mathrm{diag}(x_1, \ldots, x_d)$, but that of $R_\infty + w/2\,\mathrm{Id}$, that is, $-M$, varies in the space of skewsymmetric matrices, being governed by the differential system (3.13).

In [JMMS80][5] the Hamiltonian structure of the system satisfied by M is emphasized. This system can be interpreted as a Hamiltonian system on the space $X_d \times \mathcal{O}_{M^o}$, where $\mathcal{O}_{M^o}$ is the adjoint orbit of the skewsymmetric matrix $M^o = B_\infty - (w/2)\,\mathrm{Id}$ equipped with its usual symplectic structure. Indeed we set, for any $(\widetilde{x}, M) \in (\widetilde{X}_d \smallsetminus \Theta) \times \mathfrak{so}(d, \mathbb{C})$

$$H_i(\widetilde{x}, M) = -\sum_{j \neq i} \frac{M_{ij}^2}{x_i - x_j}$$

and we denote by $X_i(\widetilde{x}, M)$ $(i = 1, \ldots, d)$ the corresponding Hamiltonian field tangent to $\mathcal{O}_{M^o}$. We can then write Equation (3.13) as (cf. [Hit97, Th. 4.1])

$$\frac{\partial M}{\partial x_i} = X_i(\widetilde{x}, M). \qquad \square$$

3.h Relation with Schlesinger equations by Fourier transform

The universal deformation that we have constructed in §3.b by using, among others Theorem III.2.10, can also be obtained from the universal deformation constructed in §1.c for suitable initial conditions. We pass from one to the other by a partial Fourier transform, as we have done (without parameters) in §V.2.c. We will be more explicit concerning this correspondence.

We thus fix $B_0^o = \mathrm{diag}(x_1^o, \ldots, x_d^o)$ with $x^o \in X_d$ and $B_\infty \in \mathrm{M}_d(\mathbb{C})$. We will assume in the following (cf. Proposition V.2.10) that $B_\infty + k\,\mathrm{Id}$ is invertible for any $k \in \mathbb{N}$, or, if one prefers, that B_∞ has no negative integral eigenvalue.

The free $\mathbb{C}[\tau]$-module $\widehat{\mathbb{E}}^o$ of rank d equipped with the connection $\widehat{\nabla}^o$ having matrix

$$\left(\frac{B_0^o}{\tau} + B_\infty \right) \frac{d\tau}{\tau}$$

and the free $\mathbb{C}[t]$-module $\mathbb{E}^o$ of rank d equipped with the connection ∇^o having matrix

$$(B_\infty - \mathrm{Id})(t\,\mathrm{Id} - B_0^o)^{-1}\, dt = \sum_{i=1}^d \frac{(B_\infty - \mathrm{Id})D_i}{t - x_i}\, dt$$

are Fourier transforms one from the other. Let us note that the matrix $A_i = (B_\infty - \mathrm{Id})D_i$ has rank one: its unique nonzero column is the i-th column of $B_\infty - \mathrm{Id}$.

---

[5] Cf. also [Har94], [Dub96, Prop. 3.7], [Hit97, Th. 4.1] and generalizations in [Boa99].

Let $(\widehat{\mathbb{E}}, \widehat{\nabla})$ be the universal deformation of $(\widehat{\mathbb{E}}^o, \widehat{\nabla}^o)$ constructed in §3.b, parametrized by the space $X := \widetilde{X}_d \smallsetminus \Theta$. Therefore, $\widehat{\mathbb{E}}$ is a free $\mathscr{O}_X[\tau]$-module of rank d, equipped with the basis ε in which the connection matrix $\widehat{\nabla}$ is written as

$$\left(\frac{B_0(\widetilde{x})}{\tau} + B_\infty\right)\frac{d\tau}{\tau} + \frac{C(\widetilde{x})}{\tau}.$$

As in §V.2.c, let us denote by t the action of $\tau^2 \nabla_{\partial_\tau}$ and by ∂_t that of τ^{-1}, the action of x_i and ∂_{x_i} being unchanged. Then $\widehat{\mathbb{E}}$ becomes, for the same reason as in Proposition V.2.10, a $\mathscr{O}_X[t]$-module generated by ε, that we denote by $\mathbb{E}$, equipped with a connection ∇.

3.22 Lemma. *The $\mathscr{O}_X[t]$-module $\mathbb{E}$ is free of rank d, having ε as a basis.*

Let us take this lemma for granted and let us continue to examine the inverse partial Fourier transform. If we write the basis ε as a column, we have

$$\partial_{x_i}\varepsilon = {}^t C^{(i)}(\widetilde{x})\tau^{-1}\varepsilon$$
$$= {}^t C^{(i)}(\widetilde{x})(t\,\mathrm{Id} - {}^t B_0(\widetilde{x}))^{-1}({}^t B_\infty - \mathrm{Id})\varepsilon,$$

in other words, the connection matrix of ∇ in the basis ε is written as

$$\Omega = (B_\infty - \mathrm{Id})(t\,\mathrm{Id} - B_0(\widetilde{x}))^{-1}\Big(dt + \sum_i C^{(i)}(\widetilde{x})\,dx_i\Big).$$

It is not *a priori* evident that this matrix takes the form given in Theorem 1.2 (with $\psi_i(x) = x_i$). We will see that such is the case by utilizing the basis e, obtained from the base change of matrix P_0 (cf. §§3.e and 3.f). One verifies in the same way as above that the matrix Ω' of ∇ in this basis is written as

$$\Omega' = ((\Delta_\infty - \mathrm{Id}) + [\Delta_0, T])\,(t\,\mathrm{Id} - \Delta_0)^{-1}d(t\,\mathrm{Id} - \Delta_0) + [T, d\Delta_0],$$

that is also

$$\Omega' = \sum_{i=1}^d ((\Delta_\infty - \mathrm{Id}) + [\Delta_0, T])\,D_i\frac{d(t - x_i)}{t - x_i} + P_0^{-1}dP_0$$
$$= P_0^{-1}\Omega P_0 + P_0^{-1}dP_0$$

with

$$\Omega = \sum_{i=1}^d P_0\,((\Delta_\infty - \mathrm{Id}) + [\Delta_0, T])\,D_i P_0^{-1}\frac{d(t - x_i)}{t - x_i}.$$

The matrix Ω thus takes the form given in Theorem 1.2. In this way, for B_∞ fixed, the solution $(B_0(\widetilde{x}), C(\widetilde{x}))$ of Equations (2.5) with initial condition $B_0^o = x^o$ at $\widetilde{x}^o$ given by Theorem 3.7 enables one to get a solution $(A_1(\widetilde{x}), \ldots, A_d(\widetilde{x}))$ of Schlesinger equations (1.4) with initial condition $(B_\infty - \mathrm{Id})D_1, \ldots, (B_\infty - \mathrm{Id})D_d$ at $\widetilde{x}^o$.

Proof (of Lemma 3.22). As ε generates $\mathbb{E}$ over $\mathscr{O}_X[t]$, we have a surjective morphism $\mathscr{O}_X[t]^d \to \mathbb{E}$. Let $\mathbb{K}$ be its kernel. As the module $\widehat{\mathbb{E}}$ is free over $\mathscr{O}_X[\tau]$, it is also free over $\mathscr{O}_X$, hence $\mathbb{E}$ is free over $\mathscr{O}_X$. Let $\mathfrak{m}_{\widetilde{x}}$ be the maximal ideal of $\mathscr{O}_X$ at $\widetilde{x}$. We deduce that $\mathbb{K}/\mathfrak{m}_{\widetilde{x}}\mathbb{K}$ is the kernel of the induced morphism $\mathbb{C}[t]^d \to \mathbb{E}/\mathfrak{m}_{\widetilde{x}}\mathbb{E}$. We have seen in Proposition V.2.10 that it is an isomorphism. We deduce that $\mathbb{K}/\mathfrak{m}_{\widetilde{x}}\mathbb{K} = 0$ for any $\widetilde{x} \in X$, hence $\mathbb{K} = 0$ (indeed, a section of $\mathbb{K}$ consists of d polynomials in t, the coefficients of which are holomorphic functions on X which vanish at each point of X). $\square$

VII

Saito structures and Frobenius structures on a complex analytic manifold

Introduction

At the end of the seventies, K. Saito (cf. [Sai83a, Sai83b]) brought into prominence, in a conjectural way, a mixed structure on the space of parameters of the universal unfolding of any holomorphic function with an isolated critical point. This conjecture stated that there should exist a *flat "metric"* on the tangent bundle (hence an *affine structure*, see Remark 0.13.5) and a *product* compatible with the "metric", both structures being linked by relations analogous to that considered in §§VI.2.17-2.18. Therefore, the presence of an *integrable deformation* is central in such a structure, which we will call in the following a *Saito structure*. The main tool in order to exhibit such a structure is the *infinitesimal period mapping*.

This conjecture has been proved with all generality by M. Saito ([Sai89]).

One should notice that the existence of an affine structure had yet appeared in some articles, published one century earlier, on the periods of families of elliptic curves (cf. [FS82]).

More recently, B. Dubrovin (cf. [Dub96]) has shown that this structure provides solutions to some nonlinear differential equations introduced by physicists, called "WDVV equations"[1]. Moreover, he also showed that this structure appears in other contexts (*quantum cohomology* for example, see also [MM97]), thus revealing relations between mathematical domains apparently of a very different nature. An essential fact in this approach is the local existence of a potential satisfying the WDVV equations, also called *associativity equations*. This structure is also called by B. Dubrovin a *Frobenius structure* if it is presented starting from the potential, which is then called the *potential* of the Frobenius structure.

Since then, another way to produce such structures has been proposed by S. Barannikov and M. Kontsevitch [BK98] (see also [Man99b, CZ99]), making

---

[1] These are the initials of Witten, Dijkgraft, Verlinde and Verlinde.

the Frobenius structures play ([Bar99]) a prominent role in *mirror symmetry* (see for instance [Voi99, BP02]).

In this chapter we will expound the former aspect of the question. After having introduced the notion of an infinitesimal period mapping in the general framework of Saito structures, we will show the existence of universal examples of such structures, with an hypothesis of semisimplicity. We will treat some examples and we will indicate the general framework for the construction of the Frobenius structure associated to singularities of holomorphic functions by K. Saito. We will only give some indications as to the other approaches, which go far beyond the techniques developed in this book.

This chapter is partly inspired from the articles [Dub96, Hit97, Aud98a, Aud98b, Sab98] and of the book [Man99a].

1 Saito structure on a manifold

Let M be a complex analytic manifold of dimension d and let TM be its tangent bundle. We will now work with the notions introduced in §0.13. In order to emphasize the respective roles of the metric and of the associated connection, we will begin with definitions which do not involve the metric.

1.a Saito structure without a metric

1.1 Definition. A *Saito structure* on M (without a metric) consists of the following data:

(1) a *flat torsion free connection* ∇ on the tangent bundle TM,
(2) a *symmetric Higgs field* Φ on the tangent bundle TM,
(3) two global sections (vector fields) e and $\mathfrak{E}$ of Θ_M, respectively called *unit field* and *Euler field* of the structure.

These data are subject to the following conditions:

(a) the meromorphic connection[2] ∇ on the bundle π^*TM on $\mathbb{P}^1 \times M$ defined by the formula

$$\nabla = \pi^* \nabla + \frac{\pi^* \Phi}{\tau} - \left(\frac{\Phi(\mathfrak{E})}{\tau} + \nabla \mathfrak{E} \right) \frac{d\tau}{\tau}$$

is *integrable* (in other words, Relations VI.2.17 are satisfied by ∇, Φ, $R_0 = -\Phi(\mathfrak{E})$ and $R'_\infty = \nabla \mathfrak{E}$);

---

[2] Recall that $\Phi(\mathfrak{E})$ and $\nabla \mathfrak{E}$ are endomorphisms of the tangent bundle TM, i.e., Θ_M-linear endomorphisms of the sheaf Θ_M. Here, we regard them in a natural way as endomorphisms of the sheaf $\pi^*\Theta_M$. On the other hand, $\pi^*\nabla$ is a connection on this sheaf (cf. Example 0.11.10) and, similarly, $\pi^*\Phi$ is a Higgs field.

(b) the field e is $\bigtriangledown$-horizontal (i.e., $\bigtriangledown(e) = 0$) and satisfies $\Phi_e = -\operatorname{Id}$ (i.e., the product $\star$ associated to Φ has e as a *unit field*).

We will now indicate some quite immediate consequences of this definition. They will provide a better understanding of the simplest examples. We let the reader give details.

1.2 Modification of the Euler field. We can obtain Saito structures parametrized by $\lambda \in \mathbb{C}$: the field $\mathfrak{E}$ can be replaced with $\mathfrak{E} + \lambda e$. *The endomorphism R_0 is the multiplication[3] by $\mathfrak{E}$*. It is changed into $R_0 + \lambda \operatorname{Id}$. The endomorphism $\bigtriangledown \mathfrak{E}$ remains unchanged.

1.3 Relation between the Euler field and the unit field. It follows from one of the relations VI.2.17 that *the endomorphism $R'_\infty = \bigtriangledown \mathfrak{E}$ is $\bigtriangledown$-horizontal*. Moreover, *the unit field e satisfies $\bigtriangledown_e \mathfrak{E} = e$*, which is equivalent, as $\bigtriangledown$ is torsion free, to $\mathscr{L}_{\mathfrak{E}}(e) = -e$, if $\mathscr{L}_{\mathfrak{E}}$ denotes the Lie derivative relative to $\mathfrak{E}$: indeed, the relation $\bigtriangledown(R_0) = [\Phi, \bigtriangledown \mathfrak{E}] - \Phi$, applied to the pair of vectors (e, e), gives on the one hand

$$\bigtriangledown_e(R_0)(e) := \bigtriangledown_e(R_0(e)) - R_0(\bigtriangledown_e e) = \bigtriangledown_e(e \star \mathfrak{E}) = \bigtriangledown_e(\mathfrak{E})$$

and, on the other hand,

$$[\Phi_e, \bigtriangledown \mathfrak{E}](e) - \Phi_e(e) = [-\operatorname{Id}, \bigtriangledown \mathfrak{E}](e) + e \star e = e \star e = e.$$

1.4 The Euler field in flat coordinates. Any covering (in particular the universal covering) of a manifold having a Saito structure is naturally equipped with such a structure.

Let us suppose then that M is 1-connected. Still because $\bigtriangledown$ is torsion free, *there exists on M (see Theorem 0.13.4) a flat coordinate system[4] $t_1, \ldots, t_d$*, i.e., such that $\bigtriangledown(\partial_{t_i}) = 0$ for any i. One can choose it in such a way that $\partial_{t_1} = e$.

The matrix of the endomorphism $\bigtriangledown(\mathfrak{E})$ in the basis $(\partial_{t_1}, \ldots, \partial_{t_d})$ is constant. Let us assume it *semisimple*. It is then possible (and often convenient for computation) to choose these coordinates in such a way that the fields ∂_{t_i} are eigenvectors of $\bigtriangledown(\mathfrak{E})$. We can thus write, in this basis, $\bigtriangledown \mathfrak{E} = \operatorname{diag}(1, \delta_2, \ldots, \delta_d)$ with $\delta_i \in \mathbb{C}$ and $\delta_1 = 1$, as $\bigtriangledown_e \mathfrak{E} = e$. If we set $\mathfrak{E} = \sum_i a_i(t_1, \ldots, t_d) \partial_{t_i}$, the equalities $\bigtriangledown_{\partial_{t_j}} \mathfrak{E} = \delta_j \partial_{t_j}$ imply that $a_j = \delta_j t_j + r_j$ with $r_j \in \mathbb{C}$. Let us translate the flat coordinates so that $r_j = 0$ if $\delta_j \neq 0$. The Euler field can thus be written as

$$\mathfrak{E} = \sum_{j \mid \delta_j \neq 0} \delta_j t_j \partial_{t_j} + \sum_{j \mid \delta_j = 0} r_j \partial_{t_j}.$$

---

[3] Although this is not important, we will take right multiplication, in order to follow the analogy with the connection.

[4] In the étale sense, cf. Remark 0.2.4.

1.5 Covariant derivative and Lie derivative of the product.

Due to the symmetry of Φ (Definition 1.1-(2)), the relation $\nabla(\Phi) = 0$ in $\Omega^2_M \otimes_{\mathcal{O}_M} \mathrm{End}_{\mathcal{O}_M}(\Theta_M)$ amounts to the following relation, for any triple of vector fields:

$$\nabla_\xi(\eta \star \theta) - \nabla_\eta(\xi \star \theta) + \xi \star \nabla_\eta \theta - \eta \star \nabla_\xi \theta - \mathscr{L}_\xi \eta \star \theta = 0.$$

Indeed, we notice first that the left-hand term is linear in ξ, η, θ, so that it is enough to check this equality in a local basis. As ∇ is torsion free, we can consider a system of *flat* local coordinates $t_1, \ldots, t_d$. The relation then means that the expression

$$\nabla_{\partial_{t_i}}(\partial_{t_j} \star \partial_{t_k})$$

is *symmetric* in i, j, k. In order to check that this is equivalent to the equation $\nabla(\Phi) = 0$, we write $\Phi = \sum_i dt_i \otimes C^{(i)}$. The relation $\nabla(\Phi) = 0$ is equivalent to the relations $\partial C^{(i)}/\partial t_j = \partial C^{(j)}/\partial t_i$ for all i, j (see the system VI.2.17). By definition we have

$$\partial_{t_j} \star \partial_{t_k} = -\Phi_{\partial_{t_j}}(\partial_{t_k}) = -\sum_\ell C^{(j)}_{\ell,k} \partial_{t_\ell},$$

so that the symmetry of Φ is equivalent to the fact that, for any ℓ, $C^{(j)}_{\ell,k}$ is symmetric in j, k. This implies that so is $\partial C^{(j)}_{\ell,k}/\partial t_i$. Due to the flatness of the coordinates, we have

$$\nabla_{\partial_{t_i}}(\partial_{t_j} \star \partial_{t_k}) = -\sum_\ell \frac{\partial C^{(j)}_{\ell,k}}{\partial t_i} \partial_{t_\ell}.$$

Therefore, taking into account the symmetry of Φ, the symmetry in i, j, k of $\nabla_{\partial_{t_i}}(\partial_{t_j} \star \partial_{t_k})$ is equivalent to the symmetry in i, j of $\partial C^{(j)}_{\ell,k}/\partial t_i$ for all k, ℓ, in other words, to the relation $\nabla\Phi = 0$. □

The relation $\nabla(R_0) - [\Phi, \nabla\mathfrak{E}] = -\Phi$ applied to a pair (ξ, η) of vectors amounts to

$$\nabla_\xi(\eta \star \mathfrak{E}) - \nabla_\xi \eta \star \mathfrak{E} + \xi \star \nabla_\eta \mathfrak{E} - \nabla_{\xi \star \eta} \mathfrak{E} = \xi \star \eta.$$

This can be checked by the same argument as above.

Modulo the relation $\nabla(\Phi) = 0$, this is equivalent to the relation

$$\mathscr{L}_\mathfrak{E}(\xi \star \eta) - \mathscr{L}_\mathfrak{E}\xi \star \eta - \xi \star \mathscr{L}_\mathfrak{E}\eta = \xi \star \eta.$$

If we regard the product $\star$ as a section of the bundle of symmetric homomorphisms of $\Theta_M \otimes \Theta_M$ in Θ_M, on which $\mathfrak{E}$ acts by the Lie derivative, the previous relation means that[5]

$$\mathscr{L}_\mathfrak{E}(\star) = \star.$$ □

---

[5] The first relation of §1.5 also implies (cf. [Her02])

$$\mathscr{L}_{\xi \star \eta}(\star) = \xi \star \mathscr{L}_\eta(\star) + \mathscr{L}_\xi(\star) \star \eta.$$

1.6 Example (Saito structures in dimension two). We will exhibit the Saito structures on the complex plane $\mathbb{C}^2$ equipped with the usual flat connection, for which the canonical coordinates (t_1, t_2) are flat. We assume that ∂_{t_1} is the unit field. We also assume that $\nabla\mathfrak{E}$ is *semisimple* with eigenvalues 1 and δ_2 and, according to §1.4, that the coordinates are translated in such a way that the Euler field takes the form

$$\mathfrak{E} = \begin{cases} t_1\partial_{t_1} + \delta_2 t_2\partial_{t_2} & \text{if } \delta_2 \neq 0, \\ t_1\partial_{t_1} + r_2\partial_{t_2} & (r_2 \in \mathbb{C}) \ \text{if } \delta_2 = 0. \end{cases}$$

The product $\star$ will be determined by the value of $\partial_{t_2} \star \partial_{t_2}$. Let us then set

$$\partial_{t_2} \star \partial_{t_2} = \alpha_1(t_1, t_2)\partial_{t_1} + \alpha_2(t_1, t_2)\partial_{t_2}.$$

The relations that we did not use yet are

$$\nabla_{\partial_{t_1}}(\partial_{t_2} \star \partial_{t_2}) = \nabla_{\partial_{t_2}}(\partial_{t_1} \star \partial_{t_2}) = \nabla_{\partial_{t_2}}\partial_{t_2} = 0$$

and

$$\mathscr{L}_{\mathfrak{E}}(\partial_{t_2} \star \partial_{t_2}) - 2\mathscr{L}_{\mathfrak{E}}(\partial_{t_2}) \star \partial_{t_2} = \partial_{t_2} \star \partial_{t_2}.$$

The first one shows that α_1 and α_2 only depend on t_2. The second one shows that

- if $\delta_2 \neq 0$ and if we set $\delta_2 = 1/(1+m)$, we have $\alpha_1 = c_1 t_2^{2m}$ and $\alpha_2 = c_2 t_2^m$ $(c_1, c_2 \in \mathbb{C})$; therefore, such a structure exists on $\mathbb{C}^2$ if and only if $m \in \mathbb{N}$ when $c_2 \neq 0$ and $2m \in \mathbb{N}$ when $c_2 = 0$; we then have

$$\partial_{t_2} \star \partial_{t_2} = t_2^m(c_1 t_2^m \partial_{t_1} + c_2\partial_{t_2});$$

- if $\delta_2 = 0$, we have $\alpha_1 = c_1 e^{2t_2/r_2}$ and $\alpha_2 = c_2 e^{t_2/r_2}$ $(c_1, c_2 \in \mathbb{C})$; we then have

$$\partial_{t_2} \star \partial_{t_2} = e^{t_2/r_2}(c_1 e^{t_2/r_2}\partial_{t_1} + c_2\partial_{t_2}).$$

1.7 Embedding of the manifold L_Φ as a hypersurface. The Euler field $\mathfrak{E}$ defines a linear form on each fibre of the cotangent bundle, that is, a linear homomorphism of TM to the trivial bundle $M \times \mathbb{C}$. Let us denote by λ the coordinate on the factor $\mathbb{C}$. Let L_Φ be the submanifold of the cotangent bundle T^*M associated to Φ, as in §0.13.d. *If for any $x \in M$ the endomorphism $R_{0,x}$ of T_xM is regular (i.e., its minimal polynomial is equal to its characteristic polynomial), the mapping $L_\Phi \to M \times \mathbb{C}$, induced by the Euler field $\mathfrak{E}$, is a closed immersion and its image is defined by the equation $\det(\lambda\,\mathrm{Id} - R_0) = 0$.*

Indeed, it is a matter of seeing that the corresponding morphism $\mathscr{O}_M[\lambda] \to \Theta_M$ is onto. The multiplication endomorphism by $\mathfrak{E}$ is equal to R_0. It is thus regular; for any section ξ of Θ_M, the multiplication by ξ commutes with the multiplication by $\mathfrak{E}$, hence can be expressed as a polynomial in R_0 with coefficients in $\mathscr{O}_M$. As a consequence, $\xi = \xi\star e$ is a polynomial in $\mathfrak{E}$. Lastly, the kernel of $\mathscr{O}_M[\lambda] \to \Theta_M$ is generated by the minimal polynomial of R_0, which is its characteristic polynomial. $\qquad\square$

1.8 The manifold L_Φ is Lagrangian [Aud98a]. Let us suppose that M is simply connected and R_0 is regular semisimple at any point. In such a situation, the algebra structure on $T_x M$ is semisimple for any $x \in M$. The eigenvalues of R_0 define d functions $x_1, \ldots, x_d$ on M. *These functions form a system of canonical coordinates*[6] *(cf. definition 0.13.10) on M and the manifold L_Φ is nothing but the disjoint union of the graphs of the dx_i.*

Indeed, it is enough to check this locally. As in §VI.3.e, we construct (by fixing a base point $x^o \in M$) a basis e of Θ_M, according to Theorem III.2.10 and, in this basis, the matrix of R_0 is diagonal at each point of M. The expression VI.(3.12) shows that the matrix of Φ in this basis is then equal to $-dR_0$; in other words, we have, for any $i = 1, \ldots, d$, the equality

$$\Phi(e_i) = -dx_i \otimes e_i,$$

that is, for all $i, j = 1, \ldots, d$,

$$e_i \star e_j = \mathscr{L}_{e_j}(x_i) \cdot e_i.$$

As the product $\star$ is commutative, we deduce that $\mathscr{L}_{e_j}(x_i) = 0$ for $i \neq j$. According to the existence of a unit field, we can see that, as in Remark 0.13.11, the functions $\lambda_i = \mathscr{L}_{e_i}(x_i)$ do not vanish. As the Jacobian determinant $\det\big(\mathscr{L}_{e_j}(x_i)\big)$ vanishing nowhere on M, the mapping $(x_1, \ldots, x_d) : M \to \mathbb{C}^d$ has everywhere maximal rank. Let us set $e_i' = e_i/\lambda_i$; we thus have $e_i' \star e_j' = \delta_{ij} e_i'$ and $\mathscr{L}_{e_j'}(x_i) = \delta_{ij}$. We deduce that $e_i' = \partial/\partial x_i$ and that $(x_1, \ldots, x_d)$ defines in the neighbourhood of any point of M a system of canonical coordinates. The assertion on L_Φ follows from Exercise 0.13.12.   □

We deduce a holomorphic mapping, everywhere of maximal rank, from M to an open set of the manifold $\widetilde{X}_d$ of §VI.1.c.

Conversely, any holomorphic mapping everywhere of maximal rank from M to an open set of $\widetilde{X}_d$, for which $\star$, e and $\mathfrak{E}$ are as above, is obtained by canonical coordinates, hence is unique up to a permutation of these.

Theorem 4.2 will make precise these results.

Therefore, in this situation, the manifold M possesses two types of coordinate systems, one adapted to the connection, the other one to the product.

1.9 Example 1.6, continuation. In Example 1.6, when $\delta_2 \neq 0$ and $c_2 = 0$, $c_1 = 1$, we must have $2m \in \mathbb{N}$, and the matrix of R_0 is equal to

$$\begin{pmatrix} t_1 & \delta_2 t_2^{2m+1} \\ \delta_2 t_2 & t_1 \end{pmatrix}$$

so that R_0 is regular semisimple away from $\{t_2 = 0\}$. The canonical coordinates are the eigenvalues of R_0, that is,

$$x_1 = t_1 + \frac{\delta_2 t_2^{1+m}}{1+m}, \quad x_2 = t_1 - \frac{\delta_2 t_2^{1+m}}{1+m}$$

in any simply connected open set of $\{t_2 \neq 0\}$.

---

[6] In the étale sense, cf. Remark 0.2.4.

1.10 The discriminant. Let us assume that the endomorphism R_0 (multiplication by $\mathfrak{E}$) is invertible for almost all points of M. As Θ_M has no $\mathcal{O}_M$-torsion, being locally free, the endomorphism R_0 is everywhere injective and induces an isomorphism onto its image. The *discriminant* Δ of the Saito manifold M is the hypersurface of M defined by the equation $\det R_0 = 0$. This hypersurface, when it is not empty, plays an important role in the geometry of the manifold. It possesses itself a very rich geometry[7].

One can easily deduce from the Saito structure a remarkable property of this hypersurface: it is a *free divisor*. Recall the definition introduced by K. Saito [Sai80]: let D be a hypersurface of M; the subsheaf $\Theta_M \langle \log D \rangle \subset \Theta_M$ of *logarithmic vector fields* along D consists of the vector fields which are tangents to the smooth part of D; let U be an open set of M on which D is defined by the vanishing of a holomorphic function without multiple factor h, and let ξ be a vector field on U; then ξ is logarithmic along $D \cap U$ if and only if the function $\mathscr{L}_\xi(h)$ vanishes on D or also, after the Nullstellensatz, if this function is a multiple of h.

This definition clearly implies that $\Theta_M \langle \log D \rangle$ is a sheaf of $\mathcal{O}_M$-modules which coincides with Θ_M away from D. One can show that this subsheaf of Θ_M is *coherent*. One can moreover show that the sheaf $\Theta_M \langle \log D \rangle$ satisfies the following property: if Σ is a closed analytic subset of M everywhere of codimension $\geqslant 2$ and if $j : M \smallsetminus \Sigma \hookrightarrow M$ denotes the inclusion, then

$$\Theta_M \langle \log D \rangle = j_* \Theta_M \langle \log D \rangle_{|M \smallsetminus \Sigma} \cap \Theta_M \subset j_* \Theta_{M \smallsetminus \Sigma}.$$

Indeed, for any vector field ξ on an open set U of M on which $h = 0$ is an equation for D, if $\mathscr{L}_\xi(h)$ vanishes on $D \smallsetminus \Sigma$, then $\mathscr{L}_\xi(h)$ also vanishes on D.

The hypersurface is a *free divisor* (in the sense of K. Saito) if the sheaf $\Theta_M \langle \log D \rangle$ is *locally free* as a $\mathcal{O}_M$-module (hence of rank $\dim M$). One can show (by duality) that this property is equivalent to the local freeness of the sheaf of logarithmic 1-forms (cf. Remark 0.9.16-(3)).

Let us come back to the Saito manifold M and the hypersurface Δ. If R_0 is moreover regular semisimple on an open set dense of M which contains an open set dense of Δ, the divisor Δ is *free*.

Indeed, it is enough to check that the sheaf $\Theta_M \langle \log D \rangle$ is equal to the image sheaf of $R_0 : \Theta_M \to \Theta_M$ as, by assumption on R_0, the latter is locally free. The closed analytic set Σ consisting of points of Δ where R_0 is not regular semisimple has codimension $\geqslant 1$ in Δ, hence $\geqslant 2$ in M. As image R_0 is locally free, we also have

$$\text{image } R_0 = j_* (\text{image } R_0)_{|M \smallsetminus \Sigma} \cap \Theta_M,$$

which can be seen by expressing a local section of the right-hand term in some basis of image R_0 with holomorphic coefficients on the complementary set of Σ, hence with a removable singularity along Σ.

---

[7] The reader will refer to [Tei77] for that question, in the case where M is the basis of the universal unfolding of a singularity.

It is thus enough to check the equality of both subsheaves in the neighbourhood of any point of the complementary set of this closed set. There exists then, in the neighbourhood of such a point, canonical coordinates $x_1, \ldots, x_d$, and Δ is defined as the union of the hypersurfaces $x_i = 0$. In the canonical basis, the matrix of R_0 is $\mathrm{diag}(x_1, \ldots, x_d)$ and, in the neighbourhood of $x_1 = 0$ for instance, we have $x_2, \ldots, x_d \neq 0$, hence the image of R_0 is generated by the vector fields $x_1 \partial_{x_1}, \partial_{x_2}, \ldots, \partial_{x_d}$, i.e., the logarithmic fields. $\qquad\square$

1.b Saito structure with a metric

As in §0.13, we will call a *metric* on the tangent bundle Θ_M any nondegenerate symmetric $\mathscr{O}_M$-bilinear form.

1.11 Definition. A Saito structure with metric on the manifold M consists of a Saito structure $(\bigtriangledown, \Phi, e, \mathfrak{E})$ and of a metric g on the tangent bundle, satisfying the following properties:

(1) $\bigtriangledown(g) = 0$ (hence $\bigtriangledown$ is the Levi-Civita connection of g);
(2) $\Phi^* = \Phi$, i.e., for any local section ξ of Θ_M, $\Phi_\xi^* = \Phi_\xi$, where * denotes the adjoint with respect to g; in other words, we have $g(\xi_1 \star \xi_2, \xi_3) = g(\xi_1, \xi_2 \star \xi_3)$ for all vector fields ξ_1, ξ_2, ξ_3;
(3) there exists a complex number $D \in \mathbb{C}$ such that

$$\bigtriangledown(\mathfrak{E})^* + \bigtriangledown(\mathfrak{E}) = D \, \mathrm{Id}.$$

One can notice that the torsion free connection $\bigtriangledown$ can be deduced from the metric and the latter is therefore a *flat metric*. On the other hand, the existence of a Euler field gives some homogeneity to the manifold M. In the examples originally treated by K. Saito, the endomorphism $\bigtriangledown\mathfrak{E}$ is semisimple.

We will now consider some supplementary properties that one can deduce from the existence of such a metric.

1.12 Homogeneity of the metric. As $R_0 = -\Phi(\mathfrak{E})$, we have $R_0^* = R_0$. On the other hand, the skewsymmetry condition on $\bigtriangledown\mathfrak{E}$ is expressed as

$$g(\bigtriangledown_\xi \mathfrak{E}, \eta) + g(\xi, \bigtriangledown_\eta \mathfrak{E}) = D \cdot g(\xi, \eta) \quad \forall \xi, \eta$$

that is, as $\bigtriangledown(g) = 0$ and $\bigtriangledown$ is torsion free, as

$$\mathscr{L}_\mathfrak{E}(g)(\xi, \eta) := \mathscr{L}_\mathfrak{E}(g(\xi, \eta)) - g(\mathscr{L}_\mathfrak{E}\xi, \eta) - g(\xi, \mathscr{L}_\mathfrak{E}\eta) = D \cdot g(\xi, \eta) \quad \forall \xi, \eta.$$

One can write this equality as $\mathscr{L}_\mathfrak{E}(g) = D \cdot g$: the field $\mathfrak{E}$ preserves the metric, up to a multiplicative constant, by the Lie derivative; one says that it is *conformal* (recall also, for good measure, the equality $\mathscr{L}_\mathfrak{E}(\star) = \star$, seen in §1.5).

If we apply the first formula to $\xi = \eta = e$, we get $(D-2)g(e, e) = 0$, as $\bigtriangledown_e \mathfrak{E} = e$ (cf. §1.3).

1.13 The counit. Let e^* be the 1-form on M defined by $e^*(\eta) = g(e, \eta)$ for any vector field η. We say that e^* is the *counit* of the Saito structure. It follows from 1.11-(2) that, for all ξ, η, we have $g(\xi, \eta) = e^*(\xi \star \eta)$. Moreover, the form e^* is *closed*: indeed,

$$
\begin{aligned}
\partial_{z_i} e^*(\partial_{z_j}) = \partial_{z_i} g(e, \partial_{z_j}) &= g(e, \nabla_{\partial_{z_i}} \partial_{z_j}) \quad \text{(as } \nabla(g) = 0 \text{ and } \nabla e = 0) \\
&= g(e, \nabla_{\partial_{z_j}} \partial_{z_i}) \quad \text{(as } \nabla \text{ is torsion free)} \\
&= \partial_{z_j} e^*(\partial_{z_i}).
\end{aligned}
$$

If M is simply connected (or if we replace M with its universal covering), there exists a holomorphic function η such that $e^* = d\eta$ and this function (denoted originally τ by K. Saito) is unique up to the addition of a constant (cf. Exercise 0.9.8). This function η is *flat* (cf. Remark 0.13.8).

1.14 Flat coordinates adapted to the metric and to the Euler field. There exists on any simply connected open set U of M (or, better, on the universal covering space of M) a flat coordinate system $t_1, \ldots, t_d$ such that the associated vector fields $\partial_{t_1}, \ldots, \partial_{t_d}$ form a g-orthonormal basis of $T_x M$ for any $x \in U$.

In general, one cannot obtain flat g-orthonormal coordinates such that the ∂_{t_i} are eigenvectors of $\nabla \mathfrak{E}$ (as in §1.4), as for instance $g(e, e) = 0$ if $D \neq 2$. Depending on the case, we will choose one of these properties.

If $\nabla_{\partial_{t_i}} \mathfrak{E} = \delta_i \partial_{t_i}$ and $\nabla_{\partial_{t_j}} \mathfrak{E} = \delta_j \partial_{t_j}$, we have $(\delta_i + \delta_j) g(\partial_{t_i}, \partial_{t_j}) = D \cdot g(\partial_{t_i}, \partial_{t_j})$, hence ∂_{t_i} and ∂_{t_j} are g-orthogonal as soon as $\delta_i + \delta_j \neq D$.

When $g(e, e) = 0$ (if $D \neq 2$ for instance) and $\nabla \mathfrak{E}$ is semisimple, there exists a flat coordinate system $t_1, \ldots, t_d$ such that $e = \partial_{t_1}$, that $\partial_{t_1}, \ldots, \partial_{t_d}$ are eigenvectors of $\nabla \mathfrak{E}$ and that

$$
g(\partial_{t_i}, \partial_{t_j}) = \begin{cases} 1 & \text{if } i + j = d + 1, \\ 0 & \text{if } i + j \neq d + 1. \end{cases}
$$

Indeed, in any flat g-orthonormal coordinate system, the matrix of $\nabla \mathfrak{E} - D/2 \,\mathrm{Id}$ is constant and skewsymmetric; hence, for any λ, the multiplicity of an eigenvalue λ is equal to that of $-\lambda$. Let $t'_1, \ldots, t'_d$ be a flat coordinate system such that the $\partial_{t'_i}$ are eigenvectors of $\nabla \mathfrak{E}$. By the Gram-Schmidt process, we can find a constant base change (and therefore another flat coordinate system $t_1, \ldots, t_d$) such that the required properties are satisfied, with possibly only $g(\partial_{t_i}, \partial_{t_{d+1-i}}) \neq 0$. As $\partial_{t_d} \neq \partial_{t_1}$, we can, still keeping $\partial_{t_1} = e$, obtain $g(\partial_{t_i}, \partial_{t_{d+1-i}}) = 1$ by multiplying the ∂_{t_i} ($i \geqslant 2$) by constants, and modify the coordinates accordingly.

1.15 Example 1.6, continuation. Let us consider the metric g such that

$$
g(\partial_{t_1}, \partial_{t_1}) = g(\partial_{t_2}, \partial_{t_2}) = 0 \quad \text{and} \quad g(\partial_{t_1}, \partial_{t_2}) = 1.
$$

In order that this metric is compatible to the product $\star$, it is necessary and sufficient that $g(\partial_{t_2} \star \partial_{t_2}, \partial_{t_1}) = 0$, that is, $c_2 = 0$. We then have $\nabla \mathfrak{E} + (\nabla \mathfrak{E})^* = (2 + 2\delta_2) \,\mathrm{Id}$.

1.16 A potential for the metric. If we have a system (x_i) of canonical coordinates (cf. definition 0.13.10) on some open set of M (in the situation of §1.8 for instance), then we have $g(\partial_{x_i}, \partial_{x_j}) = 0$ if $i \neq j$: as a matter of fact,

$$
\begin{aligned}
g(\partial_{x_i}, \partial_{x_j}) &= g(\partial_{x_i} \star \partial_{x_i}, \partial_{x_j}) \quad \text{(canonicality)} \\
&= g(\partial_{x_i}, \partial_{x_i} \star \partial_{x_j}) \quad \text{(compatibility of g and $\star$)} \\
&= g(\partial_{x_j}, 0) = 0 \quad \text{(canonicality)}.
\end{aligned}
$$

If we set $\phi_i(x) = g(\partial_{x_i}, \partial_{x_i})$, the metric can be written as $g = \sum_i \phi_i(x)(dx_i)^2$. One can also check that the counit e^* can be written as $\sum_i \phi_i(x)\, dx_i$: indeed, by definition, $e^* = \sum_i g(e, \partial_{x_i})\, dx_i$ and the assertion follows from $e = \sum_j \partial_{x_j}$. As e^* is a closed holomorphic form (cf. §1.13), there exists on any simply connected open set of M or on its universal covering (cf. Exercise 0.9.8) a function ϕ such that $\phi_i = \partial\phi/\partial x_i$ for any $i = 1,\ldots,d$.

1.17 Exercise. Express the flatness condition for the metric g in terms of ϕ.

1.18 A tensor with four legs. Let us set $c(\xi_1, \xi_2, \xi_3) = g(\xi_1 \star \xi_2, \xi_3)$, regarded as a section of the bundle $(\Omega^1_M)^{\otimes 3}$ and let ∇c be defined as a section of the bundle $(\Omega^1_M)^{\otimes 4}$ (see for instance [GHL87, Prop. 2.58]). It is clear that c is symmetric in its arguments. We will prove that

  ∇c *is symmetric in its four arguments.*

Recall that ∇c can be computed, in any system of local coordinates $(z_1,\ldots,z_d)$, by the formula

$$
(1.19) \quad \frac{\partial c(\partial_{z_i}, \partial_{z_j}, \partial_{z_k})}{\partial z_\ell} = (\nabla_{\partial_{z_\ell}} c)(\partial_{z_i}, \partial_{z_j}, \partial_{z_k}) + c(\nabla_{\partial_{z_\ell}}\partial_{z_i}, \partial_{z_j}, \partial_{z_k})
$$
$$
+ c(\partial_{z_i}, \nabla_{\partial_{z_\ell}}\partial_{z_j}, \partial_{z_k}) + c(\partial_{z_i}, \partial_{z_j}, \nabla_{\partial_{z_\ell}}\partial_{z_k}).
$$

Let us then consider flat coordinates such that the fields $\partial_{t_1}, \ldots, \partial_{t_d}$ are g-orthonormal. The matrix of Φ can be written as $\sum_i C^{(i)} dt_i$, where $C^{(i)}$ is the matrix $C^{(i)}_{j,k}$. Then

$$
\begin{aligned}
\Phi \text{ symmetric} &\iff C^{(i)}_{k,j} = C^{(j)}_{k,i} \quad \forall i,j,k \\
\nabla\Phi = 0 &\iff \frac{\partial C^{(i)}}{\partial t_\ell} = \frac{\partial C^{(\ell)}}{\partial t_i} \quad \forall i,\ell \\
\Phi^* = \Phi &\iff C^{(i)}_{j,k} = C^{(i)}_{k,j} \quad \forall i,j,k.
\end{aligned}
$$

Lastly, we have, in flat coordinates $(t_1,\ldots,t_d)$,

$$
\nabla_{\partial_{t_\ell}}(c)(\partial_{t_i}, \partial_{t_j}, \partial_{t_k}) = \partial_{t_\ell}\big(c(\partial_{t_i}, \partial_{t_j}, \partial_{t_k})\big) = \frac{\partial C^{(i)}_{j,k}}{\partial t_\ell},
$$

which is thus symmetric.

1.20 Exercise. Write the tensor c in canonical coordinates, assuming the Saito structure is semisimple.

1.21 The homogeneity constant. In various situations (see for instance §3), the number D can be naturally expressed as $D = 2q + 2 - w$, where w is an integer and $q \in \mathbb{C}$. It is then natural to set $R_\infty = \bigtriangledown(\mathfrak{E}) - (1 + q)\,\mathrm{Id}$. The connection $\nabla = \bigtriangledown + \dfrac{\Phi}{\tau} + \left(\dfrac{R_0}{\tau} - R_\infty\right)\dfrac{d\tau}{\tau}$ is still integrable and the compatibility conditions 1.11 are equivalent to giving a Hermitian nondegenerate form G on π^*TM, which is compatible with the connection ∇ and has weight w (cf. §VI.2.b). On the other hand, the unit e is an eigenvector of R_∞, corresponding to the eigenvalue $-q$, making the meaning of this number q more precise.

2 Frobenius structure on a manifold

We will now present the definition of a Frobenius structure on a complex analytic manifold M, as given by B. Dubrovin [Dub96]. *A priori*, it does not refer to the isomonodromic relations; it rather emphasizes the properties of some tensors on the manifold.

2.a Frobenius structure

Let us then be given on TM a symmetric nondegenerate bilinear form g, an associative and commutative product $\star$ with unit e. This provides each tangent space $T_m M$ with the structure of a Frobenius algebra (in the sense given in Exercise 0.13.13). It is equivalent to be given three tensors $\ell \in \Gamma(M, \Omega^1_M)$, $g \in \Gamma(M, (\Omega^1_M)^{\otimes 2})$ and $\Gamma(M, (\Omega^1_M)^{\otimes 3})$, the last two ones being symmetric, satisfying the properties of Exercise 0.13.13.

This family of Frobenius algebras will be called a *Frobenius structure* on M if it satisfies two supplementary conditions, namely an *integrability* condition and an *homogeneity* condition. These conditions are expressed as follows:

(1) the metric g is flat and, if $\bigtriangledown$ is the associated flat torsion free connection, we have $\bigtriangledown(e) = 0$;

(2) the 4-tensor $\bigtriangledown c$ (see 1.18) is symmetric in its arguments;

(3) there exist a vector field $\mathfrak{E}$ (Euler field) and a complex number D subject to the following conditions:

   (a) the endomorphism $\bigtriangledown\mathfrak{E}$ of Θ_M is a $\bigtriangledown$-horizontal section of $\mathrm{End}_{\mathscr{O}_M}(\Theta_M)$;

   (b) we have $\mathscr{L}_\mathfrak{E}(g(\xi, \eta)) - g(\mathscr{L}_\mathfrak{E}\xi, \eta) - g(\xi, \mathscr{L}_\mathfrak{E}\eta) = D \cdot g(\xi, \eta)$ for all fields ξ, η; in other words, $\mathscr{L}_\mathfrak{E}(g) = D \cdot g$;

   (c) we have $\mathscr{L}_\mathfrak{E}(\xi \star \eta) - \mathscr{L}_\mathfrak{E}\xi \star \eta - \xi \star \mathscr{L}_\mathfrak{E}\eta = \xi \star \eta$ for all fields ξ, η; in other words, $\mathscr{L}_\mathfrak{E}(\star) = \star$.

2.1 Remarks.

(1) Given Condition (1), the horizontality (3.a) of the endomorphism $\nabla\mathfrak{E}$ is consequence of Condition (3.b). It is however important to stress upon this property. In order to check that $(1) + (3.b) \Rightarrow (3.a)$, we will work in a flat coordinate system $t_1, \ldots, t_d$ such that the fields $\partial_{t_1}, \ldots, \partial_{t_d}$ are g-orthonormal. Let us set $\mathfrak{E} = \sum_k \varphi_k(t)\partial_{t_k}$. Let us show (cf. Exercise 0.12.11(2)) that, assuming Condition (3.b), the section $\nabla_{\partial_{t_i}}\mathfrak{E}$ remains horizontal. As ∇ is torsion free (Condition (1)), it is equivalent to proving that $\mathscr{L}_{\mathfrak{E}}(\partial_{t_i}) = -\sum_k \partial_{t_i}(\varphi_k)\partial_{t_k}$ is horizontal, i.e., that the $\partial_{t_i}(\varphi_k)$ are constant for all i, k. Condition (3.b) can be written, due to the assumptions on the basis $\partial_{t_1}, \ldots, \partial_{t_d}$,

$$\partial_{t_i}(\varphi_j) + \partial_{t_j}(\varphi_i) = -D\delta_{ij} \qquad \forall\, i, j = 1, \ldots, d,$$

that is, setting $\psi_i = \varphi_i + Dt_i/2$,

$$\partial_{t_i}(\psi_j) + \partial_{t_j}(\psi_i) = 0 \qquad \forall\, i, j = 1, \ldots, d.$$

We deduce from these relations that the $\partial_{t_i}(\psi_j)$ are constant for all i, j: indeed,

$$\partial_{t_k}\partial_{t_i}(\psi_j) = -\partial_{t_k}\partial_{t_j}(\psi_i) = -\partial_{t_j}\partial_{t_k}(\psi_i) = \partial_{t_j}\partial_{t_i}(\psi_k) \quad \text{on the one hand,}$$
$$= \partial_{t_i}\partial_{t_k}(\psi_j) = -\partial_{t_i}\partial_{t_j}(\psi_k) = -\partial_{t_j}\partial_{t_i}(\psi_k) \quad \text{on the other hand.}$$

Hence $\partial_{t_k}\partial_{t_i}(\psi_j) = 0$ for any k. $\square$

(2) We deduce from Condition (3.c), taking $\xi = \eta = e$, that the unit e is an eigenvector with eigenvalue 1 of $\nabla\mathfrak{E}$, that is, $\mathscr{L}_{\mathfrak{E}}e = -e$.

(3) B. Dubrovin also imposes a semisimplicity condition for $\nabla\mathfrak{E}$. This condition is not essential for what follows.

(4) One can weaken the notion of Frobenius structure by imposing only Conditions (1) and (2), that is, without introducing the homogeneity condition provided by the existence of the Euler field. The associativity of the product (i.e., the symmetry of Φ) and Condition (2) amount then to the vanishing of the curvature of the family of connections $\nabla + \lambda\Phi$ parametrized by $\lambda \in \mathbb{C}$ (see [Dub96] or [Man99a]).

2.2 Proposition (The Saito structures are the Frobenius structures).
On any manifold M, there is an equivalence between a Saito structure with metric and a Frobenius structure.

Proof. That a Saito structure with metric gives rise to a Frobenius structure is a consequence of the properties following Definitions 1.1 and 1.11.

Conversely, given a Frobenius structure, let us set $\Phi_\xi(\eta) = -\xi \star \eta$. The commutativity and the associativity of $\star$ give the symmetry of Φ and $\Phi \wedge \Phi = 0$. The entries $C^{(i)}_{j,k}$ of Φ in some ∇-horizontal orthonormal basis are such that $\partial C^{(i)}_{j,k}/\partial t_\ell$ is a symmetric expression in i, j, k, ℓ. This implies that $\nabla(\Phi) = 0$, and hence Property (3.c) of $\mathfrak{E}$ is equivalent to the relation $\nabla(R_0) + \Phi = [\Phi, \nabla\mathfrak{E}]$, as mentioned in §1.5. $\square$

2.b The potential of the Frobenius structure and the associativity equations

Let us be given a Frobenius structure (possibly without a Euler field, as in Remark 2.1-(4)) on a manifold M that we will suppose to be 1-connected. Let us fix a flat coordinate system $(t_1,\ldots,t_d)$ on M. That the tensor $\nabla(c)$ is symmetric in its four arguments implies that there exists a holomorphic function $F : M \to \mathbb{C}$ such that we have, for all i,j,k, the relation

$$(2.3) \qquad \frac{\partial^3 F}{\partial t_i \partial t_j \partial t_k} = c(\partial_{t_i},\partial_{t_j},\partial_{t_k}) = g(\partial_{t_i} \star \partial_{t_j},\partial_{t_k}).$$

We note first that, if such a function exists, it is well determined only up to the addition of a polynomial of degree 2 in $t_1,\ldots,t_d$. Let us prove the existence of F. We will work in flat g-orthonormal coordinates for simplicity. In these coordinates, we have $c(\partial_{t_i},\partial_{t_j},\partial_{t_k}) = C^{(i)}_{j,k}$.

A first consequence of the symmetry of $\nabla(c)$ is that the 1-form $\sum_i C^{(i)}_{j,k}\,dt_i$ is closed. For any pair (j,k), there exists thus (cf. Exercise 0.9.8) a function $F_{j,k}$ (defined up to a constant) such that $dF_{j,k} = \sum_i F^{(i)}_{j,k}\,dt_i$. Still because of symmetry, we have $dF_{j,k} = dF_{k,j}$, so that we can adjust the constants (the manifold M being connected) to get moreover $F_{j,k} = F_{k,j}$ for all j,k. Still by symmetry, we see that the form $\sum_k F_{j,k}\,dt_k$ is closed, and there exists thus functions F_j such that $\partial F_j/\partial t_k = F_{j,k}$. The form $\sum_j F_j\,dt_j$ remains closed, whence the existence of F.

When $\nabla\mathfrak{E}$ is semisimple and $g(e,e) = 0$, we have, in the flat coordinates of §1.14, $g(\partial_{t_j} \star \partial_{t_j},\partial_{t_k}) = 1$ if $j+k = d+1$ and 0 otherwise, so that we can write

$$(2.4) \qquad F(t_1,\ldots,t_d) = \frac{1}{2}t_1\Big(\sum_{i=1}^d t_i t_{d+1-i}\Big) + G(t_2,\ldots,t_d).$$

2.5 Example 1.6, continuation. Let us take up the situation of Example 1.15. The only third derivatives of F which are possibly nonzero are, setting $\ell = 2m \in \mathbb{N}$,

$$\frac{\partial^3 F}{\partial t_1^2 \partial t_2} = 1 \quad \text{and} \quad \frac{\partial^3 F}{\partial t_2^3} = c_1 t_2^{2\ell} \text{ (or } c_1 e^{2t_2/r_2}).$$

We thus have, up to a polynomial of degree 2,

$$F(t_1,t_2) = \frac{1}{2}t_1^2 t_2 + c_1' t_2^{\ell+3} \quad \text{or} \quad F(t_1,t_2) = \frac{1}{2}t_1^2 t_2 + c_1' e^{2t_2/r_2}$$

for some suitable constant c_1'.

2.6 Homogeneity of the potential. Let us suppose that $\nabla\mathfrak{E}$ is semisimple. We can then choose flat coordinates as in §1.4. We have in particular $\mathscr{L}_{\mathfrak{E}}(\partial_{t_j}) = -\delta_j\partial_{t_j}$ for any j. We deduce from the conformality of $\mathfrak{E}$ with respect to the metric g (§1.12) that, for all i,j,k, the function $c(\partial_{t_i},\partial_{t_j},\partial_{t_k})$ is $\mathfrak{E}$-homogeneous of degree $D+1-\delta_i-\delta_j-\delta_k$, that is, satisfies the equation

$$\mathscr{L}_{\mathfrak{E}}c(\partial_{t_i},\partial_{t_j},\partial_{t_k}) = (D+1-\delta_i-\delta_j-\delta_k)c(\partial_{t_i},\partial_{t_j},\partial_{t_k}).$$

We deduce that $\mathscr{L}_{\mathfrak{E}}(F)-(D+1)\cdot F$ is a polynomial of degree $\leqslant 2$.

In particular, if all the δ_j are >0, this implies that F is a polynomial, as the coefficient of $t_1^{n_1}\cdots t_d^{n_d}$ in F is zero unless $\sum_j \delta_j n_j = D+1$ when $\sum_j n_j \geqslant 3$.

2.7 Associativity equations. The associativity of the product imposes constraints on (the third derivatives of) the function F. These are nonlinear partial differential equations, known as "WDVV equations".

Let us fix a flat coordinate system $(t_1,\dots,t_d)$, that we will assume to be g-orthonormal for simplicity. We thus have $\xi = \sum_m g(\xi,\partial_{t_m})\partial_{t_m}$ for any field ξ and therefore

$$\partial_{t_i}\star\partial_{t_j} = \sum_m g(\partial_{t_i}\star\partial_{t_j},\partial_{t_m})\partial_{t_m} = \sum_m \frac{\partial^3 F}{\partial t_i\partial t_j\partial t_m}\partial_{t_m}.$$

The associativity $(\partial_{t_i}\star\partial_{t_j})\star\partial_{t_k} = \partial_{t_i}\star(\partial_{t_j}\star\partial_{t_k})$ amounts to the relations[8]:

$$(2.8) \qquad \sum_m \frac{\partial^3 F}{\partial t_i\partial t_j\partial t_m}\cdot\frac{\partial^3 F}{\partial t_m\partial t_k\partial t_\ell} = \sum_m \frac{\partial^3 F}{\partial t_k\partial t_j\partial t_m}\cdot\frac{\partial^3 F}{\partial t_m\partial t_i\partial t_\ell}$$

for all $i,j,k,\ell \in \{1,\dots,d\}$, which can be expressed by the fact that

$$\sum_m \frac{\partial^3 F}{\partial t_i\partial t_j\partial t_m}\cdot\frac{\partial^3 F}{\partial t_m\partial t_k\partial t_\ell}$$

is symmetric in i,j,k,ℓ.

2.9 Exercise (Associativity equations in adapted flat coordinates). Assume that $g(e,e)=0$ and that $\nabla\mathfrak{E}$ is semisimple. Prove that, in a flat coordinate system $t_1,\dots,t_d$ as in §1.14, the associativity property can be expressed as the symmetry in i,j,k,ℓ of

$$\sum_m \frac{\partial^3 F}{\partial t_i\partial t_j\partial t_m}\cdot\frac{\partial^3 F}{\partial t_{d+1-m}\partial t_k\partial t_\ell}.$$

---

[8] Exercise: write analogous relations in flat coordinates, which are not necessarily orthonormal.

3 Infinitesimal period mapping

We describe here a procedure to construct a Frobenius manifold from a family of bundles on $\mathbb{P}^1$, equipped with a flat meromorphic connection. In order to obtain the metric of the Frobenius manifold, it will be necessary to suppose that this family possesses a nondegenerate Hermitian form. In order that the construction is defined, it is necessary that the family of bundles has a *primitive section*. The Frobenius structure is then obtained with *the infinitesimal period mapping* provided by this primitive section. In this way, we follow word for word the steps of K. Saito [Sai83a].

3.a Infinitesimal period mapping associated to a primitive section

Let us be given a bundle F of rank $d = \dim M$ on a manifold M. We deduce a bundle $E = \pi^* F$ on $\mathbb{P}^1 \times M$. We assume that it comes equipped with a flat meromorphic connection ∇ having a pole of type 1 along $\{0\} \times M$ and a logarithmic pole at infinity. It amounts to giving on F the objects ∇, Φ, R_∞, R_0 satisfying the relations of §VI.2.c. In the following, we will set $E_0 = F = i_0^* E$. Here, we have a fixed identification $E_0 = E_\infty$.

Let ω be a ∇-horizontal section of $\mathscr{E}_0$. To this section is associated an *infinitesimal period mapping*

$$\varphi_\omega : TM \longrightarrow E_0,$$

which is the bundle morphism on M defined by

$$\varphi_\omega(\xi) = -\Phi_\xi(\omega)$$

for any vector field ξ on M.

3.1 Definition (of a primitive or homogeneous section). A ∇-horizontal section ω of $\mathscr{E}_0$ is said to be

- *homogeneous* if it is an eigenvector of R_∞,
- *primitive* if φ_ω is an isomorphism of bundles.

3.b Flat connection and product on the bundle TM

If ω is a primitive section, we can carry on TM through φ_ω the structures which exist on E_0. They will be denoted with an exponent ω on the left to remind us of the dependence with respect to the primitive section. We thus have a flat connection ${}^\omega\nabla$ on TM defined by

$$^\omega\nabla(\xi) = \varphi_\omega^{-1} \nabla (\varphi_\omega(\xi)).$$

The form Φ also defines a *product*[9] $\star$ on the sections of TM by

$$\varphi_\omega(\xi \star \eta) := -\Phi_\xi(\varphi_\omega(\eta)).$$

3.2 Proposition.

(1) *The flat connection $^\omega\nabla$ on TM is torsion free or, equivalently, the section φ_ω of $\Omega^1_M \otimes \mathcal{E}_0$ satisfies $\nabla\varphi_\omega = 0$ in $\Omega^2_M \otimes \mathcal{E}_0$.*
(2) *The product $\star$ is associative, commutative and admits $e := \varphi_\omega^{-1}(\omega)$ as a unit, which is a horizontal section with respect to the flat connection $^\omega\nabla$.*

Proof.

(1) Let us locally fix a ∇-horizontal basis ε of $E_0 = E_\infty$ and local coordinates $z_1, \dots, z_d$ of M. In such a basis, the section ω has constant coefficients. We also have, taking the notation of VI.(2.2), $\varphi_\omega(\partial_{z_i}) = -C^{(i)}(z) \cdot \omega$. Then

$$\nabla_{\partial_{z_i}} \varphi_\omega(\partial_{z_j}) = -\partial_{z_i}(C^{(j)}(z)) \cdot \omega,$$

as ω is ∇-horizontal. Moreover, $\partial_{z_i}(C^{(j)}(z)) = \partial_{z_j}(C^{(i)}(z))$, after VI.(2.6). Consequently, $^\omega\nabla_{\partial_{z_i}} \partial_{z_j} = {}^\omega\nabla_{\partial_{z_j}} \partial_{z_i}$. We deduce the vanishing of the torsion of $^\omega\nabla$.

Lastly, we have by definition (cf. Exercise 0.12.1), when we regard φ_ω as a section of $\Omega^1_M \otimes \mathcal{E}_0$,

$$\nabla\varphi_\omega(\xi, \eta) = \nabla_\xi(\varphi_\omega(\eta)) - \nabla_\eta(\varphi_\omega(\xi)) - \varphi_\omega([\xi, \eta])$$

and the horizontality of φ_ω is equivalent to the vanishing of the torsion of $^\omega\nabla$. ☐

(2) The product is given by

$$\begin{aligned}
\varphi_\omega(\partial_{z_i} \star \partial_{z_j}) &= -\Phi_{\partial_{z_i}}(\varphi_\omega(\partial_{z_j})) \\
&= -C^{(i)} \cdot \varphi_\omega(\partial_{z_j}) \\
&= C^{(i)} \cdot C^{(j)} \cdot \omega,
\end{aligned}$$

and the first assertion comes from the relation $[C, C] = 0$. We have on the other hand

$$\begin{aligned}
\varphi_\omega(\xi \star e) &= -\Phi_\xi(\varphi_\omega(e)) \\
&= -\Phi_\xi(\omega) = \varphi_\omega(\xi),
\end{aligned}$$

which gives the second point (the horizontality of e is equivalent to that of ω). ☐

---

[9] Here also, we should attach an exponent ω to the notation; in the examples of §4 we will see however that this product does not depend on the chosen primitive section.

3.3 Remark. If $\mathscr{E}_0$ is equipped with a nondegenerate bilinear form g such that the relations of §VI.2.18 are satisfied, this form is carried by φ_ω to a bilinear form on TM, that we denote by ${}^\omega g$. Then ${}^\omega g$ is ${}^\omega\nabla$-horizontal, as g is ∇-horizontal and ${}^\omega\nabla$, being torsion free, is the Levi-Civita connection of the bilinear form.

3.4 Exercise. The endomorphism R_∞ of $\mathscr{E}_0$ defines a section ${}^\omega S_\infty$ of $\Omega^1_M \otimes \mathscr{E}_0$ by
$$ {}^\omega S_\infty(\xi) = R_\infty(\varphi_\omega(\xi)). $$
Prove that $\nabla({}^\omega S_\infty) = 0$ in $\Omega^2_M \otimes \mathscr{E}_0$.

3.c The Euler field

3.5 Proposition.

(1) *There exists a unique vector field ${}^\omega\mathfrak{E}$, called the* Euler field *of the primitive section ω, such that the endomorphism $\xi \mapsto \xi \star {}^\omega\mathfrak{E}$ is the endomorphism ${}^\omega R_0$ of TM.*

(2) *If ω is homogeneous of degree $-q$ (i.e., $R_\infty(\omega) = -q\omega$), we have*
$$ {}^\omega\nabla\,{}^\omega\mathfrak{E} = {}^\omega R_\infty + (1+q)\,\mathrm{Id} $$
and in particular ${}^\omega\nabla({}^\omega\nabla\,{}^\omega\mathfrak{E}) = 0$.

Proof.

(1) If the field ${}^\omega\mathfrak{E}$ exists, it must satisfy, as e is the unit of $\star$,
$$ {}^\omega\mathfrak{E} = e \star {}^\omega\mathfrak{E} = {}^\omega R_0(e) = \varphi_\omega^{-1}(R_0(\omega)) $$

Let us therefore set $\mathfrak{E}_\omega = R_0(\omega)$ and ${}^\omega\mathfrak{E} = \varphi_\omega^{-1}(\mathfrak{E}_\omega)$. We have, by assumption,
$$
\begin{aligned}
\varphi_\omega(\partial_{z_i} \star {}^\omega\mathfrak{E}) &= -\Phi_{\partial_{z_i}}(\mathfrak{E}_\omega) \\
&= -\Phi_{\partial_{z_i}} \cdot R_0(\omega) \\
&= -R_0 \cdot \Phi_{\partial_{z_i}}(\omega) \quad \text{after VI.(2.6)} \\
&= R_0(\varphi_\omega(\partial_{z_i})).
\end{aligned}
$$

(2) Let us compute in a ∇-horizontal basis $\bar\varepsilon$:
$$
\begin{aligned}
\nabla_{\partial_{z_i}}(\mathfrak{E}_\omega) &= \partial_{z_i}(B_0(\omega)) \\
&= (\partial_{z_i}(B_0))(\omega) \\
&= ([B_\infty, C^{(i)}] - C^{(i)})(\omega) \\
&= (1+q)\varphi_\omega(\partial_{z_i}) - B_\infty(\varphi_\omega(\partial_{z_i})).
\end{aligned}
$$
$\square$

We can summarize the results above:

3.6 Theorem (The infinitesimal period mapping produces a Frobenius-Saito structure).

(1) *Let M be a manifold equipped with a bundle E_0 and of data $\triangledown$, Φ, R_0, R_∞ and g satisfying the relations of §§VI.2.17 and 2.18. If $\mathscr{E}_0$ admits a primitive homogeneous section ω, the infinitesimal period mapping φ_ω equips M with the structure of a Frobenius manifold for which the field $e = \varphi_\omega^{-1}(\omega)$ is the unit field.*

(2) *Conversely, any Frobenius manifold is obtained in this way, taking as a primitive section the unit field.*   □

3.7 Remark (Construction of a primitive section). In practice, the primitive section is determined by its initial value at x^o. As a matter of fact, let $\omega^o \in E_0^o$ be an eigenvector of R_∞ (acting on E_0^o). There exists then, on any simply connected open set of M containing x^o (or on the universal covering space of M), a unique $\triangledown$-horizontal section ω of E_0, the restriction of which to x^o is ω^o. If moreover φ_ω induces an isomorphism $T_{x^o}M \to E_0^o$, there exists a hypersurface Θ_{ω^o} in M, away from which φ_ω is an isomorphism (its equation is the determinant of φ_ω in any local bases of TM and E_0). Then, when restricted to $M \smallsetminus \Theta_{\omega^o}$, ω is a primitive section.

3.d Adjunction of a variable in the infinitesimal period mapping

Let us be now given a bundle F' on a manifold M', with $\operatorname{rk} F' = \dim M' + 1$, and let us assume that F' is equipped with $\triangledown'$, R_∞', Φ', R_0' (and possibly g'), satisfying the relations of §VI.2.c (and possibly of §VI.2.18). We wish to equip the manifold $M = \mathbb{A}^1 \times M'$, where $\mathbb{A}^1$ is the affine line with coordinate τ, with the structure of a Frobenius manifold, by identifying F' with $TM_{|\{0\}\times M'} = \mathbb{C}\partial_\tau \oplus TM'$.

We will say, in such a situation, that a $\triangledown$-horizontal section ω' of $\mathscr{F}'$ is primitive if the infinitesimal period mapping

$$\psi_{\omega'} : TM_{|\{0\}\times M'} \longrightarrow F'$$

defined by $\psi_{\omega'}(\xi) = \varphi_{\omega'}(\xi) = -\Phi_\xi(\omega')$ if ξ is a section of TM' and $\psi_{\omega'}(\partial_\tau) = \omega'$, induces an isomorphism of bundles. We will denote in the same way the lifted morphism $\psi_{\omega'} : TM \to p^*F'$, if $p : M \to M'$ is the projection; in other words, we extend $\psi_{\omega'}$ by $\mathscr{O}_M$-linearity.

Let us consider on $F := p^*F'$ the data $\triangledown$, R_∞, Φ, R_0 and g defined by

$$\triangledown = p^*\triangledown', \quad R_\infty = p^*R_\infty', \quad g = p^*g', \quad \Phi = p^*\Phi' - \operatorname{Id} d\tau, \quad R_0 = p^*R_0' + \tau \operatorname{Id}.$$

We see that ω' is primitive, with the meaning given above, if and only if $\omega := 1 \otimes \omega'$ is a primitive section of F in the sense of Definition 3.1, and then $\psi_{\omega'} = \varphi_\omega$.

The Frobenius structure defined on $M = \mathbb{A}^1 \times M'$ by a homogeneous primitive section $1 \otimes \omega'$ of F, using the method of §3.b, admits as unit e the field ∂_τ.

3.8 Exercise (Adjunction of a variable to a Frobenius manifold). Let M be a Frobenius manifold. Use the procedure above to equip $\mathbb{A}^1 \times M$ with the structure of a Frobenius manifold, for which the unit field is ∂_τ, if τ is the coordinate on $\mathbb{A}^1$.

3.e Justifying the terminology

We now explain the choice of the terminology "infinitesimal period mapping" for the morphism φ_ω associated to a homogeneous primitive section ω. The explanation will be clearer for the morphism $\psi_{\omega'}$ of §3.d; this is not very restrictive, according to Exercise 3.8. We thus take up the situation of §3.d and we assume that $\widehat{R}_\infty - k\,\mathrm{Id}$ is invertible for any $k \in \mathbb{N}$.

Let us set $\widehat{\mathbb{F}} = \mathscr{O}_{M'}[\tau] \otimes_{\mathscr{O}_{M'}} F'$ (this is an algebraic variant of the bundle $p^* F'$ of §3.d). This bundle is equipped with the connection

$$\widehat{\nabla} = p^* \nabla - \left[\frac{\Phi}{\tau} + \left(\frac{R_0}{\tau} + R_\infty \right) \frac{d\tau}{\tau} \right].$$

Let $\mathbb{F}$ be the inverse partial Fourier transform of $\widehat{\mathbb{F}}$ (cf. §V.2.c and §VI.3.h). One can show, as in Lemma VI.3.22, that it is a free $\mathscr{O}_{M'}[t]$-module of rank $\dim M' + 1$ and that the connection ∇ has regular singularities (one will notice that the computation of the connection matrix done in §VI.3.h does not use that B_0 has pairwise distinct eigenvalues).

If ξ is a vector field on M' and if ω is a local section of $\mathbb{F} = \widehat{\mathbb{F}}$, we have, by definition, $\nabla_\xi \omega = \widehat{\nabla}_\xi \omega$.

Let us also consider the section $\tau\omega$. For such a field ξ, we then have

$$\nabla_\xi(\tau\omega) = \tau\widehat{\nabla}(\omega) = -\Phi_\xi(\omega).$$

We also have by definition

$$\nabla_{\partial_t}(\tau\omega) = \tau^{-1}\tau\omega = \omega.$$

Therefore, after Fourier transform, the mapping ψ_ω is nothing but the mapping

$$\Theta_M \longrightarrow \mathbb{F}, \quad \eta \longmapsto \nabla_\eta(\tau\omega).$$

The terminology that we use comes from the fact that, for the period mappings considered in algebraic geometry, the associated tangent map can often be expressed in this way.

4 Examples

4.a Universal semisimple Frobenius-Saito structures

Let $(x_1^o, \ldots, x_d^o)$ be a point of X_d (cf. §VI.1.c) and let B_∞ be a matrix such that $B_\infty - (w/2)\,\mathrm{Id}$ is *skewsymmetric*. We will associate to these data and to the choice of an eigenvector of B_∞, a Frobenius structure on the complement of a divisor in the universal covering $\widetilde{X}_d$.

Let thus w^o be an eigenvector of B_∞, with eigenvalue $\alpha \in \mathbb{C}$. We will suppose that *the coefficients w_i^o of w^o are all nonzero*. Let us then set $B_0^o = \mathrm{diag}(x_1^o, \ldots, x_d^o)$.

Proposition VI.3.8 and its complement of §VI.3.c provide us with a bundle E on $\widetilde{X}_d$, a holomorphic flat connection $\triangledown$ and a $\triangledown$-horizontal endomorphism R_∞. On the complement of a divisor $\Theta \subset \widetilde{X}_d$, we also have the holomorphic objects Φ, R_0, g and we are now in the situation described in §3.a.

As $\widetilde{X}_d$ is 1-connected, there exists a unique section ω of $\mathscr{E}$, horizontal with respect to $\triangledown$ and such that $\omega(\widetilde{x}^o) = w^o$, if $\widetilde{x}^o$ is a fixed lift of x^o in $\widetilde{X}_d$. We will restrict this section to $\widetilde{X}_d \setminus \Theta$.

Let us consider the basis e of $\mathscr{E}_{|\widetilde{X}_d \setminus \Theta}$ introduced in §VI.3.e. As the basis e is meromorphic, the entries ω_i of ω on the basis e are meromorphic functions on $\widetilde{X}_d$ with poles along Θ. Let Θ'_{w^o} be the union of the hypersurfaces $\{\omega_i = 0\} \subset \widetilde{X}_d$ and let us set $\Theta_{w^o} = \Theta'_{w^o} \cup \Theta$.

We will more precisely associate to the data (B_0^o, B_∞, w^o) a Frobenius structure on $\widetilde{X}_d \setminus \Theta_{w^o}$.

The definition of Θ_{w^o} shows that the family $u = (u_1, \ldots, u_d)$ defined at $\widetilde{x}$ by

$$u_i = \omega_i(\widetilde{x})e_i \quad (i = 1, \ldots, d)$$

is a basis of $\mathscr{E}_{|\widetilde{X}_d \setminus \Theta_{w^o}}$. The homomorphism φ_ω can then be expressed as

$$\varphi_\omega : T(\widetilde{X}_d \setminus \Theta_{w^o}) \overset{\sim}{\longrightarrow} E_{|\widetilde{X}_d \setminus \Theta_{w^o}}$$
$$\partial_{x_i} \longmapsto -\Phi_{\partial_{x_i}}(\omega) = u_i.$$

It is thus an isomorphism. It satisfies

$$\varphi_\omega(e) = \omega \quad \text{and} \quad \varphi_\omega(\mathfrak{E}) = \mathfrak{E}_\omega,$$

if we set $e = \sum_i \partial_{x_i}$ and $\mathfrak{E} = \sum_i x_i \partial_{x_i}$.

Therefore, we can apply the results of §3 to deduce a Frobenius structure on $\widetilde{X}_d \setminus \Theta_{w^o}$.

Moreover, the product $\star$ of vector fields is given by

$$\varphi_\omega(\partial_{x_i} \star \partial_{x_j}) = -\Phi_{\partial_{x_i}}(\varphi_\omega(\partial_{x_j})),$$

a formula that we extend by $\mathscr{O}$-linearity to all vector fields; one can check that

$$\varphi_\omega(\partial_{x_i} \star \partial_{x_j}) = \varphi_\omega(\delta_{ij}\partial_{x_i}),$$

hence $\partial_{x_i} \star \partial_{x_j} = \delta_{ij}\partial_{x_i}$. We deduce:

4.1 Proposition. *The product* $\star$, *the unit* e *and the Euler field* $\mathfrak{E}$ *do not depend on the chosen eigenvector* ω^o *of* B_∞ *(with the nonvanishing assumption on all of its coefficients).* □

We deduce from §1.8 and from the results above:

4.2 Theorem (Dubrovin [Dub96]). *There is a one-to-one correspondence between semisimple simply connected Frobenius manifolds (i.e., for which R_0 is regular semisimple at any point) and the quadruples $(B_0^o, B_\infty, \omega^o, U)$, where B_0^o is a regular semisimple matrix, B_∞ satisfies $B_\infty^* + B_\infty = w\,\mathrm{Id}$ with $w \in \mathbb{Z}$, ω^o is an eigenvector of B_∞, no component of which on the eigenbasis of B_0^o vanishes, and U is a simply connected (étale) open set of $\widetilde{X}_d \smallsetminus \Theta_{\omega^o}$.* □

4.b Frobenius-Saito structures of type A_d

This is the simplest example of a Saito or Frobenius structure associated to a singularity. Here, the computations are simple enough so that we do not have to use the infinitesimal period mapping, but we will take up this example with the infinitesimal period mapping in §5.c. It is remarkable that the flat coordinates can be expressed algebraically as functions of the natural coordinates, in which the product $\star$ is easily expressed.

The universal unfolding of the singularity A_d. Let us denote by M the affine space $\mathbb{C}^d$ equipped with coordinates $z = (z_0, \dots, z_{d-1})$. If u is a new variable, let us consider the subset (hypersurface) $\mathscr{H}$ of $\mathbb{C} \times M$ having equation $f(u, z) = 0$, with

$$f(u, z) = u^{d+1} + z_{d-1}u^{d-1} + \cdots + z_1 u + z_0.$$

This polynomial describes the *universal unfolding* of the function $u \mapsto u^{d+1}$, also called "singularity A_d"[10].

A *singular point* of the hypersurface $\mathscr{H}$ is a point (u^o, z^o) of $\mathscr{H}$ where the polynomial

$$f'(u, z) := \frac{\partial f}{\partial u}(u, z) = (d+1)u^d + \sum_{i=1}^{d-1} i z_i u^{i-1}$$

vanishes. The singular set $\mathscr{S}(\mathscr{H})$ is thus described by the two equations

$$(4.3) \qquad \begin{cases} f(u, z) = 0 \\ f'(u, z) = 0. \end{cases}$$

---

[10] The curious reader can refer to [AGZV88].

Its image $\Delta \subset M$ (*discriminant* set of f) by the projection $p : \mathbb{C} \times M \to M$ which forgets the coordinate u is a hypersurface of M, the equation of which is obtained by eliminating u from both equations (4.3): it is the *resultant polynomial* of f and f', also called *discriminant*[11] of f. This set Δ is the set of points $z^o \in M$ such that $f(u, z^o)$ has at least one multiple root.

One can argue similarly for f', which is a polynomial of degree d in u, with leading coefficient equal to $d + 1$. Let us notice that f' only depends on the coordinates $z' = (z_1, \ldots, z_{d-1})$. Let thus M' be the affine space $\mathbb{C}^{d-1}$ equipped with these coordinates and let Δ' be the discriminant of f': this is the *bifurcation set* of the polynomial f. On any simply connected (étale) open set $U' \subset M' \smallsetminus \Delta'$, there exist d holomorphic functions $\alpha_1, \ldots, \alpha_d$ such that $f' = (d + 1) \prod_i (u - \alpha_i)$. Let us note that $\sum_i \alpha_i \equiv 0$ and that, for all i, j such that $i \neq j$ and any $z' \in U'$, we have $\alpha_i(z') \neq \alpha_j(z')$.

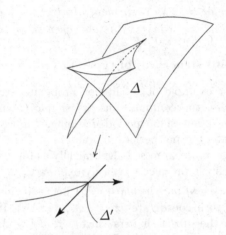

Fig. VII.1. The case $d = 3$.

Let us consider the projection $\pi : M \to M'$ which forgets the coordinate z_0. Sylvester's formula computing the discriminant shows that the equation of Δ has degree d in z_0 and that the coefficient of z_0^d is a constant. Consequently, the restriction of π to Δ is finite. Moreover, there exists a hypersurface D' of M' such that, for z' away from this hypersurface, the set $\pi^{-1}(z') \cap \Delta$ consists of d distinct points[12].

Let us consider now a simply connected (étale) open set U' contained in the complement of $D' \cup \Delta'$ and let us consider the roots $\alpha_1, \ldots, \alpha_d$ of f' on U'.

---

[11] One can find an explicit expression of the discriminant, for d small, in [Tei77] or [GKZ94, Chap. 12]. A computation with a computer, taking $d = 6$ or $d = 7$, could leave the reader puzzled... but reading [GKZ94, Chap. 12, §2] should reassure the reader.

[12] D' is the discriminant of Δ with respect to the variable z_0.

We will assume that z' stays in this open set. Then, a point $(\alpha_i(z'), z_0, z')$ is a (singular) point of $\mathscr{H}$ if and only if $f(\alpha_i(z'), z_0, z') = 0$, that is, if and only if

$$z_0 = -\left(\alpha_i(z')^{d+1} + \sum_{j=1}^{d-1} z_j \alpha_i(z')^j\right).$$

In other words, the fibre $\pi^{-1}(z') \cap \Delta$ consists of d points (assumed to be distinct)

$$z_0^{(i)}(z') := -\left(\alpha_i(z')^{d+1} + \sum_{j=1}^{d-1} z_j \alpha_i(z')^j\right) \qquad (i = 1, \ldots, d).$$

One can easily compute that, for $j = 1, \ldots, d-1$,

$$\frac{\partial z_0^{(i)}}{\partial z_j} = -\alpha_i^j - \frac{\partial \alpha_i}{\partial z_j} \cdot f'(\alpha_i(z'), z') = -\alpha_i^j.$$

Then the family of functions $x_i(z) = z_0 - z_0^{(i)}(z')$ $(i = 1, \ldots, d)$ is a coordinate system on $\pi^{-1}(U')$, as the Jacobian $(\partial x_i / \partial z_j)$ is nothing but the Vandermonde determinant on $\alpha_1, \ldots, \alpha_d$.

By definition, we have $x_i(z) = f(\alpha_i(z'), z)$. In other words, for z^o fixed in U, $x_i(z^o)$ is the *critical value* of the polynomial $f(u, z^o)$ associated to the critical point $\alpha_i(z'^o)$.

The manifold M is equipped with a free sheaf of rank $d = \dim M$: it is the sheaf of Jacobian algebras $\mathscr{O}_M[u]/f'\mathscr{O}_M[u]$, a basis of which consists of the classes modulo f' of $1, u, \ldots, u^{d-1}$. For $z^o \in M$ fixed, the fibre of this sheaf is the quotient $\mathbb{C}[u]/f'(u, z^o)\mathbb{C}[u]$. As f' only depends on z', this bundle is the pullback by π of the bundle $\mathscr{O}_{M'}[u]/f'\mathscr{O}_{M'}[u]$. Moreover, for $z'^o \in U'$, the Jacobian algebra $\mathbb{C}[u]/f'(u, z^o)\mathbb{C}[u]$ is semisimple: as a matter of fact, let us consider the polynomials $e_i(u)$, with holomorphic coefficients in $z'^o \in U'$, defined by

$$e_i(u) = \frac{f'(u)}{f''(\alpha_i)(u - \alpha_i)};$$

they clearly satisfy the equalities $e_i(\alpha_j) = \delta_{ij}$ and induce a basis of the bundle $\mathbb{C}[u]/f'(u, z^o)\mathbb{C}[u]$, as any polynomial $r(u)$ can be written in a unique way as

$$r(u) = \sum_i r(\alpha_i(z'^o))e_i(u) \mod f';$$

lastly, we have $e_i \cdot e_j = \delta_{ij}e_i \mod f'$.

The product $\star$ on Θ_M and the Euler field. The *Kodaira-Spencer map*

$$\varphi : \Theta_M \longrightarrow \mathscr{O}_M[u]/(f')$$

is the $\mathscr{O}_M$-linear morphism which associates to the vector field ∂_{z_i} the class of $u^i = \partial f/\partial z_i$ modulo f' and, more generally, which associates to the field ξ the class of $\mathscr{L}_\xi(f) = \xi(f)$.

The product $\star$ on Θ_M is obtained by carrying through φ the natural product on $\mathscr{O}_M[u]/(f')$: hence, by definition,

$$\xi \star \eta = \varphi^{-1}\big(\xi(f)\eta(f) \bmod f'\big)$$

and, for instance, $\partial_{z_i} \star \partial_{z_j} = \varphi^{-1}(u^{i+j} \bmod f')$. In particular, this product is associative, commutative, with unit ($e = \partial_{z_0}$). We have $\partial_{z_i} \star \partial_{z_j} = \partial_{z_{i+j}}$ if $i + j \leqslant d - 1$. One can also check that $\partial_{z_i} \star \partial_{z_j}$ only depends on $i + j$ and that its coefficient on ∂_{z_k} is a polynomial in z.

The *Euler field* $\mathfrak{E}$ is the pullback of the class of f by φ.

4.4 Exercise.

(1) Compute $\partial_{z_i} \star \partial_{z_j}$ for $i + j = d$.

(2) Prove that $\mathfrak{E} = \sum_{i=0}^{d-1} \rho_i z_i \partial_{z_i}$, with $\rho_i = (d + 1 - i)/(d + 1)$.

Canonical coordinates. We will check that $(x_1, \ldots, x_d)$ form a system of canonical coordinates on the open set $U := \pi^{-1}(U')$ considered above. In order to do that, let us denote by ξ_i the vector field $\varphi^{-1}(e_i)$ on U; one can check that $\xi_i \star \xi_j = \delta_{ij}$.

To show that $\xi_i = \partial_{x_i}$, it is enough to prove that

$$\frac{\partial f}{\partial x_i}(\alpha_j) = \delta_{ij},$$

which follows, on the one hand, from

$$\frac{\partial}{\partial z_j} = \sum_i \alpha_i^j \frac{\partial}{\partial x_i}$$

and, on the other hand, from $\partial f/\partial z_j(\alpha_k) = \alpha_k^j$.

The metric. At the point $z^o \in M$, let us define the bilinear form g by

(4.5) $$g(\xi, \eta) = \frac{1}{2i\pi} \int_{\Gamma_{z^o}} \frac{\xi(f)(u, z^o) \cdot \eta(f)(u, z^o)}{f'(u, z^o)}\, du,$$

where Γ_{z^o} is a circle, positively oriented, bounding a disc containing all the roots of $f'(u, z^o)$. This expression only depends on $\xi(f)$ and $\eta(f)$ modulo f', in other words only depends on $\varphi(\xi)$ and $\varphi(\eta)$. If we regard $\xi(f)(u, z^o) \cdot \eta(f)(u, z^o)\, du/f'(u, z^o)$ as a meromorphic differential form on $\mathbb{P}^1$, the residue theorem shows that

(4.6) $$g(\xi, \eta) = -\operatorname{res}_{u=\infty} \frac{\xi(f)(u, z^o) \cdot \eta(f)(u, z^o)}{f'(u, z^o)}\, du,$$

an expression where the dependence in z^o only appears in the 1-form. Making use of the coordinate v at infinity $(v = 1/u)$, we get in particular

$$g(\partial_{z_i}, \partial_{z_j}) = \operatorname{res}_{v=0} \frac{v^{d-2-i-j}}{((d+1) + (d-1)z_{d-1}v^2 + \cdots + z_1 v^d)} \, dv.$$

Then, on the one hand, $g(\partial_{z_i}, \partial_{z_j}) = 0$ if $i + j < d - 1$ and, on the other hand, for $i + j \geqslant d - 1$, $g(\partial_{z_i}, \partial_{z_j})$ is the coefficient of $v^{i+j+1-d}$ in the formal series $\frac{1}{d+1}\left(1 + \frac{1}{d+1} \sum_{i=1}^{d-1} i z_i v^{(d+1-i)}\right)^{-1}$.

4.7 Lemma (Homogeneity). *The Euler field satisfies*

$$\mathscr{L}_{\mathfrak{E}}(\star) = \star \quad \text{and} \quad \mathscr{L}_{\mathfrak{E}}(g) = \frac{d+3}{d+1} g.$$

Proof. Let us give the weight $\rho_i = (d+1-i)/(d+1)$ to the variable z_i. We will say that a polynomial $h(z_0, \ldots, z_{d-1})$ is *quasi-homogeneous* of degree δ if all its monomials have degree δ when one computes the degree of the variables z_i with the weight ρ_i; in other words, $\mathscr{L}_{\mathfrak{E}}(h) = \delta h$. We thus have $\mathscr{L}_{\mathfrak{E}}(z_i) = \rho_i z_i$. The polynomial f is homogeneous of degree 1 if one gives the variable u the weight $1/(d+1)$. The expression of $g(\partial_{z_i}, \partial_{z_j})$ mentioned above shows that it is a quasi-homogeneous polynomial of degree $(i+j+1-d)/(d+1)$. As $\mathscr{L}_{\mathfrak{E}}(\partial_{z_i}) = -\rho_i \partial_{z_i}$, we deduce that, for all i, j, we have (setting $D = (d+3)/(d+1)$)

$$\mathscr{L}_{\mathfrak{E}}(g(\partial_{z_i}, \partial_{z_j})) - g(\mathscr{L}_{\mathfrak{E}}\partial_{z_i}, \partial_{z_j}) - g(\partial_{z_i}, \mathscr{L}_{\mathfrak{E}}\partial_{z_j}) = Dg(\partial_{z_i}, \partial_{z_j}).$$

The equality $\mathscr{L}_{\mathfrak{E}}(\partial_{z_i} \star \partial_{z_j}) = (1 - \rho_i - \rho_j)\partial_{z_i} \star \partial_{z_j}$ can be shown in the same way, by induction on $i + j$, whence the first assertion. $\qquad\square$

4.8 Proposition (Flatness of the metric). *The bilinear form g is nondegenerate and flat.*

Proof. In order to prove the proposition, we will exhibit a system of flat coordinates, which was introduced in [SYS80, Th. 2.5.3]. These coordinates will appear as the coefficients of a Puiseux expansion of f when $u \to \infty$. Let us then set $h(v) = 1/f(1/v)$, that we regard as a section of the sheaf $\mathscr{O}_M[\![v]\!][v^{-1}]$ of Laurent formal series (with poles) in v with holomorphic coefficients on M:

$$h(v, \boldsymbol{z}) = v^{d+1}\left(1 + z_{d-1}v^2 + \cdots + z_0 v^{d+1}\right)^{-1}.$$

There exists a formal series $w(v, \boldsymbol{z}) = v\left(1 + a_2(\boldsymbol{z})v^2 + \cdots\right)$ such that $w(v, \boldsymbol{z})^{d+1} \equiv h(v, \boldsymbol{z})$ and $(w, \boldsymbol{z})$ is another coordinate system on $\mathbb{C} \times M$ in the neighbourhood of $v = 0$. Let us write

$$v = v(w, \boldsymbol{z}) = w\left(1 + b_2(\boldsymbol{z})w^2 + \cdots\right).$$

The function $u = 1/v$ takes the form

$$u(w, \boldsymbol{z}) = w^{-1}\left(1 - \frac{1}{d+1}\left[t_{d-1}(\boldsymbol{z})w^2 + \cdots + t_0(\boldsymbol{z})w^{d+1}\right] + \cdots\right),$$

which defines d functions $t_0(z), \ldots, t_{d-1}(z)$. Recall that, by construction, we have the relation

$$w^{d+1} f(u(w, z), z) = 1.$$

By expanding this relation and annihilating the coefficients of positive powers of w, one finds[13]

$$t_{d-1} - z_{d-1} = 0,$$
$$t_{d-2} - z_{d-2} = 0,$$

and, more generally, for $i \leqslant d - 3$,

$$t_i - z_i = q_i(t_{i+2}, \ldots, t_{d-1}, z_{i+2}, \ldots, z_{d-1}),$$

where q_i is a polynomial with constant coefficients. These formulas define two algebraic changes of coordinates inverse one to the other: $t \mapsto z$ and $z \mapsto t$. We will check that t is a *flat coordinate system* on M, with respect to the bilinear form g on Θ_M. Let us notice first that, given the triangular form of the coordinate change, we have

$$\frac{\partial}{\partial t_0} = \frac{\partial}{\partial z_0} = e.$$

Let us set $q(u, t) = f(u, z(t))$. We thus also have $q'(u, t) = f'(u, z(t))$. We wish to compute $\partial q / \partial t_i \mod q'$. We will do this in the coordinate w (and thus in the ring $\mathscr{O}_M[\![w]\!][w^{-1}]$). According to the relation

$$w^{d+1} \cdot q(u(w, t), t) \equiv 1,$$

we get

$$\frac{\partial q}{\partial t_i}(u(w, t), t) + \frac{\partial u(w, t)}{\partial t_i} \cdot q'(u(w, t), t) \equiv 0.$$

On the other hand, the expression of the series $u(w, t)$ shows that

$$\frac{\partial u(w, t)}{\partial t_i} = -\frac{1}{d+1} \left(w^{d-i} + w^{d+1} r_i(w, t) \right).$$

We can thus write, as $d - i$ and $d - j$ are $\geqslant 1$,

$$\frac{\partial q}{\partial t_i} \cdot \frac{\partial q}{\partial t_j} = \frac{q'(u(w, t), t)^2}{(d+1)^2} \left(w^{2d-(i+j)} + r_{ij}(w, t) w^{d+2} \right)$$

$$= \frac{q'(u(w, t), t)^2}{(d+1)^2} \left(v^{2d-(i+j)} + s_{ij}(v, t) v^{d+2} \right).$$

---

[13] The brave reader will do the computation.

Let us now compute the residue defining g in the variables (v, t). We find

$$g(\partial_{t_i}, \partial_{t_j}) := \mathrm{res}_{v=0} \frac{\partial q/\partial t_i \cdot \partial q/\partial t_j}{q'(u(w,t),t)} v^{-2} dv$$

$$= \mathrm{res}_{v=0} \frac{1}{d+1} \left(v^{d-2-(i+j)} + s_{ij}(v,t) \right) (1 + \cdots) dv$$

$$= \begin{cases} 0 & \text{if } i+j \neq d-1 \\ \dfrac{1}{d+1} & \text{if } i+j = d-1. \end{cases}$$

This expression also shows that the bilinear form g is everywhere nondegenerate. □

The Frobenius-Saito structure (Dubrovin, [Dub96]). We will now show that the flat metric g, the product $*$ on Θ_M and the fields $e = \partial_{z_0} = \partial_{t_0}$ and $\mathfrak{E}$ induce a Frobenius structure on the space $M = \mathbb{C}^d$. At this stage, we have checked Properties (1), (3)(b) and (3)(c) of the definition of §2.a.

Let us begin with (3)(a). We notice that, in the coordinate change $t_i = z_i + Q_i(z_{i+2}, \ldots, z_{d-1})$, the polynomial Q_i is quasi-homogeneous of degree ρ_i, so that we have, for any i,

$$\mathscr{L}_{\mathfrak{E}}(t_i) = \rho_i t_i,$$

hence $\mathfrak{E} = \sum_i \rho_i t_i \partial_{t_i}$. Therefore, in the basis given by the flat coordinates, $\nabla \mathfrak{E}$ is the constant diagonal matrix having the numbers ρ_i as diagonal entries, which implies that $\nabla(\nabla \mathfrak{E}) = 0$.

Checking 2.a-(2) is done in a roundabout way. Indeed, a computation analogous to that done for the metric shows that

$$c(\partial_{t_i}, \partial_{t_j}, \partial_{t_k}) = \mathrm{res}_{v=0} \frac{q'^2}{(d+1)^3} \left(v^{d-i} + r_i v^{d+1} \right) \left(v^{d-j} + r_j v^{d+1} \right)$$

$$\cdot \left(v^{d-k} + r_k v^{d+1} \right) \frac{dv}{v^2}$$

$$= \frac{1}{d+1} \mathrm{res}_{v=0} \left(v^{d-2-(i+j+k)} + r_i v^{d-1-(j+k)} + r_j v^{d-1-(i+k)} \right.$$

$$\left. + r_k v^{d-1-(j+i)} \right) \cdot (1 + \cdots) dv.$$

When i, j, k satisfy $i+j, j+k, k+i \leqslant d-1$, this computation gives

$$c(\partial_{t_i}, \partial_{t_j}, \partial_{t_k}) = \begin{cases} 1/(d+1) & \text{if } i+j+k = d-1 \\ 0 & \text{otherwise.} \end{cases}$$

On the other hand, when this condition is not fulfilled, this computation is not sufficient to conclude.

The idea of B. Dubrovin consists in using the canonical coordinates $x_1, \ldots, x_d$. As a matter of fact, it is enough to show the symmetry of ∇c on a dense open set of M, so that we can assume the existence of a system of canonical coordinates, according to what we have seen above.

4.9 Lemma (Darboux-Egorov condition, after Dubrovin). *Let us be given $(g, \star, e, \mathfrak{E})$ on Θ_M. If there exists a system of canonical coordinates $(x_1, \ldots, x_d)$ for the product $\star$ and if $(g, \star, e, \mathfrak{E})$ satisfy Conditions (1) and (3) of §2.a as well as the condition*

(2') there locally exists a function ϕ such that the metric g satisfies

$$g(\partial_{x_i}, \partial_{x_j}) = \delta_{ij} \frac{\partial \phi}{\partial x_i},$$

then $(g, \star, e, \mathfrak{E})$ defines a Frobenius structure on M.

This lemma applies to the unfolding of the singularity A_d. Indeed, we have $\varphi(\partial_{x_i}) = e_i(u)$ and the expression (4.5) for the metric shows that

$$g(\partial_{x_i}, \partial_{x_j}) = \begin{cases} 0 & \text{if } i \neq j \\ 1/f''(\alpha_i) & \text{if } i = j. \end{cases}$$

We will check that the function $\phi(\boldsymbol{x}) = z_{d-1}(\boldsymbol{x})/(d+1)$ satisfies the relation

$$(4.10) \qquad \frac{\partial \phi}{\partial x_i} = \frac{1}{f''(\alpha_i)}.$$

Recall, on the one hand, that the coordinate change $\boldsymbol{z} \mapsto \boldsymbol{x}$ gives the relations

$$(4.11) \qquad \frac{\partial}{\partial z_j} = \sum_{i=1}^{d} \alpha_i^j \frac{\partial}{\partial x_i} \qquad (j = 0, \ldots, d-1).$$

On the other hand, the coordinate change $\boldsymbol{z} \mapsto \boldsymbol{t}$ being triangular and the fields ∂_{t_j} being g-orthogonal, we deduce that, for any $j = 0, \ldots, d-1$,

$$g(\partial_{z_j}, \partial_{z_0}) = g(\partial_{t_j}, \partial_{t_0}) = \begin{cases} 0 & \text{if } j \neq d-1 \\ 1/(d+1) & \text{if } j = d-1. \end{cases}$$

Due to the g-orthogonality of the fields ∂_{x_i} and of (4.11), we deduce

$$(4.12) \qquad \sum_{i=1}^{d} \frac{\alpha_i^j}{f''(\alpha_i)} = \begin{cases} 0 & \text{if } j \neq d-1 \\ 1/(d+1) & \text{if } j = d-1, \end{cases}$$

which shows (4.10). $\qquad \qquad \square$

4.13 Exercise. Using (4.11), prove, for any $i = 1, \ldots, d$, the equality of polynomials

$$\sum_{k=0}^{d-1} \frac{\partial z_k}{\partial x_i} u^k = \prod_{j \neq i} \frac{u - \alpha_j}{\alpha_i - \alpha_j}$$

and deduce a proof of (4.10) not using flat coordinates.

Proof (of Lemma 4.9). The symmetry property (1.19) of ∇c is shown by using the canonical coordinates $(x_1, \ldots, x_d)$. It easily follows from the two relations

(4.14) $$g(\nabla_{\partial_{x_k}} \partial_{x_i}, \partial_{x_i}) = \frac{1}{2} \frac{\partial^2 \phi}{\partial x_i \partial x_k},$$

(4.15) $$g(\nabla_{\partial_{x_k}} \partial_{x_i}, \partial_{x_j}) = 0 \quad \text{if } i \neq j,\, j \neq k \text{ and } i \neq k.$$

The first relation is a consequence of the horizontality of g with respect to ∇ and of the vanishing of its torsion. For the second one, one moreover uses the pairwise g-orthogonality of the fields ∂_{x_i} (cf. §1.16): we have

$$g(\nabla_{\partial_{x_k}} \partial_{x_i}, \partial_{x_j}) = g(\nabla_{\partial_{x_i}} \partial_{x_k}, \partial_{x_j}) = -g(\partial_{x_k}, \nabla_{\partial_{x_i}} \partial_{x_j}) = -g(\nabla_{\partial_{x_i}} \partial_{x_j}, \partial_{x_k})$$

and, by iterating three times the cyclic permutation $i \mapsto j \mapsto k \mapsto i$, we get (4.15). $\qquad\square$

The potential F. The homogeneity property seen in §2.6 shows that, as the ρ_i are positive, the function F is a quasi-homogeneous *polynomial* of degree $D = (d+3)/(d+1)$.

4.16 Exercise (The potential of the singularities A_2 and A_3). Give details for the previous computations in the case of A_2 and A_3 singularities, and prove that the potential F is given by

$$F_{A_2}(t_0, t_1) = \frac{1}{2 \cdot 3} t_0^2 t_1 - \frac{1}{2^3 \cdot 3^3} t_1^4,$$

$$F_{A_3}(t_0, t_1, t_2) = \frac{1}{2} \left(t_0^2 t_2 + t_0 t_1^2 \right) - \frac{1}{2^6} t_1^2 t_2^2 + \frac{1}{15 \cdot 2^8} t_2^5.$$

4.17 Remark. It is interesting to compare the Frobenius structure constructed as above to the universal Frobenius structure of §4.a. Given a point z^o in the open set U considered above, let us denote by x^o its canonical coordinates (assumed to be pairwise distinct) and by t^o its flat coordinates. The weight w (cf. §1.21) is here the number of variables u, that is, 1, and the corresponding number q $(D = 2q + 2 - w)$ is thus equal to $1/(d+1)$. The matrix of $B_\infty = (q+1)\,\mathrm{Id} - \nabla\mathfrak{E}$ in the basis ∂_{t_i} is equal to $\mathrm{diag}((i+1)/(d+1)_{i=0,\ldots,d-1})$ and an explicit formula for the coordinate change $t \mapsto x$ would enable us to express this matrix in the basis of ∂_{x_i}. The primitive section ω is the unit field $\partial_{z_0} = \sum_i \partial_{x_i}$.

Therefore, in this example, the Frobenius structure of A_d restricted to U coincides with the structure constructed in §4.a if we choose as initial data at z^o the data above. One should notice that Theorem 4.2 would have only given the existence of such a structure on an open set of the universal covering space of U and only in an implicit form.

4.18 Exercise (Frobenius structure for the universal unfolding of the Laurent polynomial $u + 1/u$). We consider the unfolding $f(u, z_0, z_1) =$

$z_0 + z_1 u + 1/u$ of the Laurent polynomial $u + 1/u$ of the variable $u \in \mathbb{C}^*$. The manifold M is here the open set $z_1 \neq 0$ in the complex plane with coordinates z_0, z_1. The sheaf of Jacobian algebras is $\mathscr{O}_M[u, u^{-1}]/(f')$ and the Kodaira-Spencer mapping φ sends ∂_{z_0} to the class of 1 and ∂_{z_1} to that of u modulo f'. The metric g is defined by

$$g(\xi, \eta) = z_1 \sum_{\alpha_i | f'(u, \alpha_i) = 0} \mathrm{res}_{u = \alpha_i} \left(\frac{\varphi(\xi)\varphi(\eta)}{f'} \frac{du}{u} \right)$$

$$= -z_1 \left[\mathrm{res}_{u=0} + \mathrm{res}_{u=\infty} \right] \left(\frac{\varphi(\xi)\varphi(\eta)}{f'} \frac{du}{u} \right).$$

Prove the following results and deduce a Frobenius structure on the manifold M (more precisely on the double covering on which $\sqrt{z_1}$ is defined):

(1) The Euler field $\mathfrak{E}$ is equal to $z_0 \partial_{z_0} + 2 z_1 \partial_{z_1}$ and $D = 2$ (hence $q = 1/2$ if we choose $w = 1$). Compute the products $\partial_{z_i} \star \partial_{z_j}$.

(2) The canonical coordinates are given by

$$x_1 = z_0 + 2\sqrt{z_1}, \qquad x_2 = z_0 - 2\sqrt{z_1}$$

and the potential ϕ by $\phi(x_1, x_2) = z_0$.

(3) The flat coordinates are given by

$$t_0 = z_0, \qquad t_1 = 2\sqrt{z_1}$$

and $(\partial_{t_0}, \partial_{t_1})$ is a g-orthonormal basis of Θ_M.

(4) The potential F of the Frobenius structure is

$$F(t_0, t_1) = \frac{1}{3}t_0^3 + \frac{1}{2}t_0 t_1^2.$$

Lastly, identify this structure with the universal structure of §4.a with $B_0^o = \mathrm{diag}(1, -1)$ and $B_\infty = \mathrm{Id}/2$.

4.19 Exercise (Frobenius structure for the universal unfolding of the Laurent polynomial $u^2 + 1/u$). Same exercise as the previous one, with $f(u, z_0, z_1, z_2) = z_0 + z_1 u + z_2 u^2 + 1/u$ and $M = \mathbb{C}^3 \setminus \{z_2 = 0\}$. The Kodaira-Spencer map is given by $\varphi(\partial_{z_i}) = u^i$ $(i = 0, 1, 2)$. The metric is here defined as being $2\sqrt{z_2} \sum$ residues.

(1) Prove that $\mathfrak{E} = z_0 \partial_{z_0} + 2 z_1 \partial_{z_1} + 3 z_2 \partial_{z_2}$ and $D = 3/2$ (hence $q = 1/4$ if we choose $w = 1$). Compute the products $\partial_{z_i} \star \partial_{z_j}$.

(2) Determine the domain of existence of canonical coordinates.

(3) Show that the formulas

$$t_0 = z_0 - \frac{z_1^2}{8z_2}, \qquad t_1 = \frac{z_1}{\sqrt{z_2}}, \qquad t_2 = 2\sqrt{2}\, z_2^{1/4}$$

define (étale) flat coordinates M.

(4) Compute the potential F of this structure.

(5) Identify this structure with the universal structure of §4.a with $B_0^o = 3 \cdot 2^{-2/3} \cdot \mathrm{diag}(1, j, j^2)$ $(j^3 = 1)$ and B_∞ having eigenvalues $1/4$, $1/2$, $3/4$.

4.c Frobenius structures defined by their potential

It is a matter of exhibiting the "functions" F of d complex variables $t_1, \ldots, t_d$ satisfying the WDVV equations (2.8). By construction, the coordinates t_i are the flat coordinates on the manifold; in other words, we wish to equip an open set of $\mathbb{C}^d$, given its natural affine structure, with a structure of Frobenius manifold. A way to produce such potentials consists in expressing the WDVV relations as inductive relations on the coefficients of the Taylor expansion of F.

4.20 Exercise (Polynomial Frobenius structures in dimension 3, after S. Natanzon). We are given $\delta_1 = 1$, $\delta_2, \delta_3 \neq 0$ with $2\delta_2 = \delta_3 + 1 := D$ and we look, in flat coordinates t_1, t_2, t_3 as in §1.14, for homogeneous *polynomial* potentials F of degree $D + 1$ (up to a polynomial of degree 2) with respect to the Euler field $\mathfrak{E} = t_1 \partial_{t_1} + \delta_2 t_2 \partial_{t_2} + \delta_3 t_3 \partial_{t_3}$, satisfying the WDVV equations of Exercise 2.9. We write, as in (2.4),

$$F(t_1, t_2, t_3) = \frac{1}{2} t_1 (t_1 t_3 + t_2^2) + G(t_2, t_3).$$

(1) Prove that the only WDVV constraint not trivially satisfied is

$$\sum_{m=1}^{3} \frac{\partial^3 F}{\partial t_2 \partial t_2 \partial t_m} \cdot \frac{\partial^3 F}{\partial t_{4-m} \partial t_3 \partial t_3} = \sum_{m=1}^{3} \frac{\partial^3 F}{\partial t_2 \partial t_3 \partial t_m} \cdot \frac{\partial^3 F}{\partial t_{4-m} \partial t_2 \partial t_3},$$

which can also be written as $(G_{223})^2 = G_{333} + G_{222} G_{233}$, if we put $G_{ijk} = \partial^3 G / \partial t_i \partial t_j \partial t_k$.

(2) Express the homogeneity and polynomiality constraint on G.

(3) Deduce three types of solutions G, hence F (among which is the potential F_{A_3} of the singularity A_3).

The polynomial solutions are nevertheless quite rare. Once one has solved these inductive equations, one is left with the problem, difficult in general, of determining the domain of convergence of the Taylor series thus obtained. This is why one introduces the notion of *formal Frobenius manifold*, defined by a potential F which is a formal series in $t_1, \ldots, t_d$. The coefficients of the series, solutions of the inductive equations deduced from WDVV, are then the important objects. Here is an example:

4.21 Proposition (Potential of quantum cohomology of the projective plane). *There exists a unique sequence N_d of integers $(d \geqslant 1, N_1 = 1)$ such that the formal series*

$$G(t_2, t_3) = \sum_{d=1}^{\infty} N_d \frac{t_3^{3d-1}}{(3d-1)!} e^{dt_2}$$

is a solution of the equation $(G_{223})^2 = G_{333} + G_{222} G_{233}$. The formal potential $F(t_1, t_2, t_3) = G(t_2, t_3) + t_1 (t_1 t_3 + t_2^2)/2$ is homogeneous of degree $D + 1 = 1$ with respect to the Euler field $\mathfrak{E} = t_1 \partial_{t_1} - t_3 \partial_{t_3} + 3 \partial_{t_2}$.

Proof (Indication). One shows that G satisfies the WDVV equation above if and only if the sequence N_d satisfies the inductive equations, for $d \geqslant 2$,

$$N_d = \sum_{k+\ell=d} N_k N_\ell \, k^2 \ell \left[\ell \binom{3d-4}{3k-2} - k \binom{3d-4}{3k-1} \right]. \qquad \square$$

The remarkable fact, due to M. Kontsevitch, is that the coefficients N_d possess a nice interpretation in *enumerative geometry*: the number N_d is that of rational curves in $\mathbb{P}^2$ passing by $3d - 1$ points in a general position. That is how the Frobenius structures arises in *quantum cohomology*[14].

5 Frobenius-Saito structure associated to a singularity of function

Universal unfoldings are the source of an important family of examples of Frobenius-Saito structures. We have already analyzed with details the Frobenius-Saito structure produced by the singularity A_d (cf. §4.b). We will indicate how K. Saito has generalized such a construction to a large family of functions (cf. [Sai83b, Oda87]). We will take up the unfolding of the singularity A_d with this method in §5.c. The general case cannot be completely treated within the framework of this book, but we will indicate the main concepts, which will serve as a guide in §5.c.

5.a General sketch

One can distinguish between a local and a global setting.

- In the local setting, we start with a germ $f_o : (\mathbb{C}^{n+1}, 0) \to (\mathbb{C}, 0)$ of holomorphic function which has an *isolated critical point* at 0, that is, for some sufficiently small representative of this germ, the partial derivatives $\partial f_o / \partial u_i$ do not vanish simultaneously away from the origin.
- In the global setting, we start with a regular function f_o on a nonsingular affine variety U of dimension $n+1$ (cf. §0.10, for instance, f_o is a polynomial in the variables $u_0, \ldots, u_n$ or a Laurent polynomial in these variables) which has only isolated critical points and which "has no critical point at infinity[15]".

There are a few steps to construct such a Frobenius-Saito structure. Let us sketch them.

---

[14] Regarding this we refer to the articles and book mentioned in the introduction of this chapter. One should also note that S. Barannikov and M. Kontsevitch recently have introduced a new construction of formal Frobenius structures, giving a way to the *mirror symmetry* phenomenon in terms of such structures (cf. [BK98, Bar99, Man99b]).

[15] There exists a precise definition for this notion; see for instance [Sab06].

(1) One associates to the function f_o a meromorphic bundle with connection $(\mathbb{G}^o, \nabla)$ of rank d on the complex plane $\mathbb{C}$ (variable τ of the §V.2.c), called the *Gauss-Manin system* of f_o. One show that it has a singularity at $\tau = 0$ and $\tau = \infty$ only, the latter being regular. It comes moreover equipped with a natural lattice $\widehat{E}^o$, called the *Brieskorn lattice*[16] and which is the Fourier transform of the system of Picard-Fuchs equations classically associated to f_o (variable t of §V.2.c). The lattice $\widehat{E}^o$ is equipped with a nondegenerate Hermitian form of weight $w = n + 1$ (number of variables of the function f_o), related to the Poincaré duality on the nonsingular fibres of the function f_o.

This construction has its origin in the search for an asymptotic expansion, when τ tends to 0, for integrals like

$$I(\tau) = \int_\Gamma e^{-f_o(\boldsymbol{u})/\tau} \omega,$$

where ω is a holomorphic differential form of degree maximum and Γ is a cycle of dimension $\dim U$ (for instance, if $U = \mathbb{C}^{n+1}$, $\Gamma = \mathbb{R}^{n+1}$). This aspect is widely explained in the book [AGZV88]. We will not insist on it here.

(2) When there exist coordinates (in the étale sense) $u_0, \ldots, u_n$ on the manifold U (such is the case in the local setting, of course, and sometimes in the global setting, for instance if $U = \mathbb{C}^{n+1}$ or $U = (\mathbb{C}^*)^{n+1}$), one can consider the *Jacobian quotient* of the ring of functions on U by the ideal of partial derivatives of f_o. In various situations, one can identify in a natural way this Jacobian quotient with the quotient space $\widehat{E}^o/\tau\widehat{E}^o$ and choose functions $1, \lambda_1(\boldsymbol{u}), \ldots, \lambda_{d-1}(\boldsymbol{u})$ on U which induce, through this identification, a basis of the space $\widehat{E}^o/\tau\widehat{E}^o$ (for instance, one can choose the eigenvectors of the residue at infinity considered above; when f_o is the polynomial $u \mapsto u^{d+1}$, one recovers the family $1, u, \ldots, u^{d-1}$).

(3) One can then consider, on the product $U \times \mathbb{C}^d$ or on its germ at the origin, the function

$$f(\boldsymbol{u}, \boldsymbol{z}) = f_o(\boldsymbol{u}) + \sum_{i=0}^{d-1} z_i \lambda_i(\boldsymbol{u}),$$

that we will call "universal unfolding of f_o". The parameter space M will be a neighbourhood of the origin in $\mathbb{C}^d$ (coordinates $\boldsymbol{z}$). By assumption the Kodaira-Spencer mapping $\varphi : \Theta_M \to \mathscr{O}_{U \times M}/(\partial f/\partial \boldsymbol{u})$ is an isomorphism of locally free $\mathscr{O}_M$-modules in the neighbourhood of $0 \in M$. The natural product on the sheaf $\mathscr{O}_{U \times M}/(\partial f/\partial u_0, \ldots, \partial f/\partial u_n)$ of Jacobian algebras is carried as a product $\star$ on Θ_M. One can show that this product admits a system of canonical coordinates on some dense open set of M. The class of f in the Jacobian quotient defines, *via* φ, the Euler field $\mathfrak{E}$.

---

[16] The original construction by Brieskorn can be found in [Bri70].

(4) In order to construct the flat torsion free connection and the metric on M, one uses an infinitesimal period mapping. One first shows that the construction of the Gauss-Manin system $\mathbb{G}$ and of the Brieskorn lattice $\widehat{\mathbb{E}}$ can be done in a family, as well as that of the nondegenerate Hermitian form. *Hodge theory* provides a filtration of $\mathbb{G}^o$, which is opposite to the filtration induced by $\widehat{\mathbb{E}}^o$ and compatible with the Hermitian form, in such a way that, by the criterion of M. Saito of §IV.5.b, one gets a solution to Birkhoff's problem for $\widehat{\mathbb{E}}^o$, a solution also equipped with a nondegenerate Hermitian form. One moreover shows that the residue at infinity of the connection on the (trivial) bundle on $\mathbb{P}^1$ so constructed is *semisimple*. Its eigenvalues are nonpositive rational numbers[17]: they form the *spectrum of the singularity*[18] (local setting) or the *spectrum at infinity* of the function f_o (global setting). The corresponding characteristic polynomial is the characteristic polynomial at infinity of the Brieskorn lattice, in the sense of §III.2.b.

According to Theorem VI.2.1 and to Proposition VI.2.7 one can extend such a solution to Birkhoff's problem for $\widehat{\mathbb{E}}^o$ in a solution to Birkhoff's problem for $\widehat{\mathbb{E}}$. The bundle $F = \widehat{\mathbb{E}}/\tau\widehat{\mathbb{E}}$ on M comes therefore equipped with data $(\triangledown, \Phi, R_0, R_\infty)$ as in §VII.3.a.

It remains then to exhibit a primitive section in order to obtain a Frobenius-Saito structure on M with the associated infinitesimal period mapping, as in §VII.3. In the local setting, the existence of such a primitive section has been proved by M. Saito [Sai91]. In the global setting, the existence of such a primitive section is not known in general, but in various families of examples (cf. [Sab98, Sab06]). The reader will find more details in [Her02] and [DS03, DS04, Dou05].

As the reader will notice, the construction of such a structure involves most of the objects and techniques introduced in the previous chapters. It also requires other techniques of analysis and geometry, Hodge theory for instance.

Moreover, it is not in general possible to give an explicit description of this structure as for the singularity A_d. Let us mention however that some other singularities also give rise to explicit computations.

5.b The de Rham complex twisted by $e^{-\tau' f}$

We will indicate a general procedure to construct the Gauss-Manin system of a function. Let then U be a nonsingular affine variety (cf. §0.10) and let f be a regular function on U, depending on holomorphic parameters in some analytic manifold X. In other words, f is a section of the sheaf $\mathscr{O}_X[U] := \mathscr{O}_X \otimes_{\mathbb{C}} \mathscr{O}(U)$.

---

[17] These numbers are even negative in the local setting or if $U = \mathbb{C}^{n+1}$.

[18] For a germ of holomorphic function, the *spectrum* of a critical point has been introduced by A.N. Varchenko [Var82].

Let us denote similarly by $\Omega_X^k[U] = \mathcal{O}_X \otimes_{\mathbb{C}} \Omega^k(U)$ the sheaf of algebraic k-forms on U having holomorphic coefficients with respect to X. The *relative differential* $d : \Omega_X^k[U] \to \Omega_X^{k+1}[U]$ is thus $\mathcal{O}_X$-linear.

For simplicity, we will set below $\mathcal{O} = \mathcal{O}_X[U]$ and $\Omega^k = \Omega_X^k[U]$.

The twisted de Rham complex, variable τ'. Let τ' be a new variable (the notation is supposed to correspond to that of Chapter V). On the space $\Omega^k[\tau'] = \mathbb{C}[\tau'] \otimes_{\mathbb{C}} \Omega^k$ of polynomials in τ' with coefficients in Ω^k, the relative differential d defined by

$$d\Big(\sum_i \omega_i \tau'^i \Big) = \sum_i (d\omega_i)\tau'^i$$

can be twisted by $e^{-\tau' f}$: one notices indeed that the operator

$$d_f := e^{\tau' f} \cdot d \cdot e^{-\tau' f}$$

keeps the polynomial behaviour in τ', as we have $d_f = d - \tau' \, df\wedge$, in other words,

$$d_f \Big(\sum_i \omega_i \tau'^i \Big) = \sum_i (d\omega_i - df \wedge \omega_{i-1})\tau'^i.$$

Moreover, we have $d_f \circ d_f = 0$, as the same holds for d. We have thus defined a new complex, the *de Rham complex of U twisted by $e^{-\tau' f}$*:

$$(5.1) \qquad 0 \longrightarrow \mathcal{O}[\tau'] \xrightarrow{d_f} \Omega^1[\tau'] \xrightarrow{d_f} \cdots \xrightarrow{d_f} \Omega^{n+1}[\tau'] \longrightarrow 0.$$

The differential d_f commutes with the action of $\mathcal{O}_X[\tau']$, as it only differentiates with respect to the variables of U, so that each cohomology sheaf $\widehat{\mathrm{M}}^{(k)}$ is a $\mathcal{O}_X[\tau']$-module.

On the other hand, each sheaf $\Omega^k[\tau']$ (that we can regard as a meromorphic bundle of infinite rank on the complex line with variable τ') is equipped with the connection $\widehat{\nabla}$ obtained by twisting the natural connection (differentiation with respect to τ' only) by $e^{-\tau' f}$. We thus have

$$(5.2) \qquad\qquad \widehat{\nabla}_{\partial_{\tau'}} \omega = \frac{\partial \omega}{\partial \tau'} - f\omega.$$

This connection makes $\Omega^k[\tau']$ a module[19] over the Weyl algebra $\mathbb{C}[\tau']\langle\partial_{\tau'}\rangle$. Moreover, it commutes with the differential d_f, as the differentiation with respect to τ' commutes with the differentiation d (with respect to the variables of U). It induces thus a structure of the same kind on each cohomology sheaf $\widehat{\mathrm{M}}^{(k)}$ of the complex (5.1).

Lastly, each sheaf Ω^k of relative differential forms is equipped with an integrable connection with respect to the variables of X: this is the differentiation $d_X : \Omega^k \to \Omega_X^1 \otimes_{\mathcal{O}_X} \Omega^k$. It can be trivially extended to $\Omega^k[\tau']$

---

[19] Which does not have finite type in general!

as a partial integrable connection, still denoted by d_X. One can, in the same way, twist the latter by $e^{-\tau' f}$ to obtain a partial integrable connection $d_{X,f} = e^{\tau' f} d_X e^{-\tau' f} = d_X - \tau' d_X f \wedge$. As d and d_X commute, so do d_f and $d_{X,f}$. Therefore, each $\widehat{\mathbb{M}}^{(k)}$ is equipped with a partial integrable connection. Finally, as $d_{X,f}$ is linear with respect to τ', this partial connection can be extended as an integrable connection $\widehat{\nabla}$ on the $\mathcal{O}_X[\tau']$-module $\widehat{\mathbb{M}}^{(k)}$.

Inverse Fourier transform. We can apply a partial Fourier transform to each sheaf $\Omega^k[\tau']$ or to the cohomology sheaves $\widehat{\mathbb{M}}^{(k)}$: when we regard the sheaf $\widehat{\mathbb{M}}^{(k)}$ as module (not necessarily of finite type) over $\mathcal{O}_X[t]$, where t is the action of $-\widehat{\nabla}_{\partial_{\tau'}}$, we denote it by $\mathbb{M}^{(k)}$. It is then equipped with an integrable connection ∇ such that, for any local section $[\omega]$, we have

$$\nabla_{\partial_t}[\omega] = \tau'[\omega] \quad \text{and} \quad \nabla_{\partial_{x_i}}[\omega] = \widehat{\nabla}_{\partial_{x_i}}[\omega].$$

In particular, if $\omega \in \Omega^k$ has degree 0 in τ', the formula (5.2) shows that

$$t[\omega] = [f\omega].$$

5.3 Definition. The *Gauss-Manin systems* associated to the function f are the $\mathcal{O}_X[t]$-modules with integrable connection $\mathbb{M}^{(k)}$. The corresponding *Brieskorn lattice* $\mathbb{E}^{(k)}$ is the image of Ω^k in $\mathbb{M}^{(k)}$. It is a $\mathcal{O}_X[t]$-module.

5.c Frobenius-Saito structure of type A_d, second version

We take up the case of a family of polynomials of one variable $u \in \mathbb{C}$ whose coefficients are parametrized by the points x of a complex analytic manifold X. Let us then set

$$f(u, x) = u^{d+1} + \sum_{i=0}^{d} a_i(x) u^i,$$

where the a_i are holomorphic functions of x. We recognize an unfolding (that we do not yet assume universal) of the singularity A_d considered in §4.b. The function f defines a mapping

$$\mathbb{C} \times X \xrightarrow{\widetilde{f}} \mathbb{C} \times X$$
$$(u, x) \longmapsto (t, x) = (f(u, x), x)$$

which is *proper with finite fibres* (having cardinal $d + 1$).

Trace of functions and of 1-forms. Any function $h(u, x) \in \Gamma(W, \mathcal{O}_X[u])$ on an open set W of X has a *trace* relative to $\widetilde{f}$, which is a polynomial in t with coefficients in $\mathcal{O}_X(W)$ (the reader should check this): it is defined by the formula

$$\mathrm{tr}(h)(t, x) = \frac{1}{d+1} \sum_{(u,x) \mapsto (t,x)} h(u, x),$$

where the roots are taken with multiplicity. Let us denote by $\mathcal{O}_X[u]_{\leqslant d}$ the sheaf of polynomials of degree $\leqslant d$ in u with coefficients in $\mathcal{O}_X$ and by $\widetilde{\mathcal{O}_X[u]}_{\leqslant d}$ that of polynomials of the same kind which, moreover, have *trace zero*. The trace gives a $\mathcal{O}_X$-linear decomposition

$$\mathcal{O}_X[u] = \mathcal{O}_X[t] \oplus \widetilde{\mathcal{O}_X[u]}_{\leqslant d}[t] := \mathcal{O}_X[t] \oplus \widetilde{\mathcal{O}},$$

by identifying t and $t \circ \widetilde{f} = f$. Therefore, $\widetilde{\mathcal{O}}$ is the subsheaf of functions having trace zero. We deduce a $\mathcal{O}_X$-linear decomposition of the sheaf of relative differential 1-forms:

$$\Omega^1 := \mathcal{O}_X[u]\, du = \mathcal{O}_X[t]\, dt \oplus \widetilde{\mathcal{O}_X[u]}_{\leqslant d}[t]\, dt \oplus \mathcal{O}_X[u]_{\leqslant d-1}\, du$$

$$= \mathcal{O}_X[t]\, dt \oplus \widetilde{\mathcal{O}_X}\, dt \oplus \mathcal{O}_X[u]_{\leqslant d-1}\, du$$

$$:= \mathcal{O}_X[t]\, dt \oplus \widetilde{\Omega}^1,$$

by identifying dt and df. The projection on the summand $\mathcal{O}_X[t]\, dt$ is, by definition, the trace operator for the relative differential 1-forms. Therefore, $\widetilde{\Omega}^1$ is the subsheaf of 1-forms having trace zero.

5.4 Exercise.

(1) Identify the term $\mathcal{O}_X[u]_{\leqslant d-1}\, du$ to the sheaf $\widetilde{d\mathcal{O}_X[u]}_{\leqslant d}$, as well as to the *Jacobian quotient* $\Omega^1/\mathcal{O}_X[u]\, df \simeq \mathcal{O}_X[u]/(f')$, if f' denotes the derivative of f with respect to u.

(2) Show that the trace operator for relative 1-forms satisfies, for any $h \in \mathcal{O}_X(W)[u]$, the following properties, if d denotes the differential relative to u only:

$$\mathrm{tr}(dh) = d(\mathrm{tr}\, h)$$
$$\mathrm{tr}(h \cdot df) = \mathrm{tr}(h) \cdot dt.$$

(3) Prove that, if g has degree $\leqslant d - 1$, then, for any $k \geqslant 0$, the quotient of the division of $f^k g$ by f' has trace zero (multiply all the terms by du).

5.5 Example.
When there is no parameter, hence $f(u) = u^{d+1}$, the 1-forms having trace zero are the forms $g(u)\, du$, where g is a polynomial for which the coefficient of $u^{k(d+1)-1}$ is zero for any $k \in \mathbb{N}$.

5.6 Exercise.
Prove that

(1) $\widetilde{\Omega}^1$ is a free $\mathcal{O}_X[t]$-module of rank d;

(2) the relative differential is diagonal with respect to the decompositions

$$\mathcal{O}_X[u] = \widetilde{\mathcal{O}} \oplus \mathcal{O}_X[t] \quad \text{and} \quad \mathcal{O}_X[u]\, du = \widetilde{\Omega}^1 \oplus \mathcal{O}_X[t]\, dt$$

and $d : \widetilde{\mathcal{O}} \to \widetilde{\Omega}^1$ is *bijective* (this is the reason for introducing objects having trace zero);

(3) the quotient $\widetilde{\Omega}^1/\widetilde{\mathcal{O}} \cdot dt$ is a locally free $\mathcal{O}_X$-module of rank d.

The Gauss-Manin system and the Brieskorn lattice. Let us introduce as in §5.b a new variable τ'. The *Gauss-Manin system* is the quotient

$$
\mathbb{M} := \Omega^1[\tau'] \Big/ d_f \mathcal{O}_X[u, \tau']
$$

$$
= \mathcal{O}_X[t, \tau'] \, dt \Big/ (d - \tau' \, dt \wedge) \mathcal{O}_X[t, \tau'] \oplus \widetilde{\Omega}^1[\tau'] \Big/ (d - \tau' \, dt \wedge) \widetilde{\mathcal{O}}[\tau']
$$

$$
:= \mathcal{O}_X[t, \tau'] \, dt \Big/ (d - \tau' \, dt \wedge) \mathcal{O}_X[t, \tau'] \oplus \widetilde{\mathbb{M}},
$$

where we have set $\widetilde{\mathcal{O}}[\tau'] = \mathbb{C}[\tau'] \otimes_{\mathbb{C}} \widetilde{\mathcal{O}}$, and similarly for Ω^1 and $\widetilde{\Omega}^1$. The denominator is a $\mathcal{O}_X[\tau']$-module, but not (in a natural way) a $\mathcal{O}_X[t, \tau']$-module, as the relative differential d is not linear with respect to multiplication by t. Therefore, $\mathbb{M}$ is a $\mathcal{O}_X[\tau']$-module. The partial connection ∇_{∂_t} equips this sheaf with the structure of a $\mathcal{O}_X[t]\langle \partial_t \rangle$-module. Lastly, this module is equipped with an integrable connection ∇' with respect to the variables x_i. That ∇' commutes with the action of $\mathbb{C}[t]\langle \partial_t \rangle$ means that the connection ∇ is integrable.

We summarize the essential properties in the following exercise.

5.7 Exercise. Prove that

(1) the term $\mathcal{O}_X[t, \tau'] \, dt / (d - \tau' \, dt \wedge) \mathcal{O}_X[t, \tau']$ can be identified with the (rank one) trivial bundle $\mathcal{O}_X[t]$ equipped with its natural connection;

(2) the action of ∂_t on $\widetilde{\mathbb{M}}$ is bijective (use the bijectivity of d shown in the previous exercise);

(3) the natural mapping $\widetilde{\Omega}^1 \to \widetilde{\mathbb{M}}$ (taking the class of a degree 0 element of $\widetilde{\Omega}^1[\tau']$) is $\mathcal{O}_X[t]$-linear, injective and its image, denoted by $\widetilde{\mathbb{E}}$, is stable by ∂_t^{-1}; it is in particular a free $\mathcal{O}_X[t]$-module of rank d;

(4) $f'(t)\partial_t$ preserves $\widetilde{\mathbb{E}}$;

(5) with respect to the structure of $\mathcal{O}_X[\tau]$-module on $\widetilde{\mathbb{E}}$ for which τ acts as ∂_t^{-1} (set $\tau = \tau'^{-1}$ as in §V.2.c), there is a decomposition $\widetilde{\Omega}^1 = \tau \cdot \widetilde{\Omega}^1 \oplus \mathcal{O}_X[u]_{\leqslant d-1} \, du$; denote by $\widehat{\mathbb{E}}$ the $\mathcal{O}_X$-module $\widetilde{\mathbb{E}}$ with this structure of $\mathcal{O}_X[\tau]$-module;

(6) $\widehat{\mathbb{E}}$ is isomorphic to the free $\mathcal{O}_X[\tau]$-module (of rank d) $\mathcal{O}_X[u]_{\leqslant d-1}[\tau]$ (write any 1-form of the kind $ht^k \, dt$, with $h \in \widetilde{\mathcal{O}_X[u]}_{\leqslant d}$, as $\tau\omega$, where ω has degree $\leqslant k - 1$ in t);

(7) the quotient $\widehat{\mathbb{E}}/\tau\widehat{\mathbb{E}}$ is the free $\mathcal{O}_X$-module (of rank d) $\widetilde{\Omega}^1/\widetilde{\mathcal{O}} \cdot dt$.

In order to know the action of the connection $\widehat{\nabla}$, it is enough to define it on the sections of the $\mathcal{O}_X$-module $\mathcal{O}_X[u]_{\leqslant d-1} \, du$. Let thus $h \, du$ be such a form. Then, by definition,

$$
\tau^2 \widehat{\nabla}_{\partial_\tau} h \, du = fh \, du
$$

and we express the latter term as a function of τ: we write $fh = qf' + r$ with q of degree $\leqslant d$ and r of degree $\leqslant d - 1$; thus $fh \, du \equiv \tau q' \, du + r \, du$ holds in $\widehat{\mathbb{E}}$, as $q\,df \equiv \tau q'\,du$ in $\widehat{\mathbb{E}}$. In the same way, we have $\tau\widehat{\nabla}_{\partial_{x_i}} h\,du \equiv \tau q_i'\,du + r_i\,du$, where r_i is the remaining term in the division of $h \cdot \partial f/\partial x_i$ by f'. Let us consider then the following $\mathscr{O}_X$-linear operators on $\mathscr{O}_X[u]_{\leqslant d-1}$:

$$R_0(x): h(u,x) \longmapsto r(u,x) \qquad R_\infty(x): h(u,x) \longmapsto -q'(u,x)$$
$$\Phi_i(x): h(u,x) \longmapsto r_i(u,x) \qquad \Psi_i(x): h(u,x) \longmapsto q_i'(u,x).$$

The connection $\widehat{\nabla}$ can be written as

$$(5.8) \qquad \left(\frac{R_0(x)}{\tau} - R_\infty(x)\right)\frac{d\tau}{\tau} + \sum_i \left(\Psi_i(x) + \frac{\Phi_i(x)}{\tau}\right)dx_i.$$

We see in particular that, for any value x^o of the parameter, the restriction of the connection to $\widehat{\mathbb{E}}^o = \widehat{\mathbb{E}}/\mathfrak{m}_{x^o}\widehat{\mathbb{E}}$ takes Birkhoff's normal form. The connection $\widehat{\nabla}$, however, does not take the form VI.(2.2) in any basis of $\mathscr{O}_X[u]_{\leqslant d-1}$, as the endomorphism R_∞ may not have a constant matrix.

Theorem VI.2.1 nevertheless says that, if X is 1-connected, then for any $x^o \in X$ and any basis of the vector space $\mathbb{C}[u]_{\leqslant d-1}du$, there exists, on the complement of a hypersurface Θ_{x^o} of X, a basis of $\widehat{\mathbb{E}}$ which extends the given basis and in which the matrix of $\widehat{\nabla}$ takes Birkhoff's normal form. According to it, we can construct a flat connection ∇ on $E_0 := \widehat{\mathbb{E}}/\tau\widehat{\mathbb{E}}$. In the present situation, we can directly obtain the base change P: the *integrability* property of $\widehat{\nabla}$ implies indeed that the matrix 1-form $\Psi = \sum_i \Psi_i\,dx_i$ on X fulfills the condition

$$d\Psi + \Psi \wedge \Psi = 0;$$

it follows that the Pfaff system $dP(x) = -P(x)\Psi$ has, in the neighbourhood of x^o, a unique solution $P(x) \in \mathrm{GL}_d(\mathscr{O}_X)$ such that $P(x^o) = \mathrm{Id}$; after the base change of matrix $P(x)$, the connection takes Birkhoff's normal form. In particular, the matrices $\Phi_i(x^o)$ above are also that which come in Birkhoff's normal form.

5.9 Example 5.5, continuation. In the basis $u^\ell du$ ($\ell = 0, \ldots, d-1$), the matrix B_0^o of $R_0(x^o)$ is zero and the matrix B_∞ of $-R_\infty(x^o)$ is the diagonal matrix with $(\ell+1)/(d+1)$ as diagonal entries, with $\ell = 0, \ldots, d-1$. We recover the matrix $B_\infty = (1+q)\,\mathrm{Id} - \nabla\,\mathfrak{E}$ (with $q = 1/(d+1)$) of Remark VII.4.17.

5.10 Exercise. Prove that, for the unfolding $f(u,z) = u^{d+1} + z_{d-1}u^{d-1} + \cdots + z_0$ considered in §4.b, the form du satisfies the following property: for any z^o, the class w^o of du in $\widehat{\mathbb{E}}^o$ is an eigenvector of R_∞ and the mapping $\partial_{z_i} \mapsto \Phi^o_{\partial_{z_i}}(w^o)$ induces an isomorphism $T_{z^o}\mathbb{C}^d \to E_0^o$. Deduce, by means of Remark 3.7 and of the construction of Birkhoff's normal form above, that in the neighbourhood of any point z^o there exists a homogeneous primitive section whose restriction to z^o is equal to the class of du and thus, after §3, a Frobenius-Saito structure on this neighbourhood.

References

[AB94]   D.V. ANOSOV & A.A. BOLIBRUCH – *The Riemann-Hilbert problem*, Aspects of Mathematics, vol. 22, Vieweg, 1994.

[AGZV88]   V.I. ARNOL'D, S.M. GUSEĬN-ZADE & A.N. VARCHENKO – *Singularities of differentiable maps, Vol. I & II*, Monographs in Mathematics, vol. 82 & 83, Birkhäuser Boston Inc., Boston, MA, 1988, (Russian edition: 1982).

[AH88]   M.F. ATIYAH & N.J. HITCHIN – *The geometry and dynamics of magnetic monopoles*, Princeton University Press, Princeton, NJ, 1988.

[AM69]   M.F. ATIYAH & I.G. MACDONALD – *Introduction to commutative algebra*, Addison-Wesley, Reading, MA, 1969.

[Aud98a]   ———, "Symplectic geometry in Frobenius manifolds and quantum cohomology", *J. Geom. Phys.* **25** (1998), p. 183–204.

[Aud98b]   M. AUDIN – "An introduction to Frobenius manifolds, moduli space of stable maps and quantum cohomology", http://www-irma.u-strasbg.fr/~maudin, August 1998.

[Bar99]   S. BARANNIKOV – "Generalized periods and mirror symmetry in dimension > 3", arXiv: math.AG/9903124, 1999.

[BBRS91]   W. BALSER, B.L.J. BRAAKSMA, J.-P. RAMIS & Y. SIBUYA – "Multisummability of formal power series solutions of linear ordinary differential equations", *Asymptotic Anal.* **5** (1991), p. 27–45.

[Bd83]   L. BOUTET DE MONVEL – "𝒟-modules holonomes réguliers en une variable", in *Séminaire E.N.S. Mathématique et Physique* [BdDV83], p. 313–321.

[BdDV83]   L. BOUTET DE MONVEL, A. DOUADY & J.-L. VERDIER (eds.) – *Séminaire E.N.S. Mathématique et Physique*, Progress in Math., vol. 37, Birkhäuser, Basel, Boston, 1983.

[Bea93]   A. BEAUVILLE – "Monodromie des systèmes différentiels linéaires à pôles simples sur la sphère de Riemann (d'après A. Bolibruch)", in *Séminaire Bourbaki*, Astérisque, vol. 216, Société Mathématique de France, 1993, p. 103–119.

[Ber92]   D. BERTRAND – "Groupes algébriques et équations différentielles linéaires", in *Séminaire Bourbaki*, Astérisque, vol. 206, Société Mathématique de France, 1992, p. 183–204.

[BG91]   C. BERENSTEIN & R. GAY – *Complex Variables, an Introduction*, Graduate Texts in Math., vol. 125, Springer-Verlag, 1991.

[Bir09]   G.D. BIRKHOFF – "Singular points of ordinary linear differential equa-
           tions", *Trans. Amer. Math. Soc.* **10** (1909), p. 436–470.

[Bir13a]   _____ , "A Theorem on Matrices of Analytic Functions", *Math. Ann.* **74**
           (1913), p. 122–133.

[Bir13b]   _____ , "The generalized Riemann problem for linear differential equa-
           tions and allied problems for linear difference and q-difference equations",
           Proc. Amer. Acad. Arts and Sci. **49** (1913), p. 521–568.

[BJL79]   W. BALSER, W. JURKAT & D.A. LUTZ – "A general theory of invari-
           ants form meromorphic differential equations; Part II, Proper invariants",
           Funkcial. Ekvac. **22** (1979), p. 257–283.

[Bjö79]   J.-E. BJÖRK – *Rings of differential operators*, North Holland, Amster-
           dam, 1979.

[BK98]   S. BARANNIKOV & M. KONTSEVITCH – "Frobenius manifolds and for-
           mality of Lie algebras of polyvector fields", *Internat. Math. Res. Notices*
           4 (1998), p. 201–215.

[Boa99]   PH. BOALCH – "Symplectic geometry and isomonodromic deformations",
           Ph.D. Thesis, Oxford University, 1999.

[Bol95]   A.A. BOLIBRUKH – *The 21st Hilbert problem for linear Fuchsian sys-
           tems*, Proceedings of the Steklov Inst. of Mathematics, vol. 206, American
           Mathematical Society, Providence, RI, 1995.

[Bol98]   _____ , "On isomonodromic confluences of Fuchsian singularities", *Tr.
           Math. Inst. Steklova* **221** (1998), p. 127–142.

[Bor87]   A. BOREL (ed.) – *Algebraic 𝒟-modules*, Perspectives in Math., vol. 2,
           Boston, Academic Press, 1987.

[Bou98]   N. BOURBAKI – *Commutative algebra. Chapters 1–7*, Elements of Math-
           ematics (Berlin), Springer-Verlag, Berlin, 1989, 1998, (French edition:
           1964).

[BP02]   J. BERTIN & C. PETERS – "Variations of Hodge Structures, Calabi-
           Yau Manifolds and Mirror Symmetry", in *Introduction to Hodge theory*,
           SMF/AMS Texts and Monographs, vol. 8, American Mathematical Soci-
           ety, Providence, RI, 2002, (French edition: 1996).

[Bri70]   E. BRIESKORN – "Die Monodromie der isolierten Singularitäten von Hy-
           perflächen", *Manuscripta Math.* **2** (1970), p. 103–161.

[Bri73]   _____ , "Sur les groupes de tresses (d'après Arnold)", in *Séminaire Bour-
           baki*, Lect. Notes in Math., vol. 317, Springer-Verlag, 1973, exposé 401.

[Bru85]   A. BRUGUIÈRES – "Filtration de Harder-Narasmhan et stratification de
           Shatz", in *Modules des fibrés stables sur les courbes algébriques* [LPV85],
           p. 81–105.

[BS76]   C. BĂNICĂ & O. STĂNĂŞILĂ – *Algebraic methods in the global theory of
           complex spaces*, Editura Academiei, Bucharest, 1976.

[BV83]   D.G. BABBITT & V.S. VARADARAJAN – "Formal reduction of meromor-
           phic differential equations: a group theoretic view", *Pacific J. Math.* **109**
           (1983), p. 1–80.

[BV85]   _____ , *Deformation of nilpotent matrices over rings and reduction of an-
           alytic families of meromorphic differential equations*, Mem. Amer. Math.
           Soc., vol. 55, no. 325, American Mathematical Society, Providence, RI,
           1985.

[BV89a]   _____ , *Local moduli for meromorphic differential equations*, Astérisque,
           vol. 169-170, Société Mathématique de France, Paris, 1989.

[BV89b]   ———, "Some remarks on the asymptotic existence theorem for mero-morphic differential equations", *J. Fac. Sci. Univ. Tokyo Sect. IA Math.* **36** (1989), p. 247–262.

[Car95]   H. CARTAN – *Elementary theory of analytic functions of one or several complex variables*, Dover Publications Inc., New York, 1995, (French edition: 1961).

[Con99]   R. CONTE (ed.) – *The Painlevé Property One Century Later*, CRM Series in Mathematical Physics, Springer-Verlag, 1999.

[Cor99]   E. COREL – "Exposants, réseaux de Levelt et relations de Fuchs pour les systèmes différentiels réguliers", Thèse, Université de Strasbourg, 1999.

[Cou95]   S.C. COUTINHO – *A primer of algebraic 𝒟-modules*, London Mathematical Society Student Texts, Cambridge University Press, 1995.

[CR62]   C.W. CURTIS & I. REINER – *Representation theory of finite groups and associative algebras*, Pure and Applied Mathematics, vol. XI, Interscience Publishers, a division of John Wiley & Sons, New York-London, 1962.

[CZ99]   H.-D. CAO & J. ZHOU – "Frobenius manifold structure on Dolbeault cohomology and mirror symmetry", *Comm. Anal. Geom.* **8** (1999), no. 4, p. 795–808.

[Dek79]   W. DEKKERS – "The matrix of a connection having regular singularities on a vector bundle of rank 2 on $\mathbb{P}^1(\mathbb{C})$", in *Équations différentielles et systèmes de Pfaff dans le champ complexe*, Lect. Notes in Math., vol. 712, Springer-Verlag, 1979, p. 33–43.

[Del70]   P. DELIGNE – *Équations différentielles à points singuliers réguliers*, Lect. Notes in Math., vol. 163, Springer-Verlag, 1970.

[Dim92]   A. DIMCA – *Singularities and topology of hypersurfaces*, Universitext, Springer-Verlag, Berlin, New York, 1992.

[Dou05]   A. DOUAI – "Construction de variétés de Frobenius via les polynômes de Laurent: une autre approche", in *Singularités*, Rev. Inst. Élie Cartan, vol. 18, Univ. Nancy, Nancy, 2005, p. 105–123.

[DS03]   A. DOUAI & C. SABBAH – "Gauss-Manin systems, Brieskorn lattices and Frobenius structures (I)", *Ann. Inst. Fourier (Grenoble)* **53** (2003), no. 4, p. 1055–1116.

[DS04]   ———, "Gauss-Manin systems, Brieskorn lattices and Frobenius structures (II)", in *Frobenius manifolds (Quantum cohomology and singularities)* (C. Hertling & M. Marcolli, eds.), Aspects of Mathematics, vol. E 36, Vieweg, 2004, arXiv: math.AG/0211353, p. 1–18.

[Dub96]   B. DUBROVIN – "Geometry of 2D topological field theory", in *Integrable systems and quantum groups* (M. Francaviglia & S. Greco, eds.), Lect. Notes in Math., vol. 1260, Springer-Verlag, 1996, p. 120–348.

[Ehl87]   F. EHLERS – "Chap. V: The Weyl Algebra", in *Algebraic 𝒟-modules* [Bor87], p. 173–205.

[Ehr47]   C. EHRESMANN – "Sur les espaces fibrés différentiables", *C. R. Acad. Sci. Paris Sér. I Math.* **224** (1947), p. 1611–1612.

[Ehr51]   ———, "Les connexions infinitésimales dans un espace fibré différentiable", in *Colloque de topologie – espaces fibrés (Bruxelles, 1950)*, Georges Thone, Liège, 1951, p. 29–55.

[EV86]   H. ESNAULT & E. VIEHWEG – "Logarithmic de Rham complexes and vanishing theorems", *Invent. Math.* **86** (1986), p. 161–194.

[EV99]   ———, "Semistable bundles on curves and irreducible representations of the fundamental group", in *Algebraic geometry: Hirzebruch 70* (P. Pragacz, M. Szurek & J. Wiśniewski, eds.), Contemp. Math., vol. 241, American Mathematical Society, 1999.

[Fis76]   G. FISCHER – *Complex analytic geometry*, Lect. Notes in Math., vol. 538, Springer-Verlag, 1976.

[Fre57]   J. FRENKEL – "Cohomologie non abélienne et espaces fibrés", *Bull. Soc. Math. France* **85** (1957), p. 135–220.

[FS82]   F. FROBENIUS & L. STICKELBERGER – "Über die Differentation der elliptischen Funktionen nach den Perioden und Invarianten", *J. reine angew. Math.* **92** (1882), p. 311–327, & Complete Works, Springer-Verlag, 1968, p. 64–80.

[Fuc07]   R. FUCHS – "Über lineare homogene Differentalgleichungen zweiter Ordnung mit drei im Endlichen gelegenen wesentlich singuläre Stellen", *Math. Ann.* **63** (1907), p. 301–321.

[Gan59]   F.R. GANTMACHER – *The theory of matrices. Vols 1, 2*, Chelsea Publishing Co., New York, 1959, (Russian edition: 1942).

[GH78]   P.A. GRIFFITHS & J. HARRIS – *Principles of Algebraic Geometry*, A. Wiley-Interscience, New York, 1978.

[GHL87]   S. GALLOT, D. HULIN & J. LAFONTAINE – *Riemannian Geometry*, Universitext, Springer-Verlag, Berlin, Heidelberg, 1987.

[Giv98]   A. GIVENTAL – "Elliptic Gromov-Witten invariants and the generalized mirror conjecture", in *Integrable systems and algebraic geometry (Kobe/Kyoto, 1997)*, World Sci. Publishing, River Edge, NJ, 1998, p. 107–155.

[GKZ94]   I.M. GELFAND, M.M. KAPRANOV & A.V. ZELEVINSKY – *Discriminants, Resultants and Multidimensional Determinants*, Birkhäuser, Basel, Boston, 1994.

[GL76]   R. GÉRARD & A.H.M. LEVELT – "Sur les connexions à singularités régulières dans le cas de plusieurs variables", *Funkcial. Ekvac.* **19** (1976), p. 149–173.

[GM93]   M. GRANGER & PH. MAISONOBE – "A basic course on differential modules", in *Éléments de la théorie des systèmes différentiels* [MS93], p. 103–168.

[God64]   R. GODEMENT – *Topologie algébrique et théorie des faisceaux*, Hermann, Paris, 1964.

[God71]   C. GODBILLON – *Élements de topologie algébrique*, collection Méthodes, Hermann, Paris, 1971.

[GR65]   R.C. GUNNING & H. ROSSI – *Analytic functions of several complex variables*, Prentice-Hall Inc., Englewood Cliffs, NJ, 1965.

[GR97]   B. GRAMMATICOS & A. RAMANI – "Integrability — and how to detect it", in *Integrability of nonlinear systems (Pondicherry, 1996)* (Y. Kosmann-Schwarzbach, B. Grammaticos & K. Tamizhmani, eds.), Lect. Notes in Physics, vol. 495, Springer-Verlag, 1997, p. 30–94.

[Gra84]   A. GRAMAIN – *Topology of surfaces*, BCS Associates, Moscow, ID, 1984, (French edition: 1971).

[Gro57]   A. GROTHENDIECK – "Sur la classification des fibrés holomorphes sur la sphère de Riemann", *Amer. J. Math.* **79** (1957), p. 121–238.

[GS95]   C. Gantz & B. Steer – "Gauge fixing for logarithmic connections over curves and the Riemann-Hilbert problem", arXiv: math.AG/9504016, 1995.

[Har80]   R. Hartshorne – *Algebraic geometry*, Graduate Texts in Math., vol. 52, Springer-Verlag, 1980.

[Har94]   J. Harnad – "Dual isomonodromic deformations and moment map to loop algebra", *Commun. Math. Phys.* **166** (1994), p. 337–365.

[Her02]   C. Hertling – *Frobenius manifolds and moduli spaces for singularities*, Cambridge Tracts in Mathematics, vol. 151, Cambridge University Press, 2002.

[HH81]   G. Hector & U. Hirsch – *Introduction to the Geometry of Foliations, part A*, Aspects of Mathematics, Vieweg, 1981.

[Hit97]   N.J. Hitchin – "Frobenius manifolds (notes by D. Calderbank)", in *Gauge Theory and Symplectic Geometry (Montréal, 1995)* (J. Hurtubise & F. Lalonde, eds.), NATO ASI Series, Kluwer Academic Publishers, 1997, p. 69–112.

[Hit99]   ――――, "Riemann surfaces and integrable systems", in *Integrable Systems – Twistor, Loop Groups and Riemann Surfaces* (N.J. Hitchin, G.B. Segal & R.S. Ward, eds.), Oxford Graduate Texts in Mathematics, Clarendon Press, Oxford, 1999, p. 1–52.

[HM99]   C. Hertling & Yu.I. Manin – "Weak Frobenius manifolds", *Internat. Math. Res. Notices* (1999), p. 277–286.

[Hör73]   L. Hörmander – *An introduction to complex analysis in several variables*, North-Holland, Amsterdam, 1973, 2nd edition.

[IKSY91]   K. Iwasaki, H. Kimura, S. Shimomura & M. Yoshida – *From Gauss to Painlevé: a modern theory of special functions*, Aspects of Mathematics, vol. 16, Vieweg, Berlin, 1991.

[IN86]   A. Its & V. Novokshenov – *The Isomonodromic Deformation Method in the Theory of Painlevé Equations*, Lect. Notes in Math., vol. 1191, Springer-Verlag, 1986.

[JM81]   M. Jimbo & T. Miwa – "Monodromy preserving deformations of linear ordinary differential equations with rational coefficients II", *Physica* **2D** (1981), p. 407–448.

[JMMS80]   M. Jimbo, T. Miwa, Y. Mori & M. Sato – "Density matrix of an impenetrable Bose gas and the fifth Painlevé transcendent", *Physica* **1D** (1980), p. 80–158.

[JMU81]   M. Jimbo, T. Miwa & K. Ueno – "Monodromy preserving deformations of linear ordinary differential equations with rational coefficients I", *Physica* **2D** (1981), p. 306–352.

[JT80]   A. Jaffe & C. Taubes – *Vortices and monopoles*, Birkhäuser, Boston, 1980.

[Kas95]   M. Kashiwara – *Algebraic study of systems of partial differential equations*, Mémoires, vol. 63, Société Mathématique de France, Paris, 1995, English translation of the Master thesis, Tokyo, 1970.

[Kir76]   A.A. Kirillov – *Elements of the theory of representations*, Grundlehren der mathematischen Wissenschaften, vol. 220, Springer-Verlag, Berlin, 1976, (Russian edition: 1972).

[Kod86]   K. Kodaira – *Complex Manifolds and Deformations of Complex Structures*, Grundlehren der mathematischen Wissenschaften, vol. 283, Springer-Verlag, 1986.

[Kos90]   V.P. KOSTOV – "On the stratification and singularities of the Stokes hypersurface of one- and two-parameter families of polynomials", in *Theory of singularities and its applications* (V.I. Arnold, ed.), Advances in Soviet Mathematics, vol. 1, American Mathematical Society, Providence, RI, 1990, p. 251–271.

[KS90]   M. KASHIWARA & P. SCHAPIRA – *Sheaves on Manifolds*, Grundlehren der mathematischen Wissenschaften, vol. 292, Springer-Verlag, 1990.

[Lan65]   S. LANG – *Algebra*, Addison-Wesley, Reading, MA, 1965.

[Lev61]   A.H.M. LEVELT – "Hypergeometric functions", *Nederl. Akad. Wetensch. Proc. Ser. A* **64** (1961), p. 361–403.

[Lev75]   ———, "Jordan decomposition for a class of singular differential operators", *Arkiv för Math.* **13** (1975), p. 1–27.

[Lit40]   D.E. LITTLEWOOD – *The Theory of Group Characters and Matrix Representations of Groups*, Oxford University Press, New York, 1940.

[LPV85]   J. LE POTIER & J.-L. VERDIER (eds.) – *Modules des fibrés stables sur les courbes algébriques*, Progress in Math., vol. 54, Birkhäuser, Basel, Boston, 1985.

[LR94]   M. LODAY-RICHAUD – "Stokes phenomenon, multisummability and differential Galois groups", *Ann. Inst. Fourier (Grenoble)* **44** (1994), p. 849–906.

[LR95]   ———, "Solutions formelles des systèmes différentiels linéaires méromorphes et sommation", *Expo. Math.* **13** (1995), p. 116–162.

[LRP97]   M. LODAY-RICHAUD & G. POURCIN – "Théorèmes d'indice pour les opérateurs différentiels linéaires ordinaires", *Ann. Inst. Fourier (Grenoble)* **47** (1997), p. 1379–1424.

[LT97]   C. LAURENT-THIÉBAUT – *Théorie des fonctions holomorphes de plusieurs variables complexes*, Savoirs Actuels, CNRS Éditions & EDP Sciences, Paris, 1997.

[Mal40]   J. MALMQUIST – "Sur l'étude analytique des solutions d'un système d'équations différentielles dans le voisinage d'un point singulier d'indétermination I, II, III", *Acta Math.* **73** (1940), p. 87–129 & **74** (1941) 1–64 & 109–128.

[Mal74]   B. MALGRANGE – "Sur les points singuliers des équations différentielles", *Enseign. Math.* **20** (1974), p. 147–176.

[Mal79]   ———, "Sur la réduction formelle des équations différentielles à singularités irrégulières", Prépublication Institut Fourier, Grenoble, 1979.

[Mal83a]   ———, "Déformations de systèmes différentiels et microdifférentiels", in *Séminaire E.N.S. Mathématique et Physique* [BdDV83], p. 351–379.

[Mal83b]   ———, "La classification des connexions irrégulières à une variable", in *Séminaire E.N.S. Mathématique et Physique* [BdDV83], p. 381–399.

[Mal83c]   ———, "Sur les déformations isomonodromiques, I, II", in *Séminaire E.N.S. Mathématique et Physique* [BdDV83], p. 401–438.

[Mal86]   ———, "Deformations of differential systems, II", *J. Ramanujan Math. Soc.* **1** (1986), p. 3–15.

[Mal91]   ———, *Équations différentielles à coefficients polynomiaux*, Progress in Math., vol. 96, Birkhäuser, Basel, Boston, 1991.

[Mal94]   ———, "Connexions méromorphes", in *Congrès Singularités (Lille, 1991)*, Cambridge University Press, 1994, p. 251–261.

[Mal96]   ———, "Connexions méromorphes, II: le réseau canonique", *Invent. Math.* **124** (1996), p. 367–387.

[Man99a] YU.I. MANIN – *Frobenius manifolds, quantum cohomology and moduli spaces*, Colloquium Publ., vol. 47, American Mathematical Society, 1999.

[Man99b] ‾‾‾‾‾‾ , "Three constructions of Frobenius manifolds: a comparative study", *Asian J. Math.* **3** (1999), p. 179–220.

[Mas59] P. MASANI – "On a result of G.D. Birkhoff on linear differential equations", *Proc. Amer. Math. Soc.* **10** (1959), p. 696–698.

[Mas67] W.H. MASSEY – *Algebraic topology, an introduction*, Graduate Texts in Math., vol. 56, Springer-Verlag, 1967.

[Mat80] H. MATSUMURA – *Commutative algebra*, Benjamin/Cummings, Reading, MA, 1980, 2nd edition.

[Miw81] T. MIWA – "Painlevé property of monodromy preserving deformation equations and the analyticity of the τ function", *Publ. RIMS, Kyoto Univ.* **17** (1981), p. 703–721.

[ML71] S. MAC LANE – *Categories for the Working Mathematician*, Graduate Texts in Math., vol. 5, Springer-Verlag, 1971.

[MM97] YU.I. MANIN & S.A. MERKULOV – "Semisimple Frobenius (super) manifolds and quantum cohomology of $\mathbb{P}^r$", *Topol. Methods Nonlinear Anal.* **9** (1997), p. 107–161.

[MR92] B. MALGRANGE & J.-P. RAMIS – "Fonctions multisommables", *Ann. Inst. Fourier (Grenoble)* **42** (1992), p. 353–368.

[MS93] PH. MAISONOBE & C. SABBAH (eds.) – $\mathscr{D}$-*modules cohérents et holonomes*, Les cours du CIMPA, Travaux en cours, vol. 45, Paris, Hermann, 1993.

[Oda87] T. ODA – "K. Saito's period map for holomorphic functions with isolated singularities", in *Algebraic Geometry (Sendai, 1985)*, Advanced Studies in Pure Math., vol. 10, North-Holland, Amsterdam, 1987, p. 591–648.

[OSS80] C. OKONEK, M. SCHNEIDER & H. SPINDLER – *Vector bundles on complex projective spaces*, Progress in Math., vol. 3, Birkhäuser, Basel, Boston, 1980.

[Per95] D. PERRIN – *Géométrie algébrique*, Savoirs Actuels, CNRS Éditions & EDP Sciences, Paris, 1995.

[Ple08] J. PLEMELJ – *Problems in the sense of Riemann and Klein*, Interscience, New York, 1908.

[PS86] A. PRESSLEY & G. SEGAL – *Loop groups*, Oxford mathematical monographs, Oxford University Press, Oxford, 1986.

[Ram94] J.-P. RAMIS – *Séries divergentes et théories asymptotiques*, Panoramas & Synthèses, vol. 0, Société Mathématique de France, Paris, 1994.

[Rey89] E. REYSSAT – *Quelques aspects des surfaces de Riemann*, Progress in Math., vol. 77, Birkhäuser, Basel, Boston, 1989.

[Rob80] PH. ROBBA – "Lemmes de Hensel pour les opérateurs différentiels, application à la réduction formelle des équations différentielles", *Enseign. Math.* **26** (1980), p. 279–311.

[Sab93] C. SABBAH – "Introduction to algebraic theory of linear systems of differential equations", in *Éléments de la théorie des systèmes différentiels* [MS93], & `math.polytechnique.fr/~sabbah/`, p. 1–80.

[Sab98] ‾‾‾‾‾‾ , "Frobenius manifolds: isomonodromic deformations and infinitesimal period mappings", *Expo. Math.* **16** (1998), p. 1–58.

[Sab06] ‾‾‾‾‾‾ , "Hypergeometric periods for a tame polynomial", *Portugal. Math.* **63** (2006), no. 2, p. 173–226, arXiv: `math.AG/9805077`.

[Sai80]   K. Saito – "Theory of logarithmic differential forms and logarithmic vector fields", *J. Fac. Sci. Univ. Tokyo Sect. IA Math.* **27** (1980), p. 265–291.

[Sai83a]   ———, "The higher residue pairings $K_F^{(k)}$ for a family of hypersurfaces singular points", in *Singularities*, Proc. of Symposia in Pure Math., vol. 40, American Mathematical Society, 1983, p. 441–463.

[Sai83b]   ———, "Period mapping associated to a primitive form", *Publ. RIMS, Kyoto Univ.* **19** (1983), p. 1231–1264.

[Sai89]   M. Saito – "On the structure of Brieskorn lattices", *Ann. Inst. Fourier (Grenoble)* **39** (1989), p. 27–72.

[Sai91]   ———, "Period mapping via Brieskorn modules", *Bull. Soc. Math. France* **119** (1991), p. 141–171.

[Sch12]   L. Schlesinger – "Über eine Klasse von Differentialsystemen beliebiger Ordnung mit festen kritischen Punkten", *J. reine angew. Math.* **141** (1912), p. 96–145.

[Sch85]   P. Schapira – *Microdifferential systems in the complex domain*, Grundlehren der mathematischen Wissenschaften, vol. 269, Springer-Verlag, 1985.

[Sch94]   J.-P. Schneiders – "A coherence criterion for Fréchet Modules", in *Index theorem for elliptic pairs*, Astérisque, vol. 224, Société Mathématique de France, 1994, p. 99–113.

[Ser55]   J.-P. Serre – "Faisceaux algébriques cohérents", *Ann. of Math.* **61** (1955), p. 197–278.

[Ser56]   ———, "Géométrie algébrique et géométrie analytique", *Ann. Inst. Fourier (Grenoble)* **6** (1956), p. 1–42.

[Sib62]   Y. Sibuya – "Asymptotic solutions of a system of linear ordinary differential equations containing a parameter", *Funkcial. Ekvac.* **4** (1962), p. 83–113.

[Sib74]   ———, *Uniform simplification in a full neighborhood of a transition point*, Mem. Amer. Math. Soc., vol. 149, American Mathematical Society, Providence, RI, 1974.

[Sib90]   ———, *Linear Differential Equations in the Complex Domain: Problems of Analytic Continuation*, Translations of Mathematical Monographs, vol. 82, American Math. Society, Providence, RI, 1990, (Japanese edition: 1976).

[Sim90]   C. Simpson – "Harmonic bundles on noncompact curves", *J. Amer. Math. Soc.* **3** (1990), p. 713–770.

[Sim92]   ———, "Higgs bundles and local systems", *Publ. Math. Inst. Hautes Études Sci.* **75** (1992), p. 5–95.

[Ste51]   N. Steenrod – *The topology of fibre bundles*, Princeton University Press, Princeton, NJ, 1951.

[SvdP03]   M. Singer & M. van der Put – *Galois Theory of Linear Differential Equations*, Grundlehren der mathematischen Wissenschaften, vol. 328, Springer-Verlag, 2003.

[SYS80]   K. Saito, T. Yano & J. Sekiguchi – "On a certain generator system of the rings of invariants of a finite reflection group", *Comm. Algebra* **8** (1980), p. 373–408.

[Tei77]   B. Teissier – "The hunting of invariants in the geometry of discriminants", in *Real and Complex Singularities (Oslo, 1976)* (P. Holm, ed.), Sijthoff and Noordhoff, Alphen aan den Rijn, 1977, p. 565–678.

[Tre83] A. TREIBICH KOHN – "Un résultat de Plemelj", in *Séminaire E.N.S. Mathématique et Physique* [BdDV83], p. 307–312.

[Tur55] H. TURRITTIN – "Convergent solutions of ordinary differential equations in the neighbourhood of an irregular singular point", *Acta Math.* **93** (1955), p. 27–66.

[Tur63] _____ , "Reduction of ordinary differential equations to the Birkhoff canonical form", *Trans. Amer. Math. Soc.* **107** (1963), p. 485–507.

[Var82] A.N. VARCHENKO – "Asymptotic Hodge structure on the cohomology of the Milnor fiber", *Math. USSR Izv.* **18** (1982), p. 469–512.

[Var83] _____ , "On semicontinuity of the spectrum and an upper bound for the number of singular points of projective hypersurfaces", *Soviet Math. Dokl.* **27** (1983), p. 735–739.

[Var91] V.S. VARADARAJAN – "Meromorphic differential equations", *Expo. Math.* **9** (1991), p. 97–188.

[Var96] _____ , "Linear meromorphic differential equations: a modern point of view", *Bull. Amer. Math. Soc.* **33** (1996), p. 1–42.

[vdP98] M. VAN DER PUT – "Recent work on differential Galois theory", in *Séminaire Bourbaki*, Astérisque, vol. 252, Société Mathématique de France, 1998, p. 341–367.

[Voi99] C. VOISIN – *Mirror Symmetry*, SMF/AMS Texts and Monographs, vol. 1, American Mathematical Society, Providence, RI, 1999, (French edition: 1996).

[Was65] W. WASOW – *Asymptotic expansions for ordinary differential equations*, Interscience, New York, 1965.

[Wol64] J.A. WOLF – "Differentiable fibre spaces and mappings compatible with Riemannian metrics", *Michigan Math. J.* **11** (1964), p. 65–70.

Index of Notation

Index

Universitext

Debarre, O.: Higher-Dimensional Algebraic Geometry

Deitmar, A.: A First Course in Harmonic Analysis. 2nd edition

Demazure, M.: Bifurcations and Catastrophes

Devlin, K. J.: Fundamentals of Contemporary Set Theory

DiBenedetto, E.: Degenerate Parabolic Equations

Diener, F.; Diener, M.(Eds.): Nonstandard Analysis in Practice

Dimca, A.: Sheaves in Topology

Dimca, A.: Singularities and Topology of Hypersurfaces

DoCarmo, M. P.: Differential Forms and Applications

Duistermaat, J. J.; Kolk, J. A. C.: Lie Groups

Dumortier.: Qualitative Theory of Planar Differential Systems

Dundas, B. I.; Levine, M.; Østvaer, P. A.; Röndip, O.; Voevodsky, V.: Motivic Homotopy Theory

Edwards, R. E.: A Formal Background to Higher Mathematics Ia, and Ib

Edwards, R. E.: A Formal Background to Higher Mathematics IIa, and IIb

Emery, M.: Stochastic Calculus in Manifolds

Emmanouil, I.: Idempotent Matrices over Complex Group Algebras

Endler, O.: Valuation Theory

Engel, K.-J.; Nagel, R.: A Short Course on Operator Semigroups

Erez, B.: Galois Modules in Arithmetic

Everest, G.; Ward, T.: Heights of Polynomials and Entropy in Algebraic Dynamics

Farenick, D. R.: Algebras of Linear Transformations

Foulds, L. R.: Graph Theory Applications

Franke, J.; Hrdle, W.; Hafner, C. M.: Statistics of Financial Markets: An Introduction

Frauenthal, J. C.: Mathematical Modeling in Epidemiology

Freitag, E.; Busam, R.: Complex Analysis

Friedman, R.: Algebraic Surfaces and Holomorphic Vector Bundles

Fuks, D. B.; Rokhlin, V. A.: Beginner's Course in Topology

Fuhrmann, P. A.: A Polynomial Approach to Linear Algebra

Gallot, S.; Hulin, D.; Lafontaine, J.: Riemannian Geometry

Gardiner, C. F.: A First Course in Group Theory

Gårding, L.; Tambour, T.: Algebra for Computer Science

Godbillon, C.: Dynamical Systems on Surfaces

Godement, R.: Analysis I, and II

Goldblatt, R.: Orthogonality and Spacetime Geometry

Gouvêa, F. Q.: p-Adic Numbers

Gross, M. et al.: Calabi-Yau Manifolds and Related Geometries

Grossman, C.; Roos, H.-G.; Stynes, M.: Numerical Treatment of Partial Differential Equations

Gustafson, K. E.; Rao, D. K. M.: Numerical Range. The Field of Values of Linear Operators and Matrices

Gustafson, S. J.; Sigal, I. M.: Mathematical Concepts of Quantum Mechanics

Hahn, A. J.: Quadratic Algebras, Clifford Algebras, and Arithmetic Witt Groups

Hájek, P.; Havránek, T.: Mechanizing Hypothesis Formation

Heinonen, J.: Lectures on Analysis on Metric Spaces

Hlawka, E.; Schoißengeier, J.; Taschner, R.: Geometric and Analytic Number Theory

Holmgren, R. A.: A First Course in Discrete Dynamical Systems

Howe, R., Tan, E. Ch.: Non-Abelian Harmonic Analysis

Howes, N. R.: Modern Analysis and Topology

Hsieh, P.-F.; Sibuya, Y. (Eds.): Basic Theory of Ordinary Differential Equations

Humi, M., Miller, W.: Second Course in Ordinary Differential Equations for Scientists and Engineers

Hurwitz, A.; Kritikos, N.: Lectures on Number Theory

Huybrechts, D.: Complex Geometry: An Introduction

Isaev, A.: Introduction to Mathematical Methods in Bioinformatics

Istas, J.: Mathematical Modeling for the Life Sciences

Iversen, B.: Cohomology of Sheaves

Jacod, J.; Protter, P.: Probability Essentials

Jennings, G. A.: Modern Geometry with Applications

Jones, A.; Morris, S. A.; Pearson, K. R.: Abstract Algebra and Famous Inpossibilities

Jost, J.: Compact Riemann Surfaces

Jost, J.: Dynamical Systems. Examples of Complex Behaviour

Jost, J.: Postmodern Analysis

Jost, J.: Riemannian Geometry and Geometric Analysis

Kac, V.; Cheung, P.: Quantum Calculus

Kannan, R.; Krueger, C. K.: Advanced Analysis on the Real Line

Kelly, P.; Matthews, G.: The Non-Euclidean Hyperbolic Plane

Kempf, G.: Complex Abelian Varieties and Theta Functions

Kitchens, B. P.: Symbolic Dynamics

Klerke, A.: Probability Theory: A Comprehensive Course

Kloeden, P.; Ombach, J.; Cyganowski, S.: From Elementary Probability to Stochastic Differential Equations with MAPLE

Kloeden, P. E.; Platen; E.; Schurz, H.: Numerical Solution of SDE Through Computer Experiments

Koralov, L. B.; Sinai, Ya. G.: Theory of Probability and Random Processes. 2nd edition

Kostrikin, A. I.: Introduction to Algebra

Krasnoselskii, M. A.; Pokrovskii, A. V.: Systems with Hysteresis

Kuo, H.-H.: Introduction to Stochastic Integration

Kurzweil, H.; Stellmacher, B.: The Theory of Finite Groups. An Introduction

Kyprianou, A. E.: Introductory Lectures on Fluctuations of Lévy Processes with Applications

Lang, S.: Introduction to Differentiable Manifolds

Lefebvre, M.: Applied Stochastic Processes

Lorenz, F.: Algebra I: Fields and Galois Theory

Luecking, D. H., Rubel, L. A.: Complex Analysis. A Functional Analysis Approach

Ma, Zhi-Ming; Roeckner, M.: Introduction to the Theory of (non-symmetric) Dirichlet Forms

Mac Lane, S.; Moerdijk, I.: Sheaves in Geometry and Logic

Marcus, D. A.: Number Fields

Martinez, A.: An Introduction to Semiclassical and Microlocal Analysis

Matoušek, J.: Using the Borsuk-Ulam Theorem

Matsuki, K.: Introduction to the Mori Program

Mazzola, G.; Milmeister G.; Weissman J.: Comprehensive Mathematics for Computer Scientists 1

Mazzola, G.; Milmeister G.; Weissman J.: Comprehensive Mathematics for Computer Scientists 2

Mc Carthy, P. J.: Introduction to Arithmetical Functions

McCrimmon, K.: A Taste of Jordan Algebras

Meyer, R. M.: Essential Mathematics for Applied Field

Meyer-Nieberg, P.: Banach Lattices

Mikosch, T.: Non-Life Insurance Mathematics

Mines, R.; Richman, F.; Ruitenburg, W.: A Course in Constructive Algebra

Moise, E. E.: Introductory Problem Courses in Analysis and Topology

Montesinos-Amilibia, J. M.: Classical Tessellations and Three Manifolds

Morris, P.: Introduction to Game Theory

Mortveit, H.; Reidys, C.: An Introduction to Sequential Dynamical Systems

Nicolaescu, L.: An Invitation to Morse Theory

Nikulin, V. V.; Shafarevich, I. R.: Geometries and Groups